T0344984

Uncertainty in Industrial Practice

Uncertainty in Industrial Practice
A guide to Quantitative Uncertainty Management

Edited by

Etienne de Rocquigny

Electricité De France, France

Nicolas Devictor

Commissariat à l'Energie Atomique, France

Stefano Tarantola

*Institute for the Protection and Security of the Citizen,
Joint Research Centre of the European Commission, Italy*

John Wiley & Sons, Ltd

Other Wiley Editorial Offices

John Wiley & Sons Inc., 111 River Street, Hoboken, NJ 07030, USA

Jossey-Bass, 989 Market Street, San Francisco, CA 94103-1741, USA

Wiley-VCH Verlag GmbH, Boschstr. 12, D-69469 Weinheim, Germany

John Wiley & Sons Australia Ltd, 42 McDougall Street, Milton, Queensland 4064, Australia

John Wiley & Sons (Asia) Pte Ltd, 2 Clementi Loop #02-01, Jin Xing Distripark, Singapore 129809

John Wiley & Sons Canada Ltd, 6045 Freemont Blvd, Mississauga, ONT, L5R 4J3

Wiley also publishes its books in a variety of electronic formats. Some content that appears
in print may not be available in electronic books.

Library of Congress Cataloging-in-Publication Data

Uncertainty in industrial practice : a guide to quantitative uncertainty management / edited by
Etienne de Rocquigny, Nicolas Devictor, Stefano Tarantola.
    p. cm.
  Includes bibliographical references and index.
  ISBN 978-0-470-99447-4 (cloth)
  1. Industrial management – Mathematical models. 2.
Uncertainty – Mathematical models. 3. Risk management. I. Rocquigny, Etienne de. II.
Devictor, Nicolas. III. Tarantola, Stefano.
  HD30.25.U53 2008
  658.001 – dc22

                             2008009584

British Library Cataloguing in Publication Data

A catalogue record for this book is available from the British Library

ISBN 978-0-470-99447-4

Typeset in 10/12pt Times by Laserwords Private Limited, Chennai, India
Printed and bound in Great Britain by TJ International, Padstow, Cornwall

Contents

Preface

Uncertainty and risk management are important topics for many organizations and individuals. However, there is no generally accepted way of quantifying uncertainty, and one objective of this book is to present a number of examples of how it has been done in a selection of European organizations. We hope that this exercise will help to improve the usefulness of uncertainty and risk analysis.

This book was written by a project group of the European Safety, Reliability and Data Association (ESReDA). ESReDA is a non-profit association of European industrial and academic organizations concerned with advances in the safety and reliability field. The association always welcomes comments, contributions and ideas relating to its publications and, more generally, to development needs in the field of reliability data.

Although members of ESReDA are companies, research institutes, universities and authorities, this book would not have been possible without substantial individual effort. The members of the project group produced the text without any financial support and devoted considerable amounts of personal time to the task.

A great deal of experience fed into this publication. ESReDA is proud to present the results of the work and hopes that it will benefit the many organizations and individuals worldwide concerned with uncertainty management.

ESReDA would like to thank the authors for their contribution and also the member organizations for funding travel expenses for meetings, etc. Particular thanks are due to those organizations which allowed working group members to participate in this initiative and gave free access to their extensive in-house expertise and experience. We register our appreciation and gratitude to:

- Electricité de France R&D (EDF R&D);
- Commissariat à l'Energie Atomique – Nuclear Energy Directorate (CEA);
- European Aeronautic Defence and Space Company – Innovation Works (EADS-IW);
- Hispano-Suiza, SAFRAN Group;
- Joint Research Centre of the European Commission (JRC Ispra and Petten);

- University of Duisburg-Essen;
- Delft University of Technology, the Netherlands.

We hope that this book meets the expectations of the public and of organizations which have shown interest in the work of the group in this important field.

Henrik Kortner

President of ESReDA

Det Norsk Veritas

Contributors and Acknowledgements

This book was collectively written by the ESReDA 'Uncertainty' project group, comprising:

Etienne de Rocquigny, co-Editor, and Yannick Lefebvre, EDF

Nicolas Devictor, co-Editor, and Nadia Pérot, CEA

Stefano Tarantola, co-Editor, and William Castaings, JRC Ispra

Fabien Mangeant and Cyrille Schwob, EADS IW

Ricardo Bolado-Lavin, JRC Petten

Jean-Remi Massé, Hispano-Suiza, SAFRAN Group

Philipp Limbourg, University of Duisburg-Essen

Wim Kanning and Pieter Van Gelder, Delft University of Technology

Part II involved, in addition to the authors, the following contributors:

- CO_2 emissions: Rémi Bussac, Pascal Bailly, Philippe-Paul Thomas (EDF R&D);

- Hydrocarbon exploration: Paolo Ruffo, Alberto Consonni, Anna Corradi and Livia Bazzana (Agip);

- Electromagnetic interferences in aircraft: Hugo Canales, Michel Crokaert (Airbus), Régis Lebrun, Vincent Feuillard (EADS IW);

- Airframe maintenance contracts: Claire Laporte, Francis Pajak (Hispano-Suiza, SAFRAN Group);

- Radiological protection and maintenance: Jean-Louis Bouchet (EDF R&D);

- Aircraft fatigue modelling: Laurent Chambon (EADS IW);

- Automotive reliability in early stages: Robert Savić (ZF Friedrichshafen AG), Hans-Dieter Kochs (University of Duisburg-Essen).

Part III also involved Bertrand Iooss and Michel Marquès (CEA) and the 'Uncertainty network' of EDF R&D.

Acknowledgements go to Henrik Kortner, Terje Aven and Jon Helton, as well as to Sarah Moore, for their fruitful contributions in reviewing the book.

Introduction

Uncertainty is a fascinating subject, spanning a wide range of scientific and practical-knowledge domains. Economics, physics, decision theory and risk assessment are its traditional locations, while more recently it has also become common in modelling, numerical analysis, advanced statistics and computer science. However, the study of uncertainty has come into contact even with such subjects as epistemology, management science, psychology, public debate and theories of democracy.

The focus of this book is narrower. As the product of a working group of practitioners representing a large panel of industrial branches, it is dedicated to the better understanding of the industrial constraints and best practices associated with the study of uncertainty. It is concerned with the *quantification of uncertainties* in the presence of data, model(s) and knowledge about the problem, and aims to make a technical contribution to decision-making processes. It should illustrate the optimal trade-offs between literature-referenced methodologies and the simplified approaches often inevitable in practice, owing to data, time or budget limitations, or simply to the current formulation of regulations and the cultural habits of technical decision-makers.

The starting point of the book comprises ten *case studies*. These were chosen to reflect a broad spectrum of situations and issues encountered in practice, both *cross-industry* (covering various branches, such as power generation, oil, space and aeronautics, mechanical industries and civil structures, and various positions in the industrial cycle, from upstream research to in-service or sales support) and *cross-discipline* (including structural safety, metrology, environmental control, robust design, sensitivity analysis, etc). Their maturity varies from pioneering efforts in industrial R&D divisions (*low*) to incorporation in operational processes on a plant (*in-service*), as represented in Figure 1. They are the real source of thought throughout the book. A common methodological framework, generic to all case studies, has been derived from them, instead of being imposed as a prior theoretical setting. While much of the underlying science has already been documented in research papers or in rather domain-specific books (on, for example, reliability, environmental impact, etc), generic guidance material for non-specialists is much harder

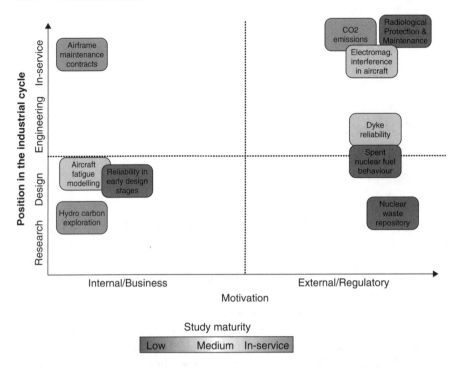

Figure 1 Case studies of the book, according to their motivation, position in the industrial cycle or to their maturity. Case studies are described in Chapter 2.

to find. Even less easy to find are international standards, of which GUM[1] may be the only representative (although this is limited to metrology and to rather simple methods). The real novelty of the book may therefore lie in its ambition to *draft from practice* a *generic* approach.

Indeed, a growing number of industrial risk studies now include some form of treatment of the numerous sources of uncertainties in order to provide more reliable results and to explore various scenarios. On one hand, appropriate uncertainty management allows for improved system designs, more accountable forecasts, better natural and industrial risk control, and more robust performance, thereby generating an 'internal/business' driver. On the other hand, many institutional bodies or regulators now demand technical justifications, including an understanding of the influence of uncertainties in industrial safety or environmental control, thereby generating an 'external/regulatory' motivation (see Figure 1).

Moreover, recent advances in information technologies make it possible to simulate several possible behaviours of products and systems. Many high-stake

---

[1] ISO Guide to the expression of Uncertainty in Measurement – see References in Section 1.5

industrial decisions are made by experts who can capitalize on a deep knowledge of their core business. Yet the explicit treatment of uncertainty on a larger scale is still weak, owing to a lack of common methods and tools or possibly due to limited cultural awareness. This book is a contribution to this vast field of engineering, with the goal of introducing more rationality and methodology into *quantitative uncertainty management*.

The focus of the book is on *quantitative uncertainty*. Yet its intent is not to deal with all types of uncertainties. It does not cover the quite challenging issues linked to the treatment of 'deep uncertainty', such as long-term futures with very damaging or irreversible consequences, which are also considered within the debate around the precautionary principle. Uncertainty in managerial, organizational or social contexts (such as organizational unpredictability, political or legal instability, versatility of public opinion, etc) is also somehow beyond the scope of this book, in so far as it is generally not practically quantifiable. Although this type of poorly-quantifiable context could be labelled a frontier domain, recommendations and examples in this book focus primarily on industrial situations in which there is *enough modelling expertise, knowledge and/or data* to support the use of *quantitative modelling of risk and uncertainty*, with probabilistic or mixed probabilistic/non-probabilistic tools.

Note that the book takes a practitioners' approach on the mathematical methods and associated theoretical groundings. Acknowledging the widespread existence of tutorials and textbooks on probability and statistics, including detailed presentation of the associated simulation and numerical methods in risk and uncertainty quantification, this book does not re-introduce them in technical detail. The reader may refer to the salient publications mentioned at the end of each chapter. Note additionally that the theoretical approach of uncertainty has involved considerable literature and significant controversies, including those associated to the rationale and conditions of the use of deterministic, probabilistic (classical or Bayesian), or extra-probabilistic settings in relation with epistemology or decision theory. While briefly recalling the context (in Chapter 1 and Chapter 14), the book does not take a strong position on one or the other competing approaches.

Indeed, existing industrial regulations, standards or codes of practices do involve a spectrum of different settings and interpretations dealing with uncertainty over the scope of the book (safety and reliability in various industrial sectors, natural risk, metrology and quality control, numerical model qualification, etc). While of course the interpretation of their results may significantly differ, it should recognised that, from a practical point of view, the implementation of most of them involves essential common features: such as the overall specification process and methodological steps, data collection and computing challenges, etc. Once again, the reader will be forwarded to some key references at the end of Chapter 1 that discuss the alternative theoretical foundations, notably regarding the use and interpretation of various probabilistic or non-probabilistic settings in representing uncertainty.

Reading guide

The book is organized as follows:

Part I introduces the common methodological framework which forms the backbone of the work; this is the essential generic structure unifying the industrial case studies (which are sketched briefly in Chapter 2), notwithstanding the peculiarities of each due to sector, physical or decision-criteria specifics.

Part II comprises ten industrial case studies recently carried out in nuclear, oil, aerospace and mechanical industries or civil structures around Europe, including a frontier case study (see Chapter 12). The chapters follow a standard format in order to facilitate comparison, in line with the common methodological framework; they illustrate a number of practical difficulties encountered or solutions found, according to the level of complexity and regulatory or financial constraints.

Part III is the scientific core of the book. It includes a review of methods which may be used in the main steps and towards the main goals of any uncertainty study. Citing a large and diverse academic literature, critical recommendations are developed from the applied point of view.

The **Appendices** gather supporting material for the practitioner. A few generic codes and standards are provided in Appendix A. Certain important websites and items of software are then reviewed with a particular focus on their operational characteristics and features in Appendix B. This is obviously a fast-moving reality for which completeness or permanence cannot be fully guaranteed. Once again, the purpose is to link the review to the salient features and key questions identified in the common methodological framework, rather than to reproduce the detailed specifications of each type of software. Appendix C gives some comments then on the promises and challenges of an extra-probabilistic setting, which enjoys growing research interest and 'frontier' industrial applications, although less mature than the core of the book recommendations.

The chapters of the book may be read independently to a certain extent, readers familiar with the subject may go directly to any case study in Part II to find illustrations of industrial practices in a given domain. Alternatively, they may refer to the more specialized chapters of Part III for a review of the methods to use in an uncertainty study. It is suggested, however, that the reader starts with Part I for an introduction to the distinctive vocabulary and approach of the book.

Notation

$\underline{x} = (x^i)_{i=1...p}$ Real-valued vector (of dimension p) of specific values for uncertain model inputs

$\underline{d}$ Real-valued vector (of dimension q) of fixed model inputs representing decision variables, conditional scenarios, or conventionally-fixed inputs

$\underline{z} = (z^l)_{l=1...r}$ Real-valued vector (of dimension r) of specific values for output variables of interest

$G(.)$ Deterministic function representing the pre-existing model linking uncertain or fixed input vectors to the vector of output variables of interest: $\underline{z} = G(\underline{x}, \underline{d})$.

$\underline{X}, \underline{Z} \ldots$ Vectors of uncertain variables (e.g. random variables in a probabilistic setting) the specific values of which are the $\underline{x}, \underline{z} \ldots$ In simplified notations, underlining is omitted: $X, Z \ldots$

$f_X(\underline{x}|\underline{\theta}_X)$ Joint probability density function of random vector $\underline{X}$, with parameters $\underline{\theta}_X$. In simplified notation, the parameter vector $\underline{\theta}_X$ is omitted, giving $f_X(\underline{x})$.

$F_X(\underline{x}), F_Z(\underline{z})$ Joint cumulative distribution function of random vector $\underline{X}$ or $\underline{Z}$

$(\underline{x}_j)_{j=1..n}$ Sample (of size n) of random vectors extracted from the cdf of $\underline{X}$.

unc_z Quantity of interest in the scalar variable of interest z representing an (absolute) dispersion around the expected value (typically a given number of standard deviations).

$\%unc_z$ Quantity of interest in the scalar variable of interest z representing relative dispersion around the expected value (typically a given number of coefficient of variations).

z_s Real value representing a threshold not to be exceeded by a scalar variable of interest z when the quantity of interest is a threshold exceedance probability.

$\underline{\theta}_X$ Vector of parameters of the measure of uncertainty of $\underline{X}$: in the probabilistic setting, it includes the parameters of the joint cdf in simplified notation, θ_X.

c_z	Quantity of interest in the variables of interest $\underline{z}$, i.e. a functional of the measure of uncertainty of $\underline{Z}$. For instance: the dispersion unc_z, a threshold exceedance probability, etc.
$varZ, E[Z]$	Variance and expectation of random variable or vector Z.
P_f	Probability of threshold exceedance (also referred to as failure probability in structural reliability), i.e. $P(G(\underline{X}) < 0)$ or $P(Z = G(\underline{X}) > z_s)$.
β	Reliability index (or Hasofer-Lind reliability index) – a way of expressing P_f often used in structural reliability, whereby $\beta = -\Phi^{-1}(P_f)$.
S_i	Sensitivity index corresponding to component X^i.
S_{Ti}	Total sensitivity index corresponding to component X^i.
Bel	Belief value of an event
Pl	Plausibility value of an event.

For the variables: superscript refers to the index within a vector (Latin letters: x^i) or the order/level of a quantile (Greek letters: x^α); subscript refers to the index within a sample (x_j), except when it denotes a specific value (e.g. penalised value: x_{pn}, threshold value: z_s); capital letters (X) denote a *random* variable (or more generally an *uncertain* variable in another setting) as opposed to a deterministic variable (or observation of a r.v.), denoted by small letters (x); underlining $(\underline{x})$ denotes a vector quantity, but this can be omitted in simplified notation. A hat $(\hat{\theta})$ denotes the estimator corresponding to an unknown deterministic quantity (θ).

Acronyms and abbreviations

bpa	Basic probability assignment (in Dempster-Shafer Theory settings)
cdf	Cumulative distribution function
c.i.	Confidence interval
CoV	Coefficient of Variation
CPU	Central Processing Unit – referring to computing time units
DP	Design Point
DST	Dempster-Shafer Theory
FAST	Fourier Amplitude Sensitivity Test – Fourier-transform based sensitivity analysis method
FORM (SORM)	First (respectively Second) Order Reliability Method
GUM	Guide to the expression of Uncertainty in Measurement – ISO standard in metrology
i.i.d.	Independent and identically distributed
LHS	Latin Hypercube Sampling

MC(S)	Monte-Carlo (Sampling): see also SRS
pdf	Probability density function
PSA	Probabilistic Safety Assessment
q.i.	Quantity of interest
r.v.	Random variable (or vector)
SRA	Structural Reliability Analysis
SRS	Simple Random Sampling
v.i.	Variable of interest

Part I
Common Methodological Framework

1

Introducing the common methodological framework

1.1 Quantitative uncertainty assessment in industrial practice: a wide variety of contexts

Quantitative uncertainty assessment in industrial practice typically involves, as is shown in Figure 1.1 below:

- a pre-existing physical or industrial system or component lying at the heart of the study, represented by a pre-existing model;
- a variety of sources of uncertainty affecting this system;
- industrial stakes and decision-making circumstances motivating the uncertainty assessment. More or less explicitly, these may include: safety and security, environmental control, process improvement, financial and economic optimization, etc. They are generally the rationale for the pre-existing model, the output and input of which help to deal with the various stakes in the decision-making process in a quantitative manner.

As will be illustrated later, mainly in Part II, these three basic features cover a very wide variety of study contexts:

The **pre-existing system** may encompass a great variety of situations, such as: a metrological chain, a mechanical structure, a maintenance process, an industrial or domestic site threatened by a natural risk, etc. In quantitative studies of uncertainties, that system will generally be modelled by a single numerical model or

Uncertainty in Industrial Practice   Edited by E. de Rocquigny, N. Devictor and S. Tarantola,
© 2008 John Wiley & Sons, Ltd

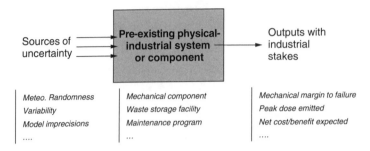

Figure 1.1 Schematic context of quantitative uncertainty assessment in industrial practice.

chain of models. The complexity of the models may vary greatly: from straightforward analytical formulae to physical models based on unsteady partial differential equations, coupled 3-D finite element models, intrinsically probabilistic models, e.g. a Boolean system reliability model predicting probabilities, etc.

The **sources of uncertainty** may include a great variety of uncertain variables of a quantitative nature, including the classical categories of: 'aleatory phenomena', 'lack of data or knowledge' or 'epistemic uncertainties', 'variability', 'measurement errors', etc. They may affect the pre-existing model in various ways: through uncertain values for model inputs, model errors or even uncertain (or incomplete) structures of the model itself, etc.

The **decision-making process** depends considerably on the **industrial stakes** involved: the study may be designed to answer regulatory requirements attached to licensing or certification of a new process or product; to ensure quality control or more robust designs; to internally optimize a technical-economic performance indicator; to feed into a larger decision model, such as a system reliability model with component uncertainties; or to help understand, in preliminary R&D stages, the importance of the various parts or parameters of the model.

1.2   Key generic features, notation and concepts

Notwithstanding the wide variety of contexts, the case studies of Part II will show that the following key generic features can be derived.

1.2.1   Pre-existing model, variables of interest and uncertain/fixed inputs

Whatever the nature or complexity of the **pre-existing model**, 'in so far as uncertainties are concerned' it may, *a priori*, be viewed conceptually as a **numerical**

function linking inputs (uncertain or fixed variables) to outputs (upon which decision criteria are established).

Formally, it is sufficient for the model to link the important **output variables of interest** (denoted $\underline{z}$) to a number of continuous or discrete inputs through a deterministic function $\underline{z} = G(\underline{x}, \underline{d})$, where some **inputs** (denoted $\underline{x}$) are **uncertain** – subject to randomness, lack of knowledge, errors or any other sources of uncertainty – while **other inputs** (denoted $\underline{d}$) are **fixed** – considered to be known – as represented in Equation 1.1 and Figure 1.2 below.

$$\underline{x}, \underline{d} \Rightarrow \underline{z} = G(\underline{x}, \underline{d}) \tag{1.1}$$

Note that $G(.)$ may represent any sort of function, including any deterministic physical or economic model (analytical or coupled 3-D finite element), or even an *intrinsically probabilistic* (or stochastic) model viewed from the perspective of a deterministic relation between some pre-existing input and output variables (e.g. failure rates at component or system level in risk analysis, or transition reaction probabilities and fluence expectation in Monte Carlo neutron physics). The computation of the pre-existing model for a given – not uncertain – point value $(\underline{x}, \underline{d})$ may hence require a very variable CPU time: from 10^{-4} s to several days for a single run, depending on the complexity of the simulation code.

Note also that the model output **variables of interest** (v.i.) are all included formally within the vector $\underline{z} = (z^l)_{l=1...r}$. Most of the time $\underline{z}$ is a *scalar* or a *small-size vector* (e.g. $r = 1$ to 5), since the decision-making process involves essentially one or few variables of interest, such as: a physical margin to failure, a net cost, a cumulated environmental dose, a failure rate in risk analysis, etc. But in some cases $\underline{z}$ may be of large dimension (e.g. predicted oil volumes at several potential well sites) or even a *function* (e.g. the mechanical margin as a function of the number of fatigue cycles, the net cost of oil production as a function of the time, etc). Vector notation will be maintained, as it is appropriate in most situations.

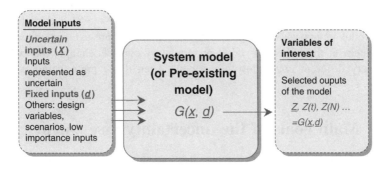

Figure 1.2 The pre-existing model and its inputs/outputs.

Regarding the **uncertain model inputs**, the vector $\underline{x} = (x^i)_{i=1...p}$ could gather formally all sources of uncertainty, whatever their nature or type (parametric, model uncertainties, etc) while in other interpretations or settings it would be restricted to some of them. The dimension of $\underline{x}$ may be very large (for example, in the case studies in Part II, p may range from 3 to several hundreds). Some components of $\underline{x}$ may be continuous, while others could be discrete or branching variables (e.g. a variable indicating the most likely model of a portfolio of uncertain models in competition with each other). It could even formally include situations where there is a spatial *field* of uncertain inputs (such as uncertain subsurface porosities) or even uncertain *functions* (such as scenarios over time); as for $\underline{z}$, the vector notation will be maintained, as it appears appropriate in most situations.

Some **model inputs** may be **fixed** – as their role is different from that of the uncertain inputs, they are given the notation $(\underline{d})$. This is the case for a number of reasons:

- some model inputs represent variables under full control: for example, the major operating conditions of an installation to be certified;
- uncertainties affecting some model inputs are considered to be negligible or of secondary importance with respect to the output variables of interest;
- for some model inputs, the decision process will conventionally fix the values despite uncertainties: for comparative purposes, it will do so by a conventional 'penalization', i.e. the choice of a fixed 'pessimistic' scenario, etc.

In industrial practice, the categorization of model inputs as 'uncertain' or 'fixed' is a matter of choice rather than theory. It can change over the course of the study and the decision-making process. Sensitivity analysis based on importance ranking and model calibration steps play key roles with respect to that choice, as will be discussed later.

Note that, for any given pre-existing system, the choice of the output variables of interest and of the appropriate chain of models to predict them depends on the industrial stakes and decision-making process. Note also that the number of model inputs varies according to the choice of pre-existing model for a given system; if the industrial stakes or regulatory controls change, this will give rise to very different uncertainty assessments even if the pre-existing system is the same.

1.2.2   Main goals of the uncertainty assessment

Industrial practice shows that the goals of any quantitative uncertainty assessment usually fall into the following four categories:

U (*Understand*): To understand the influence or rank importance of uncertainties, thereby to guide any additional measurement, modelling or R&D efforts.

A (*Accredit*): To give credit to a model or a method of measurement, i.e. to reach an acceptable quality level for its use. This may involve calibrating sensors, estimating the parameters of the model inputs, simplifying the system model physics or structure, fixing some model inputs, and finally validating according to a context-dependent level.

S (*Select*): To compare relative performance and optimize the choice of maintenance policy, operation or design of the system.

C (*Comply*): To demonstrate compliance of the system with an explicit criterion or regulatory threshold (e.g. nuclear or environmental licensing, aeronautical certification, etc).

There may be several goals in any given study and they may be combined over the course of a more-or-less elaborate decision-making process. Goals S and C refer to more advanced steps in operational decision-making, while Goals U and A concern more upstream modelling or measurement phases. Importance ranking may serve for model calibration or model simplification at an earlier stage, which becomes, after some years of research, the basis for the selection of the best designs and the final demonstration of compliance with a decision criterion. Compliance demonstration may explicitly require importance ranking as part of the process, etc.

However, as will be discussed later, the proper **identification of the most important goal(s)** of a given uncertainty assessment, as well **as of the quantities of interest** that are attached to them, are **key steps in choosing the most relevant methodologies**: this point often seems to be insufficiently appreciated in theoretical publications or prior comparisons of methodologies in official regulations or standards.

1.2.3 Measures of uncertainty and quantities of interest

The quantitative treatment of the inputs and outputs of the model may vary according to the main goal of the uncertainty assessment. However, they will, more or less explicitly, involve some '**quantities of interest**' (q.i.) in the output variables of interest. For instance, when the goal is to demonstrate compliance or to compare the results of different options, a regulation or a decision-making process involving uncertainty assessment will require the consideration of:

- percentages of error or variability in the variable(s) of interest (i.e. coefficient of variation) in measurement qualification or robust control;

- expected value of the variable of interest, such as a cost or utility in economics;

- confidence intervals of the variable(s) of interest, for instance in quality control; capabilities in robust design (i.e. ratios of a maximal acceptable range divided by 6 standard deviations);

- quantiles of the variable of interest in nuclear safety, mechanical characteristic values or corresponding to the VaR (Value at Risk) in finance;

- probabilities of exceeding a threshold or failure frequency in safety or reliability;

- ranges or simply the maximal value of the variable of interest in process control, etc.

For those quantities, an acceptable maximal value may be explicitly specified in the regulation or decision-making process, hence generating an explicit *decision criterion*. For example, for an installation, a process or a system to be licensed or certified, or for it to respect robust design objectives, the following criteria may have to be considered:

- 'there should be less than 3% uncertainty in the declared value for the output of interest';

- 'the physical margin should remain positive in spite of uncertainty, with a probability of less than 10^{-b} of being negative';

- 'the frequency of failure should be less than 10^{-b} per year, at a 95% confidence level covering the uncertainties';

- 'in spite of uncertainties, scenario A should be better (with respect to a given output variable of interest) than scenario B, to a level of confidence of at least 95%';

- 'the range of the output variable of interest should always be less than 20%' or 'the maximal value of the variable of interest should stay below a given absolute threshold', etc.

There may not be any thresholds or criteria as explicit as these, especially if the uncertainty practice is relatively recent in the given industrial field, as the case studies in Part II will demonstrate. However, there will generally be a certain *'quantity of interest'* in the output variables of interest, or more generally a *'measure of uncertainty'* on which the uncertainty assessment will issue results to be discussed in the decision process. To be more precise, what will be called a *quantity of interest* is a scalar quantity that summarizes mathematically the degree of uncertainty in the variable of interest, while the *measure of uncertainty* is the more complete mathematical distribution function comprehensively representing the uncertainty. As will be detailed in Part III, their mathematical content depends crucially on the paradigm chosen to represent uncertainty, which will hereafter be called the *uncertainty setting*. But the general structure stands, as for instance:

- in a probabilistic framework, the measure of uncertainty will be the probability measure, i.e. generally the cumulative distribution function (cdf) of the variable (or vector) of interest; the quantities of interest may be coefficients of variation, exceedance probabilities, standard deviations, or more generally any quantity derived from the cdf;

- in a non-probabilistic framework, the measure of uncertainty could be a Dempster-Shafer couple of plausibility/belief functions, while the quantity of interest might be the cumulative belief in not having exceeded a given safety threshold;

- formally, even in a deterministic framework, the measure of uncertainty could be considered the maximal range of some outputs, while the quantities of interest might be each bound of that interval.

The specification of quantities of interest and measures of uncertainty is quite natural if the final goal is of type C (*Comply*) or type S (*Select*), but it is also necessary for other types of goals, such as type U (*Understand*), or type A (*Accredit*). As will be discussed in Part III, it appears, for instance, that the importance ranking of the sources of uncertainties (Goal U) depends on the quantity of interest selected: most sensitivity analysis publications refer implicitly to variance as the quantity of interest for importance ranking, but in some cases the probability of threshold exceedance is much more relevant for industrial practice and produces very different results. Similarly, a system model can be satisfactorily calibrated (Goal A) as regards the variance of a given output of interest, but may be less acceptable regarding behaviour in the distribution tail of the output.

1.2.4 Feedback process

According to the final goal(s) motivating the uncertainty assessment, there may also be a more or less explicit *feedback process* after the initial study. Typical functions of this step might be:

- (**Goal C**) to adjust the design or the controlled variables/scenarios; to improve measurements, etc, so that the criteria can be met;

- (**Goal S**) to shift to another scenario that would further enhance performance in spite of uncertainties, e.g. by reducing the uncertainty in a critical output or by reducing costs while maintaining a given safety level, etc;

- (**Goal U/A**) to change the description of uncertainties (e.g. by removing some unimportant sources), to refine the system model to reduce uncertainties, to simplify the system model while maintaining acceptable accuracy despite inevitable uncertainties.

Of course, this feedback process can involve more than just one action, and it may be more or less strictly regulated or well defined according to the maturity of regulation or internal decision-making processes. It is generally considered essential in industrial practice, in which uncertainty assessment is often just one step in a larger or dynamic process.

1.2.5   Uncertainty modelling

Once the sources of uncertainty and corresponding input variables have been identified, there is inevitably a stage of uncertainty modelling (or quantification and characterization of the sources of uncertainty) which depends on the type of measure of uncertainty or quantities of interest chosen:

- In a probabilistic framework the uncertainty model will theoretically be a joint pdf of the vector of uncertain inputs ($\underline{x}$), although it may be specified more simply as a set of simple parametric laws for the components (e.g. Gaussian) with some independence hypotheses or approximate rank correlations.

- In an extended probabilistic framework the uncertainty model would be, for instance, a Dempster-Shafer couple of plausibility/belief functions for $\underline{x}$.

- In a deterministic framework, the maximal range of each component of $\underline{x}$.

Whatever the framework, there is always, however, a need to take into account the largest possible amount of information in order to build a satisfactory 'uncertainty model' (i.e. to choose the measure of uncertainty in the inputs). This information could include:

- direct observations of the uncertain inputs, potentially treated in a statistical way to estimate statistical models;

- expert judgement, in a more or less elaborate elicitation process, and mathematical modelling, from the straightforward choice of intervals to more elaborate Bayesian statistical modelling, expert consensus building, etc;

- physical arguments, e.g. that, however uncertain, the input should remain positive or below a known threshold for physical reasons;

- indirect observations (this is the case when the model is calibrated/validated and may involve some inverse methods under uncertainty).

As will be illustrated in Part II, uncertainty modelling may be a resource-consuming step for data collection; it appears, however, to be a crucial step to which the results of the uncertainty study may prove very sensitive, depending on the final goal and the quantities of interest involved. For instance, the choice of upper bounds or distribution tails becomes very sensitive if the quantity of interest is an exceedance probability.

1.2.6   Propagation and sensitivity analysis processes

Once an uncertainty model has been developed, the computation of the quantity (or quantities) of interest involves the well-known *uncertainty propagation* step (also known as uncertainty analysis). The uncertainty propagation step is needed to transform the measure of uncertainty in the inputs into a measure of uncertainty in the outputs of the pre-existing model. In a probabilistic setting, this implies estimating the pdf of $\underline{z} = G(\underline{x}, \underline{d})$, knowing the pdf of $\underline{x}$ and being given values of $\underline{d}$,

$G(.)$ being a numerical model. According to the quantity of interest and the system model characteristics, it may be a more or less difficult numerical step involving a wide variety of methods, such as Monte Carlo Sampling, accelerated sampling techniques, simple quadratic sum of variances, FORM/SORM or derived reliability approximations, deterministic interval computations, etc. Prior to undertaking one of these propagation methods, it may also be desirable to develop a surrogate model (equally referred to as response surface or meta-model), i.e. to replace the pre-existing system model with another which produces comparable results with respect to the output variables and quantities of interest, but which is much quicker or easier to compute.

The *sensitivity analysis* step (or *importance ranking*) refers to the computation and analysis of so-called sensitivity or importance indices of the components of the uncertain input variables $\underline{x}$ with respect to a given quantity of interest in the output $\underline{z}$. In fact, this involves a propagation step, e.g. with sampling techniques, but also a post-treatment specific to the sensitivity indices considered. This typically involves some statistical treatment of the input/output relations which control quantities of interest involving the measure of uncertainty in both the outputs and inputs (see Part II). The large variety of probabilistic sensitivity indices includes, for instance, graphical methods (scatterplots, cobwebs), screening (Morris, sequential bifurcations), regression-based techniques (Pearson, Spearman, SRC, PRCC, PCC, PRCC, etc.), non-parametric statistics (Mann-Whitney test, Smirnov test, Kruskal-Wallis test), variance-based decomposition (FAST, Sobol', correlation ratios), or local sensitivity indices of exceedance probabilities (FORM).

Note that the expression 'sensitivity analysis' is taken here in its comprehensive meaning, as encountered in the specialized uncertainty and sensitivity literature; in industrial practice the same expression may refer more generally to certain elementary treatments, such as one-at-a-time variations of the inputs of a deterministic model or partial derivatives. These two kinds of indices are usually not suitable for a consistent importance ranking, although they may be a starting point.

Part II and Part III will discuss the practical challenges involved in undertaking these two steps, for which the choice of the most efficient methods has to be carefully made. It will be shown that **it does not depend on the specificities of a physical or industrial context** as such, but rather on the **generic features** identified above: the computational cost and regularity of the system model, the principal final goal, the quantities of interest involved, the dimensions of vectors $\underline{x}$ and $\underline{z}$, etc. This is one of the important messages in this book. Having historically been designed for certain physical applications (e.g. structural reliabilistic methods to compute uncertainties in mechanics and materials), some methods do not reach their full generic potential in industrial applications when adhering to unnecessary cultural norms.

1.3   The common conceptual framework

The consideration of key generic features and concepts requires that the schematic diagram of Figure 1.2 be developed into a full conceptual framework (see

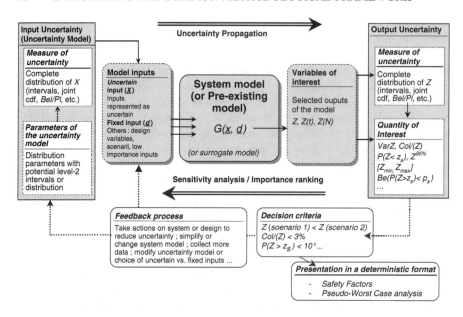

Figure 1.3 Common framework.

Figure 1.3, below) which will be used systematically throughout the case studies in Part II.

Note that the item 'presentation in a deterministic format' has been added. In industrial practice, after a more elaborate uncertainty assessment, there may be a sort of posterior process translating a probabilistic criterion into an easier-to-manipulate deterministic format, depending on the decision-making process and the operational constraints (e.g. set of partial safety factors, deterministic envelope of variation capable of guaranteeing the criteria, etc). This may be due to:

- cultural habits, when the regulation comes with more traditional deterministic margins;

- operational constraints which make it difficult to undertake systematically, on a large industrial scale, certain elaborate assessment processes;

- requirements of acceptability or facilitation of understanding.

The process may also feature when dealing with very low probabilities or highly unlikely uncertain results. Presentation in the form of mixed probabilistic/deterministic results proves, in fact, to be common in such cases, as will be demonstrated by the case studies in Part II and discussed more extensively in Chapter 19.

1.4   Using probabilistic frameworks in uncertainty quantification – preliminary comments

As already mentioned in the introduction, this book takes a practitioner's point of view on the use of mathematical settings in describing uncertainty, assuming firstly that the reader knows the basics of probability calculus, statistics and simulation methods. If necessary, textbooks such as (Bedford and Cooke, 2001; Aven, 2003; Granger Morgan and Henrion, 1990; Melchers, 1999; Rubinstein, 1981; Saltelli *et al.*, 2004) can be consulted for a refresh in risk analysis, uncertainty quantification, Monte Carlo and sensitivity analysis. The book will then focus on guiding the overall consistency and practical trade-offs to be made in choosing one of those methods in industrial practice.

Additionally, theoretical approaches of uncertainty involve a long-standing literature and significant on-going controversies, including those associated to the rationale and conditions of the use of deterministic, probabilistic (classical or Bayesian), or extra-probabilistic settings. Indeed, epistemological considerations are necessarily underlying any description of the uncertainty surrounding a system study; and there is a close link to make to decision theory paradigms (and particularly decision-making in risk analysis and management) in choosing a framework or setting to represent uncertainty. Existing regulations, standards or codes of practices do in fact refer more or less explicitly to a wide spectrum of different settings and interpretations dealing with uncertainty over the scope of the book, as will be illustrated in Part II. A very brief panel of interpretations will be introduced hereafter while the reading of useful references might be considered for deeper understanding both on founding theory and applied interpretations, including Savage, 1972; De Finetti, 1974; Kaplan and Garrick, 1981; Helton and Burmaster, 1996, and more recently, for example, Bedford and Cooke, 2001; Aven, 2003; Helton and Oberkampf, 2004.

1.4.1   Standard probabilistic setting and interpretations

Consider for instance what will be referred to as the *standard probabilistic* setting, whereby probability distributions are assigned to the components of the input $\underline{x}$ (more precisely a joint distribution on vector $\underline{x}$). Computations are then made on quantities of interest derived from the resulting distribution of the random vector $\underline{Z} = G(\underline{X}, \underline{d})$ or on sensitivity indices involving both $\underline{X}$ and $\underline{Z}$.

A first interpretation, say *frequentist* or *classical*, of that setting would consider $\underline{x}$ and $\underline{z}$ as observable realisations of uncertain (or variable) events (or properties of the system), which would occur several times independently so that, at least in theory, frequency records of both variables would allow the inference and validation of the probability distribution functions (for both inputs and output). In that context, modelling probabilistic distribution functions on the inputs may be seen as a basis for the inference of some output quantities of interest, such as a probability to exceed a regulatory threshold or an expected cost. Taking such bases for decision-making enjoys straightforward risk control interpretations and has the advantage of a potential validation through long-term observations, as daily practiced in environmental control, natural risk regulations or insurance records.

Other views, involving totally different interpretations, are however possible on the same mathematical setting (say standard probabilistic) to quantify uncertainty. A classical *subjective* interpretation of the same setting might lead to consider the probability distributions as a model of the decision-maker subjective preferences following a 'rational preference' set of axioms (such as in Savage, 1974), or degrees of belief without necessary reference to frequency observations of physical variables. A quantity of interest such as the expected utility of the output random vector $\underline{Z}$ may then be used in a decision-making process, enjoying solid decisional properties. This may not necessarily need validation by long-term observations, which, in fact, may often be impractical in industrial practice: think about such cases as the design choices of an industrial product that does not yet exist.

Alternatively, when considering global sensitivity analysis of complex physical or environmental system model in upstream model development stages, one may rely on a sort of *functional analysis* interpretation. Using probabilistic distributions on the vector of inputs $\underline{x}$ of a system and considering variance as a quantity of interest on the model output $\underline{z}$ enjoys some desirable numerical space-exploration or global averaging properties that allow well-defined sensitivity ranking procedures. However, inputs and outputs of such model may not be observable at all, as being rather abstract model parameters in upstream research processes or not corresponding to reproducible random experiments.

While being quite different, these competing interpretations of a standard probabilistic setting imply rather similar practical implementation features, as will appear later in the book, such as: the need to carefully specify the quantity of interest and select the uncertainty model with all information available; the use of numerical methods for propagation and sensitivity analysis with delicate compromise when addressing complex models G etc. Indeed, frequent uses are made of such a standard probabilistic setting in regulated practice such as metrology (ISO, 2005), pollutant discharge control or nuclear licensing without positively choosing a single theoretical interpretation.

1.4.2   More elaborate level-2 settings and interpretations

More elaborate interpretations and controversies come up when considering the issue of the lack of knowledge regarding the uncertainty description, as generated for instance by small data sets, discrepancies between experts or uncertainty in the system model. This is particularly the case in the field of risk analysis or reliability. The incorporation of observed data to infer or validate the probabilistic modelling of inputs $\underline{x}$, when done through classical statistical estimation, generates statistical fluctuation in the parameter estimates (such as the expectation as estimated from the empiric mean on a finite dataset of temperatures, or the failure rate from limited records of lifetimes); this is even more so when considering the choice of the distribution for an input x^i for which traditional hypothesis-testing techniques give at most only incomplete answers: is the Gaussian model appropriate for the distribution of x^i, as opposed to, for instance, a lognormal or beta model?

This results in a sort of 'uncertainty about the uncertainty (or probabilities)', or sometimes referred to as epistemic uncertainty about the aleatory characteristics, although this formulation is controversial in itself (see Chapter 14). Disagreement between experts (or hesitation of one expert) that are consulted to help building the uncertainty model may also be viewed as generating similar *level-2 uncertainty*, although this all depends on the way the so-called expertise *elicitation* procedure is organised and theoretically formulated (*cf.* Granger Morgan and Henrion, 1990; Cooke, 1991). A *Bayesian* setting further formalises this second probabilistic level by deliberately considering a pdf for the parameters of the pdf of an uncertain model input, representing prior (lack of) knowledge of the uncertainty model, or the 'posterior' situation after incorporation of observed data.

Various settings can be found in the literature to answer that tricky, although often inevitable, issue. Some authors may informally stick to a standard probabilistic (i.e. level 1 setting) using point estimates for the pdf parameters involved in the uncertainty model, assumed to represent the 'best estimates': the majority of the industrial case studies illustrated in Part II will evidence its practicality and pervasiveness, although seldom justified in its precise theoretical foundations. In some cases, a deterministic sensitivity analysis is undertaken on the parameters of the pdf that describes the uncertainty of a model input: let them vary in order to investigate the variation of the output quantity of interest, and possibly retain, for decision-making, its maximal value (e.g. a 'penalised' probability of exceeding a threshold, encountered in some nuclear safety studies), although the real decisional properties of such an approach is rarely discussed. Such a setting will be called *probabilistic with level-2 deterministic* in the rest of the book, as illustrated by one case study in Part II.

Probabilistic modelling of the pdf parameters of the inputs has become popular in the industry since the late 1980s (particularly in the nuclear field) and is often referred to as distinguishing epistemic and aleatory components (see in particular Helton, 1994; Helton and Burmaster, 1996; Apostolakis, 1999). Such setting will be called *double probabilistic* (instead of *probabilistic with level-2 probabilistic*); it generates not only point values for the output quantities of interest but entire distributions (or epistemic distributions of the aleatory pdf of the variable of interest). It may also be encountered in some natural risk regulations, whereby upper confidence interval estimates are preferred to the central estimate of a quantile (such as the 1000-year return intensity) although practices are not homogeneous. Note finally that recent pilot studies have introduced extra-probabilistic settings (such as DST, fuzzy logic, possibility theory etc.) to model either the level 2 on top of a probabilistic setting (i.e. a DST description of the uncertainty of the parameters of the input pdf), as will be illustrated in the frontier case study of Part II, or simply the overall uncertainty model (see Helton and Oberkampf, 2004).

Essential controversies regard the interpretation of those various settings, and even the meaning of the double probabilistic settings alone. Note firstly that, as an equivalent of the *functional analysis* interpretation mentioned earlier, one may practically consider a double probabilistic setting as just another standard probabilistic one; switch simply the definition of the pre-existing model from the one relating input to output values to an intrinsically probabilistic model relating input to output distribution parameters or probabilities. This is often the case with fault trees or event trees in reliability, where the simple deterministic functions resulting from the causal trees hide somehow the aleatory nature of the model, as illustrated by the frontier case study in Part II. Then, level-2 uncertainty analysis may be viewed as a space-exploring numerical approach in order to understand the underlying aleatory model.

In a so-called *probability of frequency approach*, which may be the most frequent approach supporting double probabilistic settings, a more formal interpretation is given. An underlying assumption is that there is a theoretically-unique uncertainty (or aleatory) model, with correct shapes of distributions for the inputs and point values for their parameters as well as true value for the quantities of interest (e.g. a true failure frequency or probability of exceeding a threshold), as could be theoretically evidenced through unlimited long-term frequency observations. However, dataset limitations or lack of expertise generate uncertainty upon those characteristics, so that a complete quantification of uncertainty includes elaborate results, such as a quantity of interest (e.g. failure frequency) and a level of confidence around it. A *classical* interpretation might handle this approach through the use of the distribution of the estimators (e.g. Gaussian distributions underlying maximal likelihood estimation confidence intervals) to model level-2 input uncertainty when data is available or more generally subjective degree of beliefs for other types of level-2 uncertainty. These level-2 distribution on parameters of the level-1 input uncertainty model are later propagated to issue level-2 estimation uncertainty of level-1 quantities of interest. A *Bayesian (or classical Bayesian)* interpretation handles it in a more formalised way by consolidating subjective preferences or expertise (incorporated in the prior uncertainty model) and observed data in a mathematical updating procedure of the level-2 pdf describing the level-1 pdf parameters. This may include mixture of distributions to account to a wider extent for the level-2 uncertainty.

A *fully Bayesian* or *predictive Bayesian* approach (as advocated by Aven, 2003) refers to a quite different interpretation. The focus is placed on observable quantities meaning the observable realisation of some physical (or economic) variables (or states, events) characterising the pre-existing system. As a simplified presentation of that approach in correspondence to the common methodology of the book, this means that inputs $\underline{x}$ and variables of interest $\underline{z}$ should correspond to (at least theoretically) observable variables (or states) of the system. The corresponding q.i, such as a system failure frequency, the expected value of the variable of interest or its probability to exceed a threshold, should be understood as decision-making intermediate tools rather than real properties of the system enjoying one true value. Hence, it is not relevant to distinguish within the results a level-2 uncertainty about the q.i. that embeds the level-1 uncertainty. Uncertainty should be eventually quantified within one single figure for the q.i., which corresponds to the expectation over the level-2 (posterior) distribution of the q.i.; or within the single distribution of the variable of interest, called the predictive distribution, which also corresponds to the level-1 distribution averaged over the level-2 uncertainty of its parameters. For instance, one would not consider a credibility interval for a system failure rate[1], but merely its expected value, consolidating all components of uncertainty into a single probabilistic figure. Note that, beyond significant differences of interpretation, it may be practically linked to the output of a Bayesian *probability of frequency approach* (e.g. taking the expectation over the level-2 distribution generated for the q.i.), provided it was specified along the same information basis on the inputs and definition of observable variables of interest.

1.5 Concluding remarks

The book does not suggest the choice of one interpretation or another recognising that, from a practical point of view, the implementation involves important common features, while of course the interpretation of the results may differ. As will be recalled in Chapter 14, the scientific literature presents a number of classifications regarding the nature of uncertainty (such as aleatory or epistemic uncertainty, variability, imprecision, etc). Some authors link the interpretation of the appropriate mathematical setting to the nature of uncertainty involved in the system studied. On this quite controversial issue the book will not propose a unique preferred interpretation; it will rather illustrate the various settings that may be found in industrial practice, in response to varying regulatory specifications and interpretations, as well as their practical implementation consequences (such as data collection, computing needs, etc). For instance, the previous section has evidenced that while there may be quite different competing interpretations, standard probabilistic and double probabilistic settings share practical implementation features, on which the book will focus. This may result in a certain incompleteness of specification of the precise words employed, such as the use of confidence intervals that would rather deserve the name of credibility intervals or prediction intervals in some cases according to the way the input uncertainty model may be theoretically interpreted and practically elicited. Once again, the reader may refer to the publications cited in the next section discussing theoretical foundations, notably on the use of probabilistic or non-probabilistic settings in representing uncertainty, and also to recent benchmarking or comparison exercises such as that reported in (Helton and Oberkampf, 2004).

---

[1] except possibly when working on a true population of similar systems running together.

References

Apostolakis, G. (1999) The distinction between aleatory and epistemic uncertainties is important; an example from the inclusion of aging effects in the PSA, in *Proceedings of PSA '99*, Washington DC.

Aven, T. (2003) *Foundations of Risk Analysis*, Chichester: John Wiley & Sons, Ltd.*

Barberà, S., Hammond, P.J. and Seidl, S. (1998) (Eds) *Handbook of Utility Theory, Volume 1 Principles*, Hingham: Kluwer Academic Publisher.

Barberà, S., Hammond, P.J. and Seidl, S. (1998) (Eds) *Handbook of Utility Theory, Volume 2 Extensions*, Hingham: Kluwer Academic Publisher.

Bedford, T. and Cooke, R. (2001) *Probabilistic Risk Analysis – Foundations and Methods*, Cambridge: Cambridge University Press.*

Cooke, R.M. (1991) *Experts in Uncertainty; Opinion and Subjective Probability in Science*, Oxford: Oxford University Press.

De Finetti, B. (1974) *Theory of Probability*, volumes I and II, New York: John Wiley & Sons, Ltd.

De Neufville, R. (2003) Architecting/Designing engineering Systems using Real Options, *MIT report* ESD-WP-2003-01.09 (http://esd.mit.edu).

De Neufville, R. (2004) Uncertainty management for engineering systems planning and design, *MIT Engineering Systems Monograph* (http://esd.mit.edu/symposium/pdfs/monograph/uncertainty.pdf).

Granger Morgan, M. and Henrion, M. (1990) *Uncertainty – A Guide to Dealing with Uncertainty in Quantitative Risk and Policy Analysis*, Cambridge: Cambridge University Press.*

Helton, J.C. (1994) Treatment of Uncertainty in Performance Assessments for Complex Systems, *Risk Analysis*, **14**, pp. 483–511.

Helton, J.C. and Burmaster, D.E. (1996) (eds.) Treatment of Aleatory and Epistemic Uncertainty, Special Issue of *Reliability Engineering & System Safety*, **54**(2–3).

Helton, J.C. and Oberkampf, W. (2004) (eds.) Alternative representations of epistemic uncertainty, Special Issue of *Reliability Engineering & System Safety*, **85**(1–3).

Helton, J.C., Cooke, R.M., McKay, M.D. and Saltelli, A. (2006) (Eds) Sensitivity Analysis of Model Output: SAMO 2004, Special Issue of *Reliability Engineering & System Safety*, **91**(10–11).

ISO (1995) *Guide to the expression of uncertainty in measurement (G.U.M.)*, European Pre-standard Env. 13005.

Kaplan, S., and Garrick, B.J. (1981) On The Quantitative Definition of Risk, *Risk Analysis*, **1**(1), 11–27.

Knight, F.H. (1921) *Risk, Uncertainty and Profit*, Hart, Schaffner & Marx.

Law, A.M. and Kelton, W.D. (2000) *Simulation Modelling and Analysis* (3rd edition), London: McGraw Hill.

Melchers, R.E (1999) *Structural reliability analysis and prediction* (2nd edition), Chichester: John Wiley & Sons, Ltd.*

Nilsen, T. and Aven, T. (2003) Models and model uncertainty in the context of risk analysis, *Reliability Engineering & System Safety*, **79**(309–317).

---

*Also suitable for beginners.

Oberkampf, W.L., DeLand, S.M., Rutherford, B.M., Diegert, K.V and Alvin, K.F (2002) Error and uncertainty in modelling and simulation, Special Issue of *Reliability Engineering & System Safety*, **75**(3), 333–357.

Paté-Cornell, M.E. (1996) Uncertainties in risk analysis; Six levels of treatment, *Reliability Engineering & System Safety*, **54**(2–3), 95–111.

Quiggin, J. (1982) A theory of anticipated utility, *Journal of Economic Behavior and Organization*, **3**, pp. 323–343.

Rubinstein, R.Y. (1981) *Simulation and the Monte-Carlo Method*, Chichester: John Wiley & Sons, Ltd.*

Saltelli, A., Tarantola, S., Campolongo, F. and Ratto, M. (2004) *Sensitivity Analysis in Practice: A Guide to Assessing Scientific Models*, Chichester: John Wiley & Sons, Ltd.*

Savage, L.H. (1954, 1972) *The Foundations of Statistics*, Dover: Dover Publication Inc.

---

*Also suitable for beginners.

2

Positioning of the case studies

2.1 Main study characteristics to be specified in line with the common framework

In order to clarify the generic characteristics of a quantitative uncertainty assessment, a number of key features have been summarized in Table 2.1. These will be systematically identified in the case studies in Part II.

2.2 Introducing the panel of case studies

Ten case studies were carried out in various contexts across Europe in the energy sector (nuclear, thermal power, oil), the aerospace and automotive industries, and in civil structures. They form the substance of Part II of the book. Apart from covering a range of industrial sectors, they are intended to illustrate the variety of study characteristics encountered in practice, such as (*cf.* Table 2.2):

- the phenomenological domain implied in the pre-existing model: mechanics, geosciences, radiology, electromagnetism, metrology, reliability engineering, engineering economics;

- the position in the industrial cycle and associated goals and decision criteria, ranging from upstream ranking of R&D priorities, where no explicit criteria exist and final goals are rather of type U or A, to more operationally regulated or decision contexts, with clear regulatory guidelines (type C) or internal decision rules (type S);

Uncertainty in Industrial Practice   Edited by E. de Rocquigny, N. Devictor and S. Tarantola,
© 2008 John Wiley & Sons, Ltd

Table 2.1   Table of study characteristics.

Final goal of the uncertainty study	One or more of the four main goals (Understand, Accredit, Select or Comply) – goals may change over the duration of the industrial process.
Variables of interest (v.i.)	Variables in which the uncertainty is to be quantified. For example, maximal dose over time, or full function of dose over the time period, or maximal dose + first time at which dose exceeds a level, etc.
Quantity of interest (q.i.)	A scalar quantity that summarizes mathematically the uncertainty in the variable of interest. For example, quantile, coefficient of variation, bilateral confidence interval, DST interval of the expectation, or capability (in robust design), etc.
Decision criterion (criteria)	If any, an explicit mathematical relation involving the quantity of interest and used in the decision-making process. For example, exceedance probability should be below 10^{-b}, coefficient of variation below x%, intervals of uncertainty should not overlap to ensure significantly more reliability, capability greater than 1,2, etc.
System model or pre-existing model	A numerical function describing the system considered, the output and input of which help to deal with the various stakes in a quantitative decision-making process. For example, deterministic slope stability calculation, final-element electromagnetic model, cash flow projection, etc.
Uncertainty setting	Mathematical paradigm chosen to represent uncertainty. For example, standard probabilistic framework, mixed deterministic-probabilistic, non-probabilistic such as Dempster-Shafer.
Model inputs, and uncertainty model developed	Inputs of the system model, either uncertain ($\underline{x}$) or fixed ($\underline{d}$). Various sources and types of uncertainty may be represented in $\underline{x}$, according to uncertainty setting. For example, point value for a temperature, material property as a 2-D field of values, etc.
Propagation method(s) chosen	Numerical methods to transform efficiently the input uncertainty model into the output measure of uncertainty, according to the quantity of interest selected.

Table 2.1 (*continued*)

	For example, Monte Carlo, Quasi-Monte Carlo, FORM, etc.
Sensitivity analysis method(s) chosen	Statistical/numerical methods to analyse the influence of the different uncertain inputs on the quantity of interest. For example, rank correlations, FORM sensitivity indices, Sobol' indices, etc.
Feedback process	Set of actions that can be taken after an uncertainty study, such as: change of design, simplification or calibration of models, reduction of uncertainty, collection of more data, etc: possible actions depend on the reducible or irreducible nature of uncertain inputs.

- more technically, the uncertainty setting and quantities of interest involved: these features strongly influence the relevance of the panel of uncertainty methods, as will be explained in Part III.

Most of the case studies refer to a more or less probabilistic setting, although with notable differences (level 1 or 2, mixed to a certain extent with deterministic approaches): this can be seen as a reflection of more recent industrial practice and regulatory guidance, although it could also be considered rather as a case for making further improvements, since in many sectors and domains standards and practices remain either informal or more implicitly deterministic.

Conversely, the last case study, concerned with Reliability Modelling in Early Design Stages, has a particular status. Although it was developed on a real example in the automotive industry, it constitutes a frontier case study in being based on an extension of the probabilistic framework related to Dempster-Shafer Theory (DST), briefly introduced in Appendix 3. While considerable theoretical literature exists on that paradigm, it still raises important controversies as compared to probabilistic setting in uncertainty quantification and lacks sufficient large-scale industrial record. It should not be considered as standard industrial practice.

It would appear that, notwithstanding the many variable features, the overall procedure in uncertainty treatment is quite consistent. These case studies also illustrate the many practical challenges and opportunities offered by real data situations and the actual availability of expertise: this constitutes the basis of the recommendations in Part III.

Table 2.2 classifies the case studies according to the phenomenological domain, industry branch and main features such as goals, and settings.

Table 2.2   Main features of case studies.

Name (including *shortened name*)	Industrial SECTOR and physical domain	Final goal and decision criteria	Setting and quantity of interest
CO_2 *emissions*: estimating uncertainties in practice for power plants	THERMAL POWER Metrology/ chemistry	Goal C (+ S & U) – European regulatory criterion (coefficient of variation)	Probabilistic with level-2 deterministic Relative uncertainty (multiple of coefficient of variation)
Hydrocarbon exploration: decision-support through uncertainty treatment	OIL Geosciences	Goal S (+ U and A)	Standard probabilistic Variances volume (Goal U) + downstream probabilities of being > threshold (economical benefit – Goal C)
Risk due to Personal Electronic Devices: carried out on radio-communication systems aboard aircraft (*Electromagnetic interferences in aircraft*)	AEROSPACE Electro-magnetism	Goals U and C – regulatory criteria in preparation	Standard probabilistic Exceedance probability
Safety assessment of a radioactive high-level *Nuclear waste repository* – comparison of dose and peak dose	NUCLEAR Geosciences (soil chemistry, hydrogeology)	Goal U – no formal criteria	Standard probabilistic Mean, variance, quantiles

Table 2.2   *(continued)*

Name (including *shortened name*)	Industrial SECTOR and physical domain	Final goal and decision criteria	Setting and quantity of interest
Airframe maintenance contracts	AEROSPACE Maintenance and engineering economics	Goal U – no formal criteria	Double probabilistic Variance of the upper bound
Uncertainty and reliability study of a creep law to assess the fuel cladding behaviour of PWR spent fuel assemblies during interim dry storage (*Spent nuclear fuel behaviour*)	NUCLEAR Material physics and mechanics	Goal U – no formal criteria	Standard probabilistic Variance and exceedance probability, histogram
Radioprotection and maintenance	NUCLEAR Nuclear radiology	Goal S (+ C and U)	Standard probabilistic Expectation, variance and 95% c.i. on each scenario. Discrete pdf of the optimal scenario
Partial safety factors and stability of river dykes (*Dyke reliability*)	CIVIL INFRA-STRUC-TURES Soil mechanics and civil engineering	Goal C – national regulatory criterion (threshold probability)	Standard probabilistic Exceedance probability (failure probability)
Probabilistic assessment of fatigue life – *Aircraft fatigue modelling*	AEROSPACE Material physics and mechanics	Goal C – engineering criterion (confidence interval)	Standard probabilistic Fractile (curves) 95%, 5%, 50%

(continued overleaf)

Table 2.2   (*continued*)

Name (including *shortened name*)	Industrial SECTOR and physical domain	Final goal and decision criteria	Setting and quantity of interest
Reliability Modelling *in early* design *stages* using the Dempster-Shafer Theory of Evidence	AUTOMO-TIVE Reliability engineering	Goal U (and C downstream)	Probabilistic with level-2 non-probabilistic (DST) Be/Pl. of the system P_f/hour < threshold

The following figure gathers the main technical elements that characterize an uncertainty study: the complexity of the pre-existing model, the complexity of the uncertainty setting and the quality of the available information. The difficulty and thus risk associated with the realization of a study increase with the complexity of the uncertainty setting and the pre-existing model. High quality information (data or expertise) may come at a substantial cost, which should increase with the complexity of the given pre-existing model and setting. A study begun in a complex uncertainty setting, with a complex pre-existing model, and combined with poor

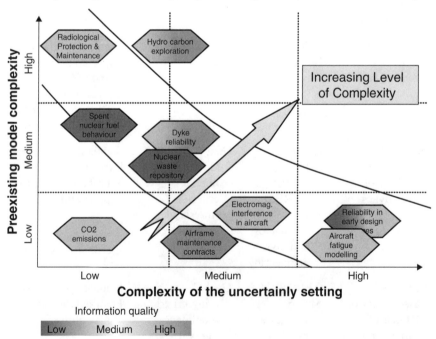

Figure 2.1 Case studies of the book, according to their complexity.

information on the input variables, may require a large investment of resources in order to achieve the goal of the study. The budget therefore has to be strategically allocated between those three axes.

2.3   Case study abstracts

This section presents all the case studies through short abstracts summarizing their context and specificities. A full description of each case study will be given in the individual chapters of Part II.

CO_2 emissions: estimating uncertainties in practice for power plants (Chapter 3)

The overall context is the CO_2 reduction planned within the Kyoto agreements: every power producer has to declare an annual level of emissions, and thereafter may trade on an emission permit market. The final goal of the mandatory uncertainty study is to quantify, within a declared and accountable process, the relative uncertainty in the annual emission measurement and to demonstrate compliance with a criterion (Goal C). This criterion specifies explicit target levels for a quantity of interest called the 'reference relative uncertainty' for two variables of interest: annual CO_2 emissions and declared fuel consumption. The two variables of interest are predicted through pre-existing models that represent analytically the alternative metrological chains involved.

For this kind of metrological exercise, a classical uncertainty methodology is derived from the Guide to the expression of Uncertainty in Measurement (GUM) through simple and accountable steps: sources of uncertainty, mainly metrological in nature, are modelled, principally through expert judgement and statistical estimates from calibration/supplier data; propagation is carried out on the analytical pre-existing models through rather simple methods (Monte Carlo simulation or quadratic variance approximation); ranking of the importance of uncertainty sources through rank correlation is an essential sensitivity analysis step to optimize the choice of a metrological chain. The main feed-back processes, after a first uncertainty and sensitivity study, are the selection of the most robust existing metrological chain, and potentially the subsequent re-calibration or up-grading of the most critical measurement components to reduce uncertainty.

Hydrocarbon exploration: decision-support through uncertainty treatment (Chapter 4)

The evaluation of risk in hydrocarbon exploration, which is related to the probability of the existence of a hydrocarbon reservoir, is a fundamental component of any drilling decision in the oil industry. The main goal of the risk evaluation, i.e. of uncertainty treatment, is to find traps where oil/gas has accumulated and been retained in greater volumes than in other traps in order to optimize the drilling decision (Goal S). Supporting goals are the better understanding of the model and identification of the main contributors to uncertainty (Goal U), as well as model simplification (Goal A).

The two fundamental components of the pre-existing model involved in oil exploration are: Basin Modelling (BM), which gives a 4-D description (in space and time) of the basin status and evolution, and Petroleum System Modelling (PSM), which produces the history of the geological processes that led to generation and accumulation of hydrocarbons in the current traps. In this case study, both BM and PSM are compounded into a single coherent uncertainty and sensitivity framework, under a standard probabilistic setting. Uncertainty propagation involves Quasi-Monte Carlo sampling. Global sensitivity analysis is carried out using state dependent parameter modelling (SDP) and by applying the method of Sobol' to a neural network-based surrogate model. Finally, empirical distributions of oil and gas volumes for each trap of the basin are estimated through Monte Carlo analysis.

Determination of the risk due to personal electronic devices (PEDs), carried out on radio-navigation systems aboard aircraft (Chapter 5)

The advance of commercial electronics has led to an increasing use of Personal Electronic Devices (PEDs) aboard aircraft. Such devices represent a new and evolving source of electromagnetic radiation which carry a risk of interference with onboard electronic equipment. Different international aeronautical committees are currently working on this subject (EUROCAE WG 58 and RTCA SC-202). Several scenarios of electromagnetic coupling phenomena have been identified and described by these committees. One such scenario is analysed in the case study: the possible effects that spurious emissions from PEDs (mainly laptops and mobile phones) may have on radio navigation systems through radiation in the ILS (Instrumental Landing System) band (108 MHz – 112 MHz). Different stakes arise:

(i) sources of electromagnetic emissions (PEDs) are controlled by neither the airlines nor the aircraft manufacturers; therefore, their control in terms of the power realistically emitted from their initial frequency band of use is very difficult;

(ii) very little data is available on their behaviour out of their initial frequency band of use;

(iii) the models representing the electromagnetic coupling (installation parameters, definition of model (granularity)) are representative but not always very accurate;

(iv) the physical value which makes it possible to detect a functional adverse effect is represented by the susceptibility of the antenna, but this value is by nature very difficult to determine;

(v) the scenario describing the (*a priori* uncertain) number of PEDs to be taken into account is an agreement between various people.

The main goal of this case study, which is inspired by safety considerations, is to define indicators that could drive future research studies (Goal U). A model-based approach is used to represent the global electromagnetic problem. The sources of uncertainty are modelled with probability density functions. Propagation of uncertainties is undertaken under several assumptions with the First-Order Reliability Method (FORM), coupled with a Monte Carlo approach for validation purposes. Finally, a sensitivity analysis is performed to rank uncertainty sources in order of importance and thus to identify which model inputs would need to be improved in order to obtain more accurate model predictions.

Safety assessment of a radioactive high-level waste repository – comparison of dose and peak dose (Chapter 6)

A Performance Assessment (PA) aims to demonstrate the safety of a radioactive High-Level Waste Repository (HLW). Developing a PA for an HLW repository requires the design of a system model and the implementation of the corresponding computer code representing the release of pollutants from the facility, their transport through the host geological medium and their spread over the biosphere. All scenarios which could jeopardize the safety of the repository, as well as the uncertainty associated with all model inputs, are considered in the calculation of the eventual dose received by the exposed population.

The dose over time is the main variable of interest in most studies performed so far, though other ancillary variables of interest, such as the peak dose over time, could also be considered. The main goal of this case study is to identify the most relevant uncertain inputs and to improve the understanding of the model (Goal U). Some aspects of the outcomes obtained for both variables of interest are compared: the estimability of quantities of interest and the applicability of different sensitivity analysis techniques (variance decomposition, regression techniques and graphical methods). Several quantities of interest are considered for each variable of interest: the mean, the variance, several quantiles (50%, 95% and 99%) and the associated confidence intervals.

A cash flow statistical model for airframe accessory maintenance contracts (Chapter 7)

This case study deals with a cash flow statistical model for airframe engine maintenance contracts: the final goal is to understand the relative importance of uncertain input factors (Goal U). The pre-existing model plots the contract seller's cash-flow forecast's cumulative curve during the contract period in the mean, $Z(t)$, and with confidence percentiles. These are the variables of interest. The model inputs for the sensitivity analysis are related to logistic support, supply chain costs, and other accessory reliability inputs. This latter kind of uncertainty is of level 2, since the pre-existing model is already intrinsically probabilistic in its reliability component. The quantity of interest used for sensitivity analysis is the variance of confidence percentiles. The final goal is to rank the importance of the uncertain inputs. This is important for the feedback decision criterion in logistic and supply chain management.

For this purpose, two sensitivity analysis methods are used: a linear adjusted model based on complete two-level design of experiments (DoE), and the estimation of Sobol' sensitivity indices by means of Monte Carlo simulations. The two approaches produce the same results on specific applications. The linear model happens to be sufficient to determine the main uncertainty sources for cash flow confidence interval variability with much more practicable computation times. The order of magnitude of the quantity of interest can be computed analytically through the linear model, which can therefore be seen as a surrogate model. The application to a maintenance contract for a fleet of electronic control units shows that predominant impact rankings correspond to the turnaround time and the mean time between unscheduled removals.

Uncertainty and reliability study of a creep law to assess the fuel cladding behaviour of PWR spent fuel assemblies during interim dry storage (Chapter 8)

In the 1990s a research project called PRECCI (French acronym for 'Research Programme on the long-term Evolution of the Irradiated Fuel Assemblies') was launched, with the aim of predicting the long-term evolution of spent fuel in storage and disposal conditions and the potential release of radionuclides in nominal and incidental scenarios. Part of the study focussed on the determination of the rate of rod failures according to time under dry storage conditions. Indeed, during dry storage, fuel cladding undergoes a deformation by creep under the effect of the internal pressure and temperature, which can lead to a cladding rupture and consequently a release of radionuclides in the container.

The aim of the case study is to illustrate the interest in uncertainty and sensitivity analysis in the framework of a research project, especially to identify the main sources of uncertainty which merit further R&D efforts (Goal U). The focus is on methodology, since the approach is being applied to more up-to-date and complete models in industry. The study is based on the three following parts: statistical analysis of model responses based on Monte Carlo simulations, sensitivity analysis of the variables of interest according to uncertain model inputs, and reliability analysis, in order to determine which uncertainty sources need to be better understood or controlled.

Radiological protection and maintenance (Chapter 9)

Maintenance actions in nuclear power plants may expose operators to ionizing radiation. At Electricité de France (EDF), the choice of the optimal radioprotection solution depends on the results of a piece of numerical simulation software called PANTHERE. Nevertheless, dose estimates are affected by numerous sources of uncertainty, which put in question the robustness of the final decision. A probabilistic framework is used in this exploratory case study to model the uncertainty of PANTHERE inputs and to assess the resulting uncertainty in the dose estimate. The uncertainty model relies mainly on expert judgement, but an inverse method is also implemented in order to benefit from dose rate measurements carried out in the field. Monte Carlo simulations are then used to obtain an estimate of the uncertainty of PANTHERE outputs and to choose the optimal radioprotection scenario, with a significant but acceptable CPU cost (a few hundreds of runs of the model). The major sources of uncertainty are also identified by correlation analysis.

Partial safety factors to deal with uncertainties in slope stability of river dykes (Chapter 10)

Dyke safety is a crucial issue for flood protection. It involves designing engineering structures to cover uncertain threats. Slope instability of the inner slope is one of the failure mechanisms of river dykes. The goal of this case study is to give a first derivation of partial safety factors, ensuring dyke reliability through compliance with a threshold defined with respect to all the uncertainties involved. The first step is to identify all sources of uncertainty. The safety format defines which sources are treated as uncertain and which as deterministic. The cohesion, friction angle, water level and phreatic line are considered as uncertain inputs in the model. The second step is to calculate the partial safety factors, involving uncertainty propagation.

A two-step strategy is applied for the uncertainty analysis. FORM analyses and FORM corrections are carried out to propagate the uncertainties. These results are used for the downstream derivation of the partial safety factors. The final result of this study is a set of partial safety factors with respect to slope stability of the inner slope of river dykes. The partial safety factors are in fact the transcription of a probabilistic uncertainty treatment presented in user-friendly format to avoid problems with the implementation. This study is a first step towards deriving a theoretical background for the partial safety factors. Additional research would be necessary to derive an optimal set of partial safety factors for Dutch dyke systems.

Probabilistic assessment of fatigue life (Chapter 11)

The general purpose of the study is to secure the mechanical integrity of an aeronautical structure through compliance with fatigue limits (Goal C). Current industrial practice involves stress *vs.* life curves (the so-called SN curves). These materialize the fatigue limits used in conjunction with the safety factors which account for the scatter inherent in fatigue experiments. The outcomes of the study are twofold. The first is to evaluate the ability of stochastic fatigue criteria to predict scatter at a structural level, based on the material scatter.

In the field of classical fatigue, it is customary to assume, through the use of fatigue criteria, that this prediction is valid for the average behaviour. It is of interest to confirm whether this assumption is also valid for the scatter in a stochastic model. The fatigue model considered in this study was developed by EADS Innovation Works, and has been shown to provide relatively robust estimates of the mean fatigue life of various coupon geometries, but the deterministic criterion needs further validation. The structural configuration under study is a specimen with a residual dent, modelled with 3-D finite elements. The second outcome of the study is to generate, numerically, pSN curves with confidence intervals (95% or 99% probability with 95% confidence) for structures based on coupons data, in order to provide designers with a fair estimate of the expected scatter on structures early in the design process.

At the material level, a single uncertainty source is considered (the 'material fatigue resistance' of the material β). The cumulative density function (cdf) of the uncertain input is evaluated by a procedure based on statistical estimation. Manufacturing, geometry and loads are implicitly assumed consistent from test to test. This is debatable but achievable to the first order if (great) care is taken. Additional uncertainty sources should be introduced in order to be able properly to characterize complex specimens (which for simplicity's sake is not done here, even for the structural specimen). Model uncertainty is not explicitly taken into account. The confidence in the fatigue criterion is based on comparisons with (average) experimental data. A Monte Carlo approach is used in the first instance to propagate the uncertainties. As the computation time for a full 3-D finite element model is expected to be too long (especially for a non-linear problem), a cheaper yet sufficiently accurate approach has been devised.

Reliability modelling in early design stages using the Dempster-Shafer Theory of Evidence (Chapter 12) – a frontier case study using an extra-probabilistic setting

The Dempster-Shafer Theory of Evidence (DST) enjoys increasing research interest and pilot industrial applications as an alternative to probabilistic modelling. In early design stages, where expert estimates of a system's component failure probabilities replace field data, the extent of ignorance or expert disagreement may be very large, and very few accurate test results may be available. Therefore, a 'conservative' uncertainty treatment, in the sense that the full extent of ignorance is expressed on the output, is interesting to support reliable and safe design. DST, merging interval-based and probabilistic uncertainty modelling, provides interesting functions for the representation, aggregation and propagation of uncertainty from model inputs to model outputs. Monte Carlo sampling is used for the propagation of uncertainty through a fault tree model of the system. The system investigated in this *research* case study, an automatic transmission from the ZF AS Tronic series, is still in the development stage. Aggregation of expert estimates and propagation of the resulting mass function through the system model are undertaken. The results are used to predict whether the system complies with a given target failure measure (Goal C).

Part II
Case Studies

3

CO_2 emissions: estimating uncertainties in practice for power plants

3.1 Introduction and study context

The overall context of the study is the CO_2 reduction planned within the Kyoto agreement; every power producer has to declare an annual level of emissions and thereafter may trade on an emissions permit market. The goal of the mandatory uncertainty study is to quantify precisely, within an explicit and accountable process, the relative uncertainty in the annual emission measurements. Note that this type of study is concerned solely with industrial emission uncertainty, which is, of course, a very limited aspect of the broad issue of uncertainty in climate change.

The first European CO_2 emissions allowance trading system (ETS) started on 1 January 2005, as established by Directive 2003/87/EC of the European Parliament (EC, 2003) and of the Council of 13 October 2003 (referred to below as 'the Directive'). CO_2 is by far the most important Greenhouse Gas (GHG), accounting for 82% of total EU GHG emissions in 2002. The energy industries (public electricity and heat production) generate around 29% of total CO_2 emissions in the EU (based on data from 2002, see EEA 2004). The EC Decision of 29 January 2004 (EC, 2004, referred to below as the 'EU Guidelines') set out detailed criteria for the monitoring, reporting and verification of CO_2 emissions resulting from various activities, such as combustion installations; specifications were also contained in earlier documents such as IPCC 1996, GHG 2003 and WBSD-WRI, 2003. The EU Guidelines dictate that the only data which operators have to determine with a specific uncertainty target in the case of calculation-based

Uncertainty in Industrial Practice   Edited by E. de Rocquigny, N. Devictor and S. Tarantola,
© 2008 John Wiley & Sons, Ltd

methodology is *fuel consumption metering*. However, it is also specified that *'the operator shall have an understanding of the impact of uncertainty on the overall accuracy of his reported emission data'* and that *'the operator shall manage and reduce the remaining uncertainties of the emissions data in his emissions report'*. Note that this concerns the CO$_2$ emissions inventory at site level. Greenhouse gas emission inventories at the scale of entire industrial sectors or countries involve other specific approaches (see e.g. Salway, 2002) for the estimation of uncertainties as required by the United Nations Framework Convention on Climate Change (UNFCC) or the Kyoto protocol.

In practice, uncertainty information has to be reported by an operator for two variables of interest: annual fuel consumption and annual CO$_2$ emissions. The guidelines set target values for the *relative uncertainty*. Those targets depend on the order of magnitude of the emissions and on the type of fuel involved. For instance, if emissions are larger than 500 kt/yr, which is the case for large thermal power plants, the target for the relative uncertainty in 2006 was a maximum of ±2.5% of annual fuel consumption if natural gas or oil was burnt, and ±3% for coal. The quantity of interest referred to as *relative uncertainty* can be defined either as twice the coefficient of variation of the distribution of the variable of interest, or as the ratio of half a 95% confidence interval to the expected value of the same distribution (see Section 3.3.1): both definitions are considered equally acceptable in the guidelines.

This chapter presents an example of CO$_2$ emission uncertainty estimation as practised on a 2*600 MWe pulverised coal-fired plant at *Electricité de France* (EDF). Only CO$_2$ resulting from coal combustion is considered.

3.2   The study model and methodology

3.2.1   Three metrological options: common features in the pre-existing models

Direct measurement of CO$_2$ emissions is not recommended by the guidelines, as it may be quite unstable. There are three practical metrological options available, each of them involving a number of individual measurements and an analytical combination of the results to estimate both fuel consumption and the associated CO$_2$ emissions:

- Measurement by coal weighing before burning

- Mass balance (annual inventories of coal)

- Heat rate (followed by inversion of the physical model)

Hence, there are three corresponding pre-existing deterministic models: each needs specific uncertainty treatment. All three are based on analytical formulae to aggregate the alternative metrological options leading from individual measurement results to the variables of interest (see Figure 3.1 for details). Annual fuel consumption Q_{fuel} is evaluated first, but differently according to the three metrological

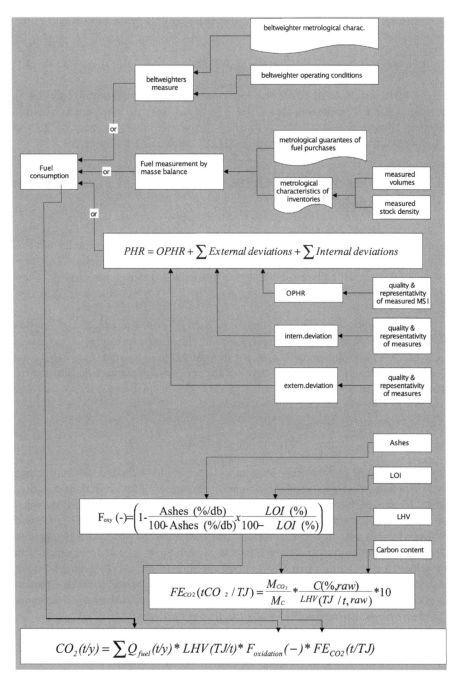

Figure 3.1 The three alternative metrological chains.

possibilities. Then CO_2 emissions are estimated according to the following formula, which is common to the three models:

$$CO_2(t/y) = \sum Q_{fuel}(t/y)^* LHV(TJ/t)^* F_{oxidation}(-)^* FE_{CO2}(t/TJ) \qquad (3.1)$$

where *LHV* stands for the fuel raw low heat value in TJ/t (TeraJoules per tonn), FE_{CO2} is the CO_2 emission factor and $F_{oxidation}$ is the oxidation factor. This last factor is computed as a simplified function of the original ash content of the coal (*Ashes*) and the rate of unburned carbon (*LOI*) as detailed in Figure 3.1.

3.2.2   Differentiating elements of the fuel consumption models

In the *first option* fuel consumption is directly metered with a *beltweigher* with integrator. There is no model as such, and the fuel consumption uncertainty is directly dependent on the operating characteristics of the beltweigher. This type of device is commercially available with a guaranteed metering uncertainty of less than ±0.5%. However, this uncertainty generally depends on the operating load of the beltweigher.

In the *second option* a *mass balance approach* is used. The annual consumption is calculated from the purchased fuel, the difference in the quantities of stock held over the year, and 'other fuel', used for transportation or re-sold. Coal stocks are determined through a volumetric survey of the coal stockpiles and a density survey. The model equation follows:

$$Q_{combusted} = Q_{purchased} + (stocks_{start} - stocks_{end}) - Q_{other} \; (in \; tons)$$

$$stock_{date}(t) = volume \; (m^3)^* density \; (t/m^3) \qquad (3.2)$$

The purchased fuel metering uncertainty depends on the type of metering device used. For most devices, suppliers can easily guarantee an uncertainty of less than ±2%. Moreover, the quality control of these devices is often already implemented on site, since they are used for the purchase of the fuel. Note that the purchased fuel quantity is usually more accurately metered than the stocked fuel quantity. Therefore, the minimum uncertainty associated with the mass balance approach can be achieved with a minimum fuel stock at the time of the surveys, namely at the beginning and the end of the year. In the *third option* the annual fuel consumption is calculated from the plant's gross *heat rate* (*PHR*) and electric gross power output (*W*, which is measured directly) using the following formula:

$$Q_{combusted}(t) = \frac{W \, (MWh)^* PHR(kJ/kWh)}{HHV(kJ/kg)} \qquad (3.3)$$

where *HHV* is the mean high heating value of the fuels burned during the period, weighted by the consumption of each of these fuels. *PHR* is the mean plant gross heat rate during the period, weighted by the electric gross power output. In practice the *PHR* is determined monthly through the following formula (determination of heat rate by efficiency losses):

$$PHR = OPHR + \sum External\ deviations + \sum Internal\ deviations \quad (3.4)$$

The Optimal Plant Heat Rate (*OPHR*) was determined at the time of plant commissioning. The external deviations include the efficiency losses due to the external operating conditions, such as fuel characteristics including hydrogen content, cold source temperature, and atmospheric conditions. Internal deviations include efficiency losses due to the plant components or the operators, such as damage or ageing of the components, burner adjustments or oxygen level in the flue gas. A more detailed analytical model relates external and internal deviations to a number of individual measurements for which sources of uncertainty can eventually be estimated.

3.3 Underlying framework of the uncertainty study

3.3.1 Specification of the uncertainty study

The final goal of this case study is primarily the demonstration of *compliance* with *uncertainty criteria* embodied by target values in the guidelines (Goal C-COMPLY). Three different metrological chains are compared to investigate which complies best with the criteria (Goal S-SELECT). Through the establishment of industrial emission practices, it also appears that *understanding the importance* of the various sources of uncertainty will become even more important in improving the metrological options in the long term (Goal U-UNDERSTAND).

Two variables of interest are considered (see Figure 3.2): annual fuel consumption Q_{fuel} and annual CO$_2$ emissions t_{CO2} (both in tons per year). The target values specified by the European/national regulations can be seen as *uncertainty decision criteria*: a *quantity of interest* representing the 'relative uncertainty' is compared to a maximal percentage of relative uncertainty. The quantity of interest 'relative uncertainty' can be understood either as a relative 95% confidence interval or as twice the coefficient of variation, which are identical under Gaussian assumptions, but assumed to be acceptably similar otherwise. Mathematically:

$$\%unc_z = \frac{1}{2E(Z)}(z^{97.5\%} - z^{2.5\%})$$

$$\%unc_z \approx k\frac{\sqrt{var(Z)}}{E(Z)} \quad (3.5)$$

where $z^\beta = F_Z^{-1}(\beta)$ stands for the β-th quantile of a random variable Z, and k is about equal to 2. The overall uncertainty *setting* is *probabilistic* for all sources

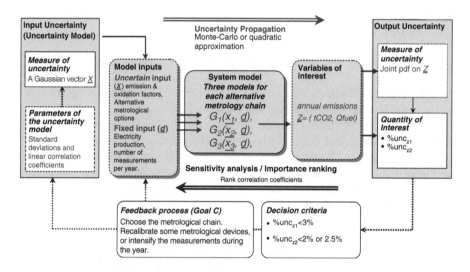

Figure 3.2 Overall uncertainty problem in the common framework.

of uncertainty, however heterogeneous their natures: measurement errors, intrinsic coal variability or variability of the atmospheric conditions, lack of knowledge of some operational conditions or physical characteristics, etc.

However, the imprecisely known standard deviations of the distributions of some model inputs are treated deterministically, with the application of an enlargement factor. Indeed, according to the international standard in metrological uncertainty, GUM (Guide to the expression of Uncertainty in Measurement, ISO, 1995), an 'uncertainty enlargement factor' k has to be applied to the 'raw' standard deviation estimates taken to model the sources prior to propagation: this is a simple and classical way to account for some of the epistemic uncertainty affecting the estimation of measurement deviations in a limited sample/information. The setting is therefore essentially a standard probabilistic one, supplemented however by a *level-2 deterministic* treatment of some sources of uncertainty (which are penalized with an enlargement factor). Note finally that *model uncertainty* is not very relevant to this example owing to the basic cumulative approaches that underlie the physical phenomena involved in the three metrological chains.

3.3.2   Description and modelling of the sources of uncertainty

Sources of uncertainty vary in number according to the three competing metrological options and corresponding measurement aggregation models, including six uncertain model inputs for the common terms (LHV, emission and oxidation factors): in total, the number of uncertain inputs varies from fewer than ten (beltweigher option) to a few dozen (heat rate option), summed over 12 monthly measurements for some of them.

• Simplified practices, according to the GUM metrological standard

Owing to the metrological nature of the exercise, practices recommended by the GUM standard are implemented. The general simplified assumption is that of *non-biased, normal distributions*, since all the uncertainties pertain to the metrological domain, for which this simplified assumption is ubiquitous (and indispensable due to the low budget of this type of study).

However, for some uncertain inputs, uniform distributions are chosen because, for physical or operational reasons, their domain is definitively bounded: a uniform distribution is chosen if there is no reason to think that some part of the domain is more probable (triangular distributions are an alternative if there is a reasonable expectation of greater likelihood somewhere in the uncertainty domain).

If data are not directly available, standard deviations are estimated through both statistical analysis and informal expert judgement; expert judgement is also employed to estimate the bounds of uniform distributions or rank correlation coefficients between some uncertain inputs reckoned to be significantly dependent. In many cases, this information comes from supplier data on each metrological sensor.

Rank correlations, elicited through expert judgement, are taken to model the most important dependencies, essentially month-to-month measurement error correlations when calibration is not repeated at the end of each month. Consider, for instance, the source of uncertainty associated with the emission factor FE_{CO2}. The operators are compelled to use accepted national or international standards for the sampling and analysis of the fuels. For one specific type of coal, uncertainty is estimated by conventional expert judgement to be ±2.3%: however, this value has to be adjusted to the variability associated with accounting for the different qualities of the fuels burned and stocked on site. The mean CO$_2$ emission factor of the coals burned in the plant during the year is therefore calculated from the coals' individual emission factors, weighted by their consumptions. The statistical analysis of the coals delivered for the last three years shows that the emission factors of these fuels are in the range of 90.3 to 97.4 tCO_2/TJ with a mean value of 94.55 tCO_2/TJ. If the plant has a fuel management system which is perfectly effective (best case), the uncertainty attached to the mean emission factor is the same as that attached to the fuel analysis, namely ±2.3%. If the plant has no fuel management system (worst case), a uniform distribution is assumed for the coal emission factor, between minimal and maximal values. The corresponding uncertainty attached to the mean emission factor (combined with the uncertainty attached to the coal analysis) is then estimated to be ±4.9%.

• Validity and practical difficulties

As a general rule, it should be noted that the relatively low level of sophistication of these practices is explained by the type of industrial study under consideration.

The quantity of interest and decision criteria involve essentially a measure of the central variability of the variable of interest and not an exceedance probability: consequently, the shape of the distribution of the uncertain inputs and of the variable of interest is less of an issue. The stakes involved are thus moderate: focus is placed on guaranteeing accuracy in the declaration of the average emissions, not on safety-driven considerations that would include threshold exceedance phenomena.

Naturally, both the choice of Gaussian or uniform distributions and the precise values of their parameters can be improved through additional operational practices and through the improved qualification of measurement chains, especially regarding the sources of uncertainty with the highest importance ranks.

Even within this simple framework, some difficulties are encountered in practice. Each model/data chain involves a large number of individual metrological steps, for which information on the sources of uncertainty is quite poor (mainly expert judgement, often coming from supplier data). Besides this, sensor representativity uncertainty is generally hard to estimate, and in real situations outside the supplier laboratory benchmarks a large number of poorly known influential physical phenomena (such as outside temperature, electro-magnetic noise, dust, etc.) can affect metrological performance, so that supplier estimates for standard deviation may sometimes be underestimated.

Secondly, the assignment of the deterministic enlargement coefficient k to quantify the second-level epistemic uncertainty proves tricky and, ultimately, quite sensitive to the results of the analysis. The difficulty in assigning such enlargement coefficients is due, in practice, to the rare availability of original measurement deviation statistics; this information is often kept by the supplier of the measurement equipment. Interpreting the underlying qualification process is thus difficult for the plant operator: in some cases, it is quite unclear whether the elementary figures for sensor uncertainty already include enlargement factors of $k = 2, 3$, or not.

3.3.3 Uncertainty propagation and sensitivity analysis

Uncertainty is propagated from input to output either by standard Monte Carlo sampling (MCS, Rubinstein, 1981) or by using the Taylor quadratic approximation for the estimation of the output variance (see the following formula):

$$E(Z) \approx G(E(\underline{X}))$$

$$Var Z \approx \sum_{i=1}^{p} \left(\frac{\partial G}{\partial X^i} \right)^2 Var X^i + \sum_{i1 \neq i2}^{p} \left(\frac{\partial G}{\partial X^{i1}} \right) \left(\frac{\partial G}{\partial X^{i2}} \right) \cdot Cov(X^{i1}, X^{i2}) \quad (3.6)$$

In fact, although MCS does not represent a computing challenge because of the simplicity of the model, the Taylor quadratic approximation (see ISO, 1995), enabling the prediction of the coefficient of variation, is quite an attractive option for the routine calculations that have to be performed annually by each power plant through dedicated spreadsheets, beyond the initial reference studies.

As mentioned above, both quantities of interest for the assessment of output uncertainty (i.e. twice the coefficient of variation of the output distribution or a relative half-95% confidence interval around the output expectation) are acceptable to the regulator. To estimate the output distribution, and the related quantities of interest, some 20000 Monte Carlo trials were performed (see Figure 3.3). The results show a distribution of the output that can be viewed as close to Gaussian (see, for example, skewness and kurtosis, close to 0 and 3 respectively). Both formulations of the quantity of interest lead to similar results. Propagation-induced uncertainty, i.e. inaccuracy in predicting a given quantity of interest, was controlled with repeated series of 20000 trials and compared against the Taylor approximation, which showed limited differences in both cases.

For the three options computed, accuracy in predicting the quantity of interest with the approximated Taylor propagation method is at the level of a few percents, which is deemed quite acceptable for this kind of study. Moreover, higher accuracy would make only limited sense, since the accuracy of the (input) uncertainty model would then become much more important.

Using Monte Carlo sampling, sensitivity measures can be obtained by estimating rank correlations between sampled values x of the uncertain model inputs and the corresponding values of the variable of interest z (Saltelli *et al.*, 2000).

Statistics	Value
Trials	20 000
Mean	3 683 906 428
Median	3 683 975 312
Standard	
Deviation	51 644 341
Variance	2,67E+15
Skewness	0.03
Kurtosis	3.02

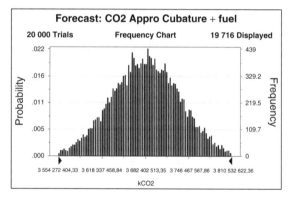

Figure 3.3 Example of CO₂ uncertainty analysis results using standard Monte Carlo sampling.

If the Taylor quadratic approximation is used to estimate the output variance, the following measure of sensitivity can also be used:

$$S_i = \left(\frac{\partial G}{\partial X^i} \right)^2 \frac{var\, X^i}{var\, Z} \tag{3.7}$$

The sensitivity results for the main uncertain input are comparable between rank correlations and the S_i. In the case that the decision criteria are not fulfilled, the sensitivity analysis helps the analyst to identify the most important inputs, on which more knowledge would be needed in order to reduce the uncertainty of the output to acceptable levels.

3.3.4   Feedback process

The main feedback process downstream of the uncertainty treatment is the selection of the best measurement chain, taking into account both that the decision criteria for the two quantities of interest have been fulfilled and that not too much importance has been attached to less controllable uncertainty sources. Besides this, a decision to upgrade the measurement chain could be taken in order to reduce the output uncertainty. This can be achieved either by re-calibrating the chain more frequently or even by acting on some of the most influential inputs, with a standard cost-benefit approach. Table 3.1 below summarizes the study characteristics.

3.4   Practical implementation and results

Table 3.2, below, gives the main results. Regarding the *first metrological option*, fuel consumption uncertainty is related to the operating characteristics of the beltweighers, for which commercial devices come up with theoretically guaranteed metering uncertainty of less than ±0.5%. However, this uncertainty generally depends on the operating load of the beltweigher, which may be outside the optimal range. In the case of the power plant considered, the metering uncertainty could be less than the target ±2.5% only for a beltweigher load greater than 8%, corresponding to a coal flow rate of 200 t/h. This condition is not consistently fulfilled in practice, resulting in an estimated uncertainty of ±3.3%, based on the observed operating conditions of the beltweigher during the year 2003. This method is therefore not recommended for the power plant to determine its annual CO$_2$ emissions.

Regarding the *second option* (fuel mass balance approach), the stock survey is guaranteed with an uncertainty of less than 5% and the coal stock represented approximately 18% of the fuel purchased in 2003. Note also that fuel purchased has less uncertainty due to fair shipping and supplier control. The resulting overall uncertainty associated with the fuel consumption is ±2.0%. This metrological option is preferred, both because it complies with the target, and because of its relative simplicity.

Table 3.1   Main study characteristics.

Final goal of the uncertainty study	Goal C – compliance with a threshold Some ancillary goals, the importance ranking of the sources (goal U) and the optimization of the metrological chain (goal S)
Variables of interest	Vector: annual CO$_2$ emissions and fuel consumption
Quantity of interest	Relative uncertainty (multiple of coefficient of variation or relative half-95% confidence interval around the expectation)
Decision criterion	Relative uncertainty below specified x%
Pre-existing model	Three analytical models representing the alternative metrological chains
Uncertainty setting	Probabilistic setting with level-2 deterministic treatment
Model inputs and uncertainty model developed	Metrological sources: metrological error or variability per component of the metrological chains. Gaussian distributions modelled principally through expert judgement plus some statistical estimates; penalised standard deviation applied through enlargement coefficients.
Propagation method(s)	Monte Carlo and Taylor quadratic approximation
Sensitivity analysis method(s)	Rank correlations and Taylor sensitivity indices
Feedback process	Selection of the most robust existing metrological chain; potential re-calibration or up-grading of the most critical measurement components to reduce uncertainty.

Regarding the *third option* (plant heat rate approach), fuel consumption uncertainty was estimated to be ±2.5%, assuming that the optimal plant heat rate is known with an uncertainty of ±2.0% (best case). The *OPHR* accounts for 66% of the fuel consumption uncertainty with this method; having been estimated long ago, at commissioning time, it should be checked before being used with this method in the context of the EU Guidelines. However, this method may be complementary to the mass balance approach when the fuel stocks surveys do not correspond to the declaration period.

Table 3.2 Results for the power plant considered.

	1 – Beltweigher		2 – Mass balance		3 – Heat rate	
	CO$_2$ uncertainty	Fuel consumption uncertainty	CO$_2$ uncertainty	Fuel consumption uncertainty	CO$_2$ uncertainty	Fuel consumption uncertainty
Target values (coal)	±3%	±2.5%	±3%	±2% (on purchased)	±3%	±2.5%
Results for the power plant considered (coal)	±3.8%	±3.3%	±2.8%	Coal ±2.0% Purchased coal ±1.5%	±3.1%	±2.5
Importance ranking – sensitivity indices						
Coal consumption	73.5%	100%	28.0%	57.0%	–	–
Carbon content	26.0%	–	50.9%		41.7%	–
OPHR	–	–	–	–	38.3%	66.1%
HHV	–	–	–	–	9.8%	17.1%
W	–	–	–	–	7.0%	11.6%
Hydrogen content	–	–	–	–	1.6%	2.8%
Stock (beginning)	–	–	11.3%	22.1%	–	–
Stock (end)	–	–	9.7%	20.8%	–	–

At the end of the study, EdF communicated to the institutional bodies the 'reference relative uncertainty' as specified by the regulations; internally, EdF also used the results of the sensitivity analysis in order to optimize the choice of the measurement chain; a dedicated simplified spreadsheet tool was developed to facilitate the transmission to operational plant services.

3.5   Conclusions

Regarding the implementation of the EU guidelines, the recommended method for the annual CO_2 emissions determination in the power plant considered was that based on the fuel mass balance: this choice was made both because the approach allows for compliance with a threshold and because the measurement chain is simple enough to be maintained and reported to external audits. After the initial study linked to formal criteria, it proved desirable to have a greater understanding of the impact of uncertainty on the overall reported CO_2 emissions data; the challenge in the medium term is to improve the measurement chains.

Regarding the uncertainty methodology, this kind of study is well covered by the GUM guidelines, according to which it is essential to keep the methodology quite simple, owing to the limited stakes and budgets involved. The setting adopted in the study is standard probabilistic, supplemented by a level-2 deterministic treatment of the epistemic uncertainty generated by the limitations of the data needed to estimate the parameters of the uncertainty model. Considerable expert judgement and reliable supplier information are required, making the quantification of the uncertainty sources the most difficult step. Two options for uncertainty propagation and sensitivity analysis are acceptable for this kind of problem: the full Monte Carlo approach for reference studies, and the Taylor quadratic approximation for repetitive operational studies. This latter option, which uses simple spreadsheet tools, appears much easier to disseminate to operators.

References

EEA (2004) *Annual European Community GHG inventory 1990–2002 and inventory report 2004*, submission to the UNFCC secretariat. European Environment Agency. (Available online at: www.eea.eu.int)

EC (2003) *Directive 2003/87/EC of the European Parliament and of the Council of 13 October 2003 establishing a scheme for greenhouse gas emission allowance trading within the Community*. Brussels, Official Journal of the European Union, http://eurlex.europa.eu

EC (2004) *Commission Decision of 29 January 2004 (C(2004) 130) establishing guidelines for the monitoring and reporting of greenhouse gas emissions pursuant to Directive 2003/87/EC of the European Parliament and of the Council*. Brussels, Official Journal of the European Union, http://eurlex.europa.eu

GHG (2003) *Protocol guidance on uncertainty assessment in GHG inventories and calculating statistical parameter uncertainty* (September) (Available online at: www.ghgprotocol.org)

IPCC (1996) *Revised 1996 IPCC Guidelines for National Greenhouse Gas Inventories – Reference Manual (Volume 3) Energy*. Intergovernmental Panel on Climate Change. (Available online at: www.ipcc.ch)

ISO (1995) *Guide to the expression of uncertainty in measurement (G.U.M.)*, EUROPEAN PRESTANDARD ENV 13005 (published in 1999).

Rubinstein, R.Y. (1981) *Simulation and the Monte Carlo Method*, Chichester: John Wiley & Sons, Ltd.

Saltelli, A., Chan, K. and Scott, E.M. (2000) (Eds) *Sensitivity Analysis*, Chichester: John Wiley & Sons, Ltd.

Salway, A. G. (2002) *Treatment of Uncertainties for National Estimates of GHG Emissions*, AEA Technologies, UK. (Available online at: www.naei.org.uk)

WBCSD-WRI (2003) *The Greenhouse Gas Protocol – A corporate accounting and reporting standard*, version 2 (December 2003). World Business Council for Sustainable Development – World Resources Institute. (Available online at: www.ghgprotocol.org)

4

Hydrocarbon exploration: decision-support through uncertainty treatment

4.1 Introduction and study context

In hydrocarbon exploration the main goal is to find traps where oil and/or gas have accumulated and been retained in quantities that are larger than a given economic threshold. Regardless of all the economic variables that come into play, the basis of every drilling decision is associated with the probable presence (or absence) of hydrocarbons (hydrocarbon risk) in the potential traps of the basin. The evaluation of this probability is the joint effort of a team of geologists, geochemists and geophysicists, collaborating to gain the best understanding of the drilling prospects in a particular location.

Modelling geological processes is subject to uncertainty because input data are scarce and imprecise and also because the modelling algorithms are approximations of true geological processes. This fact is the rationale for this paper, which recommends that uncertainty treatment and sensitivity analysis be used in tandem to support drilling decisions.

The decision to drill a given trap is taken on the basis of comparisons of estimated volumes of oil/gas for any pair of traps. The model produces pdfs of the volumes, so the comparisons are made between pairs of distributions with different shapes and statistical moments. While different hypothesis tests can be employed for the comparisons, this issue is not treated in the present chapter. The study can be split schematically into two components: the *geometric* and the *fluid*. The first is the object of Basin Modelling (BM) activity, which gives a 4-D description (in space

Uncertainty in Industrial Practice   Edited by E. de Rocquigny, N. Devictor and S. Tarantola,
© 2008 John Wiley & Sons, Ltd

and time) of the basin status and evolution. The *fluid* is the object of Petroleum System Modelling (PSM), which produces the history of the geological processes that led to generation and accumulation of hydrocarbons in the current traps.

4.2   The study model and methodology

4.2.1   Basin and petroleum system modelling

A general introduction to BM and PSM techniques can be found in Welte *et al.* (1997). The modelling steps and the related workflow are described below (Figure 4.1). The Basin Modelling considered in this study encompasses the definition of shape of the geological structures, the spatial distribution of the geological properties for each geological layer, the structural evolution of the basin during geological time (hundreds of millions of years), and the definition of the history of the heat flow at the basis of the sediments.

Petroleum System Modelling comprehends the description of the evolution of the Pressure and Temperature (P&T) fields in the sediments; the history of the Generation and Expulsion (G&E) of the hydrocarbons from source rocks; the Migration of hydrocarbons from source rocks to reservoirs and the preservation of Trapping conditions of the hydrocarbons throughout the evolution of the basin (M&T).

The Basin Depth Model (see Figure 4.1) is built with a *layer-cake* vertical depth conversion method from interpreted seismic time maps. This consists in the transformation of the 'time thickness' into a depth thickness using the corresponding seismic layer velocity. In this way, the Basin Depth Model is constructed layer by layer, from top to bottom:

$$D_i = D_{i-1} + (T_i - T_{i-1}) * V_i \tag{4.1}$$

where D_i is the depth of reflector i, which is the bottom of geological layer i, D_{i-1} is the depth of reflector $i-1$, which is the top of geological layer i, T_i is the seismic reflection time from reflector i, T_{i-1} is the seismic reflection time from reflector $i-1$, $T_i - T_{i-1}$ is the 'time thickness' of layer i, V_i is the interval seismic velocity of geological layer i.

Velocity maps are obtained via geostatistical techniques, which produce optimal velocity maps integrating different data sources (called *a priori information*). In this study such techniques are used to produce a Basin Depth Model compound of seven interfaces and six layers (see the vertical section in Figure 4.2 left. Layers 1 and 4 are the hydrocarbons' source rocks ('Source 1' and 'Source 2', respectively), while layer 3 is the carrier for the hydrocarbons' migration into potential traps (in the same layer). The top surface of layer 3 and section AA' are shown in Figure 4.2 right together with the nine major traps.

The description of the resulting Basin Model is completed with (uncertain) information about the layers' characteristics (porosity, permeability, thermal conductivity), derived from wells, literature and sedimentological studies. The sediment compaction is modelled according to the assumption that the main rock deformation

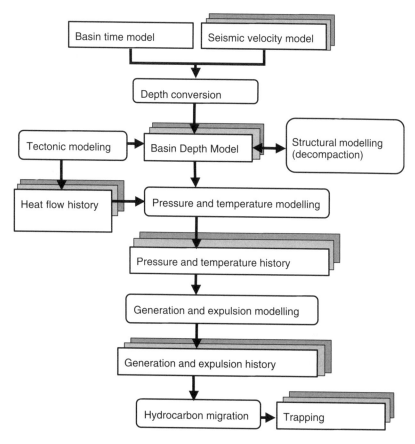

Figure 4.1 Workflow of BM and PSM. Boxes with rounded corners are modelling steps. Rectangular boxes are data. Repeated shadowed rectangular boxes are multiple realizations/simulations of those data.

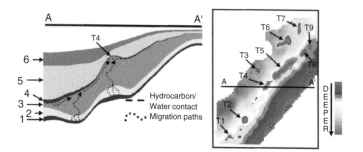

Figure 4.2 (left) Vertical section of the basin model carrier and reservoir layer 3 (right) Top surface of layer 3, section AA' and the nine major traps. Note the position of trap T4 in both graphs. Reproduced with permission from Elsevier.

takes place in a vertical direction and that the sediment porosity depends on the depth. Applying the conservation of the solid thickness through time produces the following equation:

$$\frac{d}{dt}[(1 - \phi(t))\, s(t)] = 0 \tag{4.2}$$

where $\phi(t)$ is the porosity at geological time t and $s(t)$ is the thickness of a layer at geological time t. This equation can be constrained using present time value (thickness and porosity). Solving the compaction equation makes it possible to recover the geometrical evolution of the basin through time in terms of thickness and porosity of each layer of the basin. The result is the 4-D Basin Model, which can be seen as a set of 3-D models, one for each selected geological time step, describing the evolution of the 3-D Basin Model through geological time. According to the 4-D Basin Model, a heat flow history can be established from tectonic modelling and well data (Podladchikov et al., 2002), resulting in a set of heat flow maps, one for each selected geological time step. The heat flow maps influence pressure and temperature evolution in the basin through convective and conductive heat transport in the geological basin. The convective element is controlled mainly by the heat transported by fluid (water) in the basin and is driven by the overpressure (i.e., the deviation of the pressure from hydrostatic pressure) gradient. The first step in computing the thermal evolution of the sedimentary basin is the computation of the overpressure regime of the basin, which takes into account the sediment compaction. The conservation of mass is combined with Darcy's law under the hypothesis of slightly compressible fluid, to obtain the pressure equation:

$$\nabla \cdot \left(\frac{K}{\mu_W} \nabla(\delta p) \right) = \phi \beta_{pw} \frac{d_s(\delta p)}{dt} + \frac{d_s(\varepsilon)}{dt} + \phi \beta_{pw} \frac{d_s P_i}{dt} \tag{4.3}$$

where $\frac{d_s(\cdot)}{dt} = \frac{\partial(\cdot)}{\partial t} + \vec{v}_s \cdot \nabla(\cdot)$, $\vec{v}_s$ is the velocity of the sedimentation, K is the rock permeability, μ_W is the water viscosity, $\delta p = P - P_i$ is the overpressure, P_i is the hydrostatic pressure defined by $P_i = P_0 + \int_z^0 \rho_w g dz$, P_0 is the hydrostatic pressure at sea level, ρ_w is the density of water, g is the acceleration of gravity, ϕ is the porosity, β_{pw} is the water compressibility, and ε is the deformation of the porous matrix.

The term governing the pressure equation is the deformation of the porous matrix ε which is computed according to the porosity variation through time. The dependence of the layer permeability K on porosity is calculated through time according to the Kozeny-Carman's law:

$$K = K_A \left(\frac{\phi}{\phi_A} \right)^3 \left(\frac{1 - \phi_A}{1 - \phi} \right)^2 \tag{4.4}$$

where K is the permeability at geological time T, K_A is the present-time permeability, ϕ is the porosity at geological time T, ϕ_A is the present-time porosity. The searched fluid velocity v_W is computed from the overpressure distribution $v_W = -(K/\mu_W)\nabla(\delta p)$. The thermal history of the basin is modelled taking into account both the thermal conductivity distribution in three-dimensional space and the fluid flow convective effects. The thermal conductivity distribution is caused by the sediment distribution with different petrophysical properties, while the fluid flow convective effects come from the solution of the pressure equation. The resulting temperature equation is:

$$(\rho C)\frac{d_s T}{dt} = -\rho_W C_W \vec{v}_W \cdot \nabla T + \nabla \cdot (\Lambda \nabla T) \tag{4.5}$$

where T is the temperature, ρ_W is the density of water, C_W is the specific heat of water, v_W is the velocity of water, (ρC) is the product of density by specific heat for the water-sediment system: $\rho C = \rho_W C_W \phi + (1 - \phi)\rho_s C_s$, ϕ is the porosity, ρ_s is the density of sediment, C_S is the sediment specific heat, Λ is the thermal conductivity, defined by $\Lambda = \Lambda_S^{1-\phi}\Lambda_W^{\phi}$, Λ_S is the sediment thermal conductivity, Λ_W is the water thermal conductivity.

The temperature equation is solved taking into account the boundary conditions: the temperature at the top of the basin and the heat flux at the base of sediment layers. The 'source rock' is a part of a low permeability layer (the *source rock layer*) with an abundance of organic matter, which degrades through a number of parallel reactions (*primary* cracking) into hydrocarbons, which are more stable components (oil and gases). This is followed by *secondary* cracking, which transforms oils and wet gases into lighter components (dry gases) and coke.

Both *primary* and *secondary* reactions are assumed to be independent of pressure, and their temperature dependence is given by an Arrhenius-type equation, which relates the generated quantity of type-i oil and type-i gas to temperature and activation energy:

$$\frac{d}{dt}(w_{oi}) = A_i e^{-\frac{E_i}{RT}}\left(F_{oi} - w_{oi} - w_{gi}\right) - A_g e^{-\frac{E_g}{RT}}w_{oi}; \quad \frac{d}{dt}(w_{gi}) = A_g e^{-\frac{E_g}{RT}}w_{oi} \tag{4.6}$$

where w_{oi} is the quantity of type-i oil generated, w_{gi} is the quantity of type-i gas generated, E_i is the activation energy for type-i oil, E_g is the activation energy for type-i gas, A_g, A_i, F_{oi} are calibration constants derived from laboratory measurements, R is the gas constant, T is the temperature.

Expulsion of the generated hydrocarbons from the source rock is a very complex and still poorly understood process. It is modelled as a multiphase flow (hydrocarbons and water) within a low permeable porous medium (the source rock itself), controlled by the relative permeability of each fluid phase. In practice, hydrocarbon flow is driven by the overpressure resulting from the compound effect of the sediment compaction and the hydrocarbons generated (by the transformation itself). The result of this modelling step is a set of expulsion maps, one for each selected geological time step and for each hydrocarbon component (oil and gas in the simplest case). These maps represent the result of the expulsion out of the *source rock layer*.

Once the hydrocarbons are expelled, they move along permeable rocks (in 'secondary migration' or simply 'migration') until they reach a trap. If all the layers were permeable, hydrocarbon would reach the surface of the earth, while as soon as an impermeable layer is encountered, overlaying a permeable one, an hydrocarbon accumulation may occur in the so-called 'traps' at the top of the permeable layer. The simplest approach to model hydrocarbon migration is based on the so-called 'ray-tracing technique', which assumes that hydrocarbons move just at the top of a permeable layer overlaid by an impermeable one (acting as a seal) and following the steepest path. Subsequent processes, like spillage and leakage, may cause hydrocarbons to migrate out of traps. All the processes are modelled over the geological time scale, considering the evolving basin geometry and changing properties of rocks and fluids. The final results of the whole PSM workflow are the total amounts of hydrocarbons (oil and gas) filling each trap of the basin area at the present time.

4.3   Underlying framework of the uncertainty study

To account for the large uncertainties in hydrocarbon exploration, a probabilistic approach is adopted which makes it possible to compute the 'hydrocarbon risk' for each trap in a basin. In basin modelling, geostatistical techniques can be used to model the uncertainties of the geological layer geometry. Geometric uncertainty is related to the process of depth conversion. Interpreted seismic time reflections are transformed into depth reflectors taking into account the uncertainty of seismic propagation velocities.

In petroleum system modelling, multiple 'geological realizations' of the basin are modelled as if they were scenarios (Ruffo *et al.*, 2001). All possible combinations of scenarios are considered and, for each scenario, as many realizations as needed are produced for uncertainty and sensitivity analyses. Every phase of PSM (including BM) contributes to the prediction uncertainty and can, in principle, be scrutinized using sensitivity analysis. Yet, besides the large amount of CPU time needed, the main difficulty in applying uncertainty and sensitivity analyses to the entire PSM workflow arises from the complex management of the workflow. Therefore, uncertainty is usually evaluated only for some of the phases of PSM (Ruffo, 2002): modelling of 'temperature and pressure', or 'generation and expulsion' or 'migration and entrapment'.

The modelling of geological processes may involve non-linear mathematical formulations: for example, thresholds for some quantities may trigger a specific event, such as the generation of gas; for this reason, the regression/correlation techniques of sensitivity analysis are not adequate.

4.3.1   Specification of the uncertainty study

The objective of the study is to propagate scenario and model input uncertainties through the BM and PSM models. The uncertain predictions are subsequently used to decide (Goal S-SELECT) whether it is economically advantageous to drill a new trap and to assess the risk of drilling a trap with relatively high uncertainty in volumes.

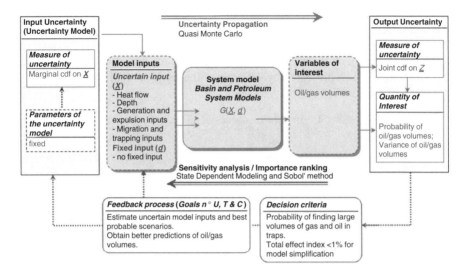

Figure 4.3 Overall uncertainty problem in the common framework.

The ancillary goals are: understanding the model, importance ranking and model simplification (Goals U-UNDERSTAND and A-ACCREDIT). Sensitivity analysis is particularly useful for Goal A as it helps in finding non-influential uncertain model inputs which could be fixed in subsequent simulations. This use of sensitivity analysis serves to build parsimonious models (i.e. models in which all uncertain inputs somehow contribute to the predictions), in contrast to the tendency to produce models with many unimportant inputs.

The goals of the study are approached here essentially through model simulations. Once the first identified trap has been explored, experimental data will be available to conduct additional analyses that could help to refine the strategy for subsequent drillings (i.e. to improve the predictions of oil and gas volumes). Indeed, based on the combination of observed data and model predictions, the modeller will be able to update the probability distributions for the uncertain model inputs and, in particular, for the scenarios, so as to identify the most probable combination of geometry and heat flow histories. A common practice consists in comparing model predictions and available observations to see whether they fit and how well. In practice, a measure of goodness of fit can be defined and calculated for each model prediction.

Correspondingly, each sampled point over the input space is associated with the measure of goodness of fit. The goodness of fit function can be considered as the posterior joint probability distribution function for model parameters and scenarios, from which posterior scenario probabilities can be derived. In turn, an additional set of runs with updated pdfs for the scenarios can give better predictions for the non-observed traps. This procedure is called Generalized Likelihood Uncertainty Estimation (GLUE) and is attributed to Beven and Binley, 1992. Unfortunately, for this exercise observed data were not available, hence the uncertainty analysis with the posterior joint probability distribution functions was not carried out. The core of the model is represented by the code of the G&E modelling phase, namely $\underline{Z} = G(\underline{X}, \underline{d})$, where:

- vector $\underline{X}$ contains 15 uncertain sources (13 model inputs + 2 scenarios) (see Section 4.3.2);

- vector $\underline{Z}$ contains 2 variables of interest: the oil and gas volumes, available at 9 traps of the basin, i.e. 18 values;

- vector $\underline{d}$ contains all model inputs that are not considered uncertain ($\underline{d}$ may include model inputs with negligible importance, as evidenced after achieving Goal A).

4.3.2   Description and modelling of the sources of uncertainty

The main sources of uncertainty of a basin depth model are the interpretation of seismic times and the estimation of seismic velocities. Disregarding the radical error of having misinterpreted seismic reflection travel times, the largest uncertainty is associated with seismic velocities, as they are indirectly estimated from seismic signal coherency (Yilmaz, 1987).

A specific geostatistical technique for probabilistic depth conversion (Ruffo and Scola, 1992) was used to simulate 100 equally probable velocity maps for the layers of the basin, and hence the basin depth models. From these, eight representative depth models were selected, which were representative of the broad range of situations in the area of interest. A similar approach was used to select four heat flow maps, as these were representative of the possible effects on the PSM results.

The depth models and the heat flow maps were both treated as scenarios and all their combinations considered, producing 32 different 4-D basin models as input to the PSM. The P&T phase was actually run 32 times – once for each scenario – and for each P&T run, the subsequent phases (G&E and M&T) were also run 32 times. This produced 1024 runs, in which all (15) uncertain inputs associated with Basin Model, P&T, G&E and M&T (see below) were considered. Such an uncertainty approach extends that of Tarantola *et al.* (2001), in which the geometry of the Basin Model was kept fixed (rather than having eight scenarios), there were eight 3-D P&T runs (instead of 32), and 32 G&E runs (instead of 1024).

Fifteen sources of uncertainty were considered, grouped in four subsets: (1) Basin Model (2 model inputs): 8 model geometries combined with 4 heat flow maps to get 32 P&T scenarios; (2) G&E (6 model inputs): Total Organic Carbon, Porosity-Stress Curve, Water Threshold Saturation, for Source 1 (layer 1) and Source 2 (layer 4); (3) Migration (4 model inputs): Expulsion Efficiency from Source 1 and 2, Leakage of Gas and Leakage of Oil from a trap; (4) Trap (3 model inputs): Net to Gross ratio; Water Irreducible Saturation, Thickness of the reservoir layer.

4.3.3   Uncertainty propagation and sensitivity analysis

To address Goal S, a quasi-random number generator is used to generate a 15-input sample of size 1024; the estimated computational time to execute the set of simulations is around 72 hours. The 1024 model simulations are run to obtain the probability distributions of the trapped oil and gas volumes. On the basis of such probability distributions and of a given economic threshold, the optimal traps to drill may be decided upon (see next section).

For Goals U and T a sensitivity analysis technique based on State Dependent Parameter (SDP) modelling (Ratto et al., 2004) was adopted, which estimates first-order sensitivity indices for the 15 uncertain inputs, based on the 1024 model predictions already calculated.

In addition to SDP, the method of Sobol' was employed (Sobol', 1993), producing higher order and total effect sensitivity indices. These are useful in identifying both interactions among the inputs as well as inputs that do not contribute at all to the output variance (and which could therefore be fixed). However, the method of Sobol' is computationally unaffordable in this case as it would require a much larger computational cost than SDP. In order to circumvent this problem, a surrogate model was developed and validated by modelling the global PSM workflow through a neural network (NN). To approximate the original model $\underline{Z} = G(\underline{X}, \underline{d})$ adequately, the neural network was trained on the 1024 quasi-random simulations, using the 15 model inputs $\underline{X}$ and the 18 output predictions $\underline{Z}$. The validation of the surrogate model was performed by comparing the first-order indices of Table 4.2 with the first-order indices obtained with the Sobol' method applied to the NN simulations (see Table 4.3).

4.3.4   Feedback process

The sensitivity analysis helps to find those uncertain model inputs that could be fixed to their nominal values without any appreciable reduction in the variance of the variables of interest. It is then possible to run subsequent simulations on a simplified model, without loss of information and with a significant reduction in the computational time. Sensitivity analysis also helps to identify the scenarios and model inputs that are most responsible for the uncertainty in the variables of interest. In addition, once observations of oil/gas volumes have been gathered at traps, the uncertain distributions for both uncertain inputs and scenarios can be updated and uncertainty predictions of oil/gas volumes further refined.

Table 4.1   Main study characteristics.

Final goal of the uncertainty study	To predict volumes of gas and oil under uncertainty and decide which trap to drill (Goal S). Supporting goals are model-understanding and importance ranking (goal U) and model simplification (Goal A).
Variables of interest	Volumes of oil and gas in each trap (with their uncertainty bounds).
Quantity of interest	Probability of oil and gas volumes in the traps and variance of oil and gas volumes.
Decision criteria	Decision criteria linked to probability of finding large volumes of gas and oil in traps. First-order and total effect indices for model understanding and importance ranking (Goal U). Total effect indices below a given threshold (e.g. 0.1) for model simplification (Goal A).
Pre-existing model	Basin Model and Petroleum System Model
Uncertainty setting	Standard probabilistic framework
Model inputs and uncertainty model developed	Model inputs are affected by epistemic uncertainty. Some of them are set up to represent scenario uncertainty (i.e. different geometries and alternative heat flow scenarios). These latter inputs are set up using discrete uniform distributions.
Propagation method(s)	Quasi-Monte Carlo sampling, as it samples the input space with good regularity.
Sensitivity analysis method(s)	SDP (State Dependent Parameter) technique for first-order indices on the system model; Sobol' indices on the surrogate model for first-order, higher order and total effect indices.
Feedback process	To fix non-influential model inputs. To estimate uncertain model inputs and most probable scenarios using experimental data. To improve uncertainty predictions of oil and gas volumes using updated probability distributions for the uncertain inputs.

Table 4.2   First-order sensitivity indices of the 15 inputs for oil and gas volumes for traps T4, T5 and T9. The sensitivity analysis was conducted using the SDP method on 1024 simulations.

	Output	T4 – oil	T4 – gas	T5 – oil	T5 – gas	T9 – oil	T9 – gas
Basin Model	Geometry (scenario)	**0.51**	0.22	0.20	0.15	**0.05**	**0.55**
	Heat Flow (scenario)	0.01	0.00	0.00	0.00	0.00	0.00
G&E	Total Organic Carbon layer 1	0.01	0.00	0.03	0.07	0.01	0.06
	Porosity-stress layer 1	0.00	0.00	0.12	0.03	0.02	**0.11**
	Water thresh. saturat. layer 1	0.00	0.00	0.00	0.00	0.00	0.00
	Total Organic Carbon layer 2	0.00	0.00	0.03	0.00	0.00	0.00
	Porosity-stress layer 2	0.03	0.00	**0.39**	0.01	0.01	0.00
	Water thresh. saturat. layer 2	0.00	0.00	0.00	0.00	0.00	0.00
Migration	Explusion efficiency layer 1	0.00	0.00	0.00	0.01	0.00	0.00
	Explusion efficiency layer 2	0.00	0.00	0.06	0.00	0.00	0.00
	Leakage of oil from trap	0.00	0.00	0.00	0.00	0.01	0.00
	Leakage of gas from trap	0.09	0.12	0.00	**0.32**	0.00	0.00
Trap	Net to Gross Ratio	0.09	**0.25**	0.00	0.11	0.00	0.03
	Water Irreducible saturation	0.02	0.07	0.00	0.02	0.00	0.01
	Thickness of the reservoir	0.04	0.08	0.00	0.00	0.00	0.00

4.4   Practical implementation and results

4.4.1   Uncertainty analysis

The uncertainty analysis shows that oil accumulation is detected only in traps T4 and T5, while gas tends to accumulate in all traps. The results of the uncertainty analysis for trap T5 for both oil and gas volumes are depicted in Figure 4.4 (each bar corresponds to different spill-in scenarios from other traps). It is found that trap T5 contains the maximum quantity of oil and gas and hence is the most favourable trap to drill provided that the economic threshold (which fluctuates with the market) allows it. The spilling from one trap to another or to outer regions of the modelled

Table 4.3   First-order sensitivity indices of the 15 inputs for oil and gas volumes for traps T4, T5 and T9. The sensitivity analysis was conducted using the Sobol' method on 131072 neural-network simulations.

	Output	T4 – oil	T4 – gas	T5 – oil	T5 – gas	T9 – oil	T9 – gas
Basin Model	Geometry (scenario)	**0.58**	**0.25**	**0.20**	**0.17**	**0.37**	**0.60**
	Heat Flow (scenario)	0.00	0.00	0.00	0.00	0.01	0.00
G&E	Total Organic Carbon layer 1	0.01	0.00	0.03	0.08	0.02	0.10
	Porosity-stress layer 1	0.00	0.00	0.14	0.04	0.15	**0.13**
	Water thresh. saturat. layer 1	0.00	0.00	0.00	0.00	0.01	0.00
	Total Organic Carbon layer 2	0.00	0.00	0.04	0.00	0.00	0.00
	Porosity-stress layer 2	0.03	0.00	**0.41**	0.01	0.12	0.00
	Water thresh. saturat. layer 2	0.00	0.00	0.00	0.00	0.00	0.00
Migration	Explusion efficiency layer 1	0.00	0.00	0.00	0.01	0.00	0.00
	Explusion efficiency layer 2	0.00	0.00	0.06	0.00	0.00	0.00
	Leakage of oil from trap	0.00	0.00	0.00	0.00	0.00	0.00
	Leakage of gas from trap	0.10	0.15	0.00	**0.37**	0.00	0.00
Trap	Net to Gross Ratio	0.12	**0.29**	0.00	0.14	0.00	0.04
	Water Irreducible saturation	0.02	0.09	0.00	0.03	0.01	0.01
	Thickness of the reservoir	0.04	0.10	0.00	0.01	0.01	0.00

area (see double arrows in Figure 4.5), is quite important for the correct evaluation of the trapped hydrocarbon volumes. By repeating the calculations of volume statistics for 30 different depth models, the probabilities of spill-over from one trap to another are obtained. The spilling paths from a single trap are mutually exclusive events. 'No spillage' could also be an option. Considering the hydrocarbon quantities that may have filled a trap, as shown for trap T5 in Figure 4.4, the total amount resulting from all the simulations summarizes the contributions due to the different spilling-in scenarios. For example, the case 'no spill-in' represents all the simulations where trap T5 was filled directly from the drainage area alone, while the case 'T4' is the set of all the simulations in which trap T5 was filled from the spilling of T4.

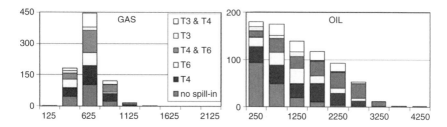

Figure 4.4 Trap T5 – Empirical distributions of gas (left) and oil (right) volumes with breakdown information on alternative spill-in scenarios.

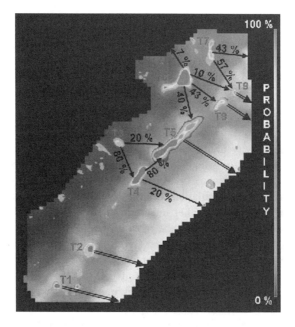

Figure 4.5 Top of layer 3 in grey – probability of spilling scenarios is indicated by arrows. Reproduced with permission from Elsevier.

The study shows that T5 can be filled in only from the six alternative cases listed in Figure 4.4. The scatterplot of the volume of gas for trap T5 against the model input 'gas leakage' (Figure 4.6 left) yields information on the non-linear relation linking these quantities. Another example is the use of a cobweb plot (Kurowicka and Cooke, 2006) on three quantities (see Figure 4.6 right), showing that, for the third model geometry of the basin model, trap T4 never contains oil, while all the other model geometries 'convey' both oil and gas to trap T4 (in Figure 4.6 right they are in light grey). The cobweb visual tool can be extremely useful to practitioners in identifying dependencies and conditional dependencies between model inputs and output responses.

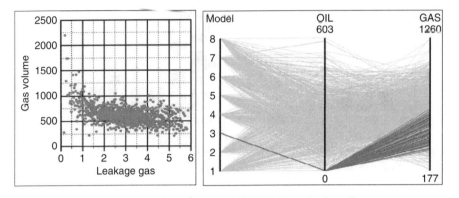

Figure 4.6 (left) Non-linear relationship between gas volume and 'leakage of gas' for trap T5; (right) Cobweb plot on three quantities: model geometry, oil and gas in trap T4. In the third scenario of model geometry no oil is conveyed to the trap. Reproduced with permission from Elsevier.

4.4.2   Sensitivity analysis

The first-order sensitivity indices computed with the SDP method (see Table 4.2) show that for all traps for which no oil accumulation is expected (i.e. all traps except T4 and T5), the sum of first-order indices is less than 0.40. This means that the improbable oil accumulations are the result of strong interactions between geometries, heat flows and the other uncertain inputs.

As the first-order indices show, the most critical input for the gas-prone traps (i.e. T1, T2, T3, T6, T7, T8, T9) is Model Geometry. For traps T4 and T5 the most critical input for gas accumulation is related to Net/Gross ratio and Gas Leakage, respectively. Concerning the accumulation of oil, the most important inputs are Model Geometry for trap T4, and source-rock quantities (Porosity-Stress curve of Source 2) for trap T5. Model Geometry plays an important role in the different modelling phases; its effect on results is non-linear and it may also introduce discontinuities in the space of results. These considerations call for a more systematic use of Model Geometry uncertainty, which implies the ability to manage at least 32 model geometry simulations.

In order to characterize interactions, it is useful to compute the second-order and the total effect sensitivity indices. In brief, the second-order sensitivity indices (not displayed here because there are 105 combinations of such indices) show that Model Geometry interacts with source-rock model inputs (TOC, porosity-stress curve), migration inputs (expulsion efficiency, gas leakage) and trapping inputs (net/gross, reservoir thickness). In addition, the total sensitivity indices of all the variables of interest are below 0.10 with respect to nine inputs (as shown in Table 4.4). This indicates that these inputs could be fixed to their nominal values without any appreciable reduction in the variance (i.e. loss of information) in the variables of interest. Therefore, it would be possible to run subsequent simulations

Table 4.4   Total effect sensitivity indices of the 15 inputs for oil and gas volumes for traps T4, T5 and T9. The sensitivity analysis was conducted using the Sobol' method on 131072 neural-network simulations. Nine inputs (in bold) are not influential, thus the model can be consistently simplified.

	Output	T4 – oil	T4 – gas	T5 – oil	T5 – gas	T9 – oil	T9 – gas
Basin Model	Geometry (scenario)	0.65	0.32	0.26	0.22	0.57	0.68
	Heat Flow (scenario)	**0.00**	**0.01**	**0.00**	**0.00**	**0.02**	**0.00**
G&E	Total Organic Carbon layer 1	0.02	0.02	0.07	0.14	0.11	0.13
	Porosity-stress layer 1	0.02	0.03	0.20	0.08	0.33	0.18
	Water thresh. saturat. layer 1	**0.00**	**0.00**	**0.00**	**0.00**	**0.01**	**0.00**
	Total Organic Carbon layer 2	**0.01**	**0.01**	**0.05**	**0.01**	**0.01**	**0.00**
	Porosity-stress layer 2	0.04	0.02	0.46	0.03	0.19	0.03
	Water thresh. saturat. layer 2	**0.00**	**0.01**	**0.00**	**0.00**	**0.01**	**0.00**
Migration	Explusion efficiency layer 1	**0.01**	**0.01**	**0.01**	**0.01**	**0.01**	**0.01**
	Explusion efficiency layer 2	**0.01**	**0.01**	**0.08**	**0.02**	**0.01**	**0.01**
	Leakage of oil from trap	**0.00**	**0.00**	**0.00**	**0.00**	**0.00**	**0.00**
	Leakage of gas from trap	0.11	0.18	0.00	0.44	0.01	0.00
Trap	Net to Gross Ratio	0.15	0.31	0.02	0.15	0.03	0.05
	Water Irreducible saturation	**0.03**	**0.10**	**0.01**	**0.03**	**0.02**	**0.02**
	Thickness of the reservoir	**0.04**	**0.11**	**0.01**	**0.01**	**0.03**	**0.01**

on a simplified model, with a significant reduction in the computational cost and in the analysis time.

4.5   Conclusions

The current study was possible thanks to the porting of the G&E modelling phase on a parallel platform, in practice a cluster of 24 Linux workstations. The CPU time needed to perform the study, with this setting, was about 84 hours (72 hours for the 1024 runs of the G&E phase). As the sensitivity study has suggested, Model Geometry plays an important role in the different modelling phases; its effect on the results is non-linear and it may also introduce discontinuities in the space of the variables of interest. One should focus on the uncertainty of Model Geometry in a more systematic way; this would involve the ability to manage an increasing number of model geometry scenarios, which in turn would require the establishment of a semi-automated procedure for use in the future.

The scenario approach is an appropriate way to manage uncertainty in basin modelling. Figure 4.4 shows the outcome of the uncertainty analysis in terms of volumes of oil and gas for trap T5. Moreover, if observations were to become available for a given trap, or groups of traps (in the form of volumes of oil or gas derived from drilling operations), this would allow for the analysis of the goodness of fit between observations and model predictions, which in turn would enable the estimation of the most probable scenarios.

References

Beven, K. and Binley, A. (1992) The future of distributed models: model calibration and predictive uncertainty, *Hydrological Processes*, **6**(3), 279–298.

Kurowicka, D. and Cooke, R.M. (2006) *Uncertainty Analysis with High Dimensional Dependence Modelling*, Chichester: John Wiley & Sons, Ltd.

Podladchikov, Y.Y., Schmalholz, S.M., Cavazza, D. and Ruffo, P. (2002) *Palaeo-Heat Flow Assessment Using Automatic Inverse Modelling of Sedimentary Basin Formation*. Abstract in the Proceedings of the EAGE Workshop on 'Multidimensional Geological Modelling: from Cutting-Edge Research to Application in the Field', Firenze.

Ratto, M., Tarantola, S., Saltelli, A. and Young, P.C. (2004) 'Accelerated estimation of sensitivity indices using State Dependent Parameter models', in *Proceedings of the 4th International Symposium on Sensitivity Analysis of Model Output (SAMO) 2004*, Santa Fe.

Ruffo, P., Bazzana, L., Consonni, A., Corradi, A., Saltelli, A. and Tarantola, S. (2006) *Hydrocarbon exploration risk evaluation through uncertainty and sensitivity analyses techniques*, published in R.E.S.S. Volume **91** (2006) pp 1155–1162.

Ruffo, P. (2002) *Probabilistic Petroleum System Modelling: An Overview*. Abstract in the Proceedings of the EAGE Workshop on 'Multidimensional Geological Modelling: from Cutting Edge Research to Application in the Field', Firenze.

Ruffo, P. and Scola, V. (1992) *Structural Model Uncertainty of a Prospect – A Geostatistical Approach*. Abstract in the Proceedings of the 54th EAGE Meeting, Paris.

Ruffo, P., Tarantola, S., Corradi, A. and Grigo, D. (2001) *Uncertainties Evaluation in Petroleum System Modelling.'* Abstract in the Proceedings of IFP International workshop on 'Quantitative Basin Evaluation, a Tool for Lowering Exploration Risks', Paris.

Sobol', I. (1993) Sensitivity analysis for non-linear mathematical models, *Mathematical Modelling and Computational Experiment*, **1**, 407–414.

Tarantola, S., Corradi, A., Ruffo, P. and Saltelli, A. (2001) *Global sensitivity analysis techniques for the analysis of the oil potential of sedimentary basins*. Long Abstract in the Proceedings of the 3rd International Symposium on Sensitivity Analysis of Model Output (SAMO) 2001, Madrid.

Welte, D.H., Horsfield, B. and Baker, D.R. (1997) (Eds) *Petroleum and Basin Evolution*, Berlin: Springer-Verlag.

Yilmaz, O. (1987) Seismic Data Processing. *Investigations in Geophysics*, **2**. SEG, Tulsa.

5

Determination of the risk due to personal electronic devices (PEDs) carried out on radio-navigation systems aboard aircraft

5.1 Introduction and study context

The advance of commercial electronic devices has led to an increasing use of Portable Electronic Devices (PEDs) on board aircraft, by flight crew, cabin crew and passengers, and for various purposes, including entertainment. Such devices are not typically designed or tested with respect to aviation standards and therefore represent a new and evolving source of electromagnetic radiation which may interfere with aircraft systems. One of the associated issues is that, although qualification tests for the electromagnetic immunity of airborne equipment have become progressively more rigorous over the years, many aircraft, including of recent manufacture, operate with electronic equipment and systems conforming to earlier standards. It must be established whether the use of PEDs on board aircraft is safe and whether the operation of these devices will have any adverse effects on aeroplane performance. As a result of Electromagnetic Compatibility and Safety analysis, the various regulatory guidelines now state that the use of PEDs during critical phases of flight is forbidden. This problem is a multi-industry topic, as it concerns the commercial electronics industry (definition of the emission parameters), aircraft manufacturers (definition of factors of the aircraft's physical features), aeronautical suppliers (equipment requirements in terms of electromagnetic susceptibility) and

Uncertainty in Industrial Practice Edited by E. de Rocquigny, N. Devictor and S. Tarantola,
© 2008 John Wiley & Sons, Ltd

airlines (carrying the passengers and legally responsible for the correct use of PEDs inside an aircraft).

Even though the field is regulated, many questions remain unresolved and future needs have already been forecast: to survey the new PED technologies; to scale/score the risk due to PEDs in relation to other risks; to manage technological developments and the responsibilities of the various industrial sectors involved; and to design policy to control passengers' behaviour. Various international aeronautical committees are currently working in the area (EUROCAE WG 58 & RTCA SC-202, for example). Different scenarios of electromagnetic coupling phenomena have been defined and described by these committees.

The goal of this paper is both to use and to deepen the common understanding of the PED problem in order to link the analysis to a more general framework in uncertainty studies (Mangeant *et al.*, 2006). First, the common physical and technological understanding of this problem, shared by the different standards committees, will be recapped: the definition of the failure modes of the Instrumental Learning System (ILS) antenna, the spurious electromagnetic emissions of the Personal Electronic Devices, the coupling mechanisms occurring in the ILS frequency band. Then the main regulatory and industrial stakes (responsibilities, risk sharing and risk control among the various industrial players) will be reviewed.

Different methods and tools to quantify the risk for one of these potential risky scenarios will be compared: the effects of spurious emissions of PEDs (mainly laptops and mobile phones) on radio navigation systems (the ILS antenna) through radiation and conducting effects in the ILS frequency band (108 MHz–112 MHz).

From an uncertainty management point of view, this is a good example of an uncertainty study model in an industrial context. One has to cope with having only a small amount of data, with difficulties in validating the robustness and accuracy of the models and with gaining access to pertinent data. The main goals of the study are: (1) to identify the main contributors and drive future R&T studies and data collections; and (2) to define a downstream process with the different actors of the study.

5.2   The study model and methodology

5.2.1   Electromagnetic compatibility modelling and analysis

The basic principle of Electromagnetic Compatibility analysis is the following. For each potential interference issue, the emission levels of the source are compared to the susceptibility level of the potentially impaired equipment. This comparison is made through coupling values that characterize the coupling path between the emitting source (in this study, the Personal Electronic Devices) and the aircraft's 'victim' equipment (the ILS antenna). Therefore, the aim of the electromagnetic compatibility (EMC) analyses will be to ensure that there will be no adverse effect on the aircraft's systems and equipment from any PEDs that might be brought on board (standard JAR OPS 1.110).

5.2.2   Setting the EMC problem

- **Understanding the physical problem**

Within the ILS frequency band (108–112 MHz), several electromagnetic sources (the PEDs) emit electromagnetic fields with different potential signal shapes from the interior of the aircraft. These electromagnetic fields propagate inside the aircraft, pass through the apertures of the aircraft (windows, slots around doors and/or canopy) before reaching the antenna. The antenna collects the sum of these signals (superposition of all the electromagnetic fields), in addition to the useful signal coming from the airport. The problem is thus to evaluate the electromagnetic constraints (cumulative noise) at the entry point of the ILS antenna which could provoke adverse functional effects (i.e. false signals or lack of useful signal for the pilot). At this point, one should understand the three sub-problems characterizing the global problem: definition of the failure mode, characterization of the spurious emissions and definition of the coupling mechanisms.

- **Definition of the failure modes**

The ILS system is designed to receive useful signals coming from the airport that enable the pilot to guide the aircraft towards the ground and the airport for landing. This signal makes it possible to compute the glide and slopes to reach the airport safely. The useful signal of the ILS at the levels of the onboard electronic equipments is called the 'ddm'. The 'ddm' is an analogic signal and is proportional to the difference between two signal levels of two side-band frequencies around the ILS carrier frequency. Two main failure modes have been identified which could potentially induce a false signal:

  - *Jamming effect*: The level of noise (undesired signal) is so high in the frequency band of interest that the useful (desired) signal can not be received and understood by the ILS system. This mechanism is also known as the desensitization of the antenna.

  - *Side-band effects*: A pernicious signal is added to the useful signal at the same frequency, directly conveying false information to the pilot.

  This paper focuses on the first failure mode (the jamming effect).

- **Definition of the spurious emissions of Portable Electronic Devices (PEDs)**

Portable electronic devices are typically lightweight consumer electronic devices, individually owned and operated by passengers or crew-members, which have functional capability for communications, entertainment, data processing, and/or other operations. These PEDs could be part of an aircraft's installed system (examples are pico-cell and wireless local area network managing cells) or provided by the operator in order to help the flight or cabin crew in their tasks.

Non-intentional (spurious) emissions, resulting from the internal electrical operation of the devices, represent energy that is wasted from the devices. This energy can be either radiated (*radiated spurious emissions*) or conducted (*conducted spurious emissions*) along a connection such as a power network or a local area network. PEDs without intentional transmitting capability include, but are not limited to, electric razors, basic laptops, basic electronic games, CD players, radios, etc. The following recaps the main characteristics of spurious emissions due to PEDs.

- *Frequency domain*: unpredictable, even if it is observed that most of the spurious emissions from the PEDs are composed of narrow frequency bands;

- *Frequency bandwidth*: unpredictable – may be either noise-like or narrow band;

- *Waveform*: unpredictable without measurement;

- *Emitted level*: generally very low level. The standard limits, which are not all harmonized, give values that the emissions must not exceed. Yet, in the case of faulty devices (due to ageing or malfunctioning), this limit value could be exceeded.

Intentionally generated emissions are useful signals with well-defined characteristics, emitted for communication and command purposes. PEDs with intentional transmitting capability include, but are not limited to, cellular communication, wireless networking technology, hand-held radio transceivers, and transmitters that control devices such as toys.

- **Coupling paths**

Generally, in an ILS system two main coupling mechanisms between electronic equipment and a Portable Electronic Device operated by a passenger are possible:

- **Coupling to receiver antenna, also called 'front-door effects'**. This coupling is due to the incident radiated electromagnetic energy received at the level of the antenna.

- **Coupling to cables of the ILS system, also called 'back-door effects'**. This coupling represents the radiated electromagnetic energy coupled to the cables and then transmitted by these cables to the equipment units.

5.2.3   A model-based approach

As aircraft test measurements are very expensive and cannot be performed on the whole fleet of aircraft, a model-based approach is preferred to assess the effects of the variety of configurations due to PED locations, aircraft installations, intrinsic antenna mode, etc.

Concerning notation, let us consider N PEDs located among one of the N_T possible locations inside the aircraft. Of course, the number of possible configurations is equal to the number of combinations $C_{N_T}^N$. Let us denote with $(j_1, \ldots, j_N)$ one of these $C_{N_T}^N$ configurations.

Concerning signal processing, a given signal S at the level of the antenna has to be computed. This signal is the sum of the useful signal coming from the airport and the noise due to the N PEDs located inside the aircraft cabin.

$$S = S_0 + \varepsilon_N \tag{5.1}$$

where S is the signal at the level of the antenna, S_0 is the signal of reference received from the airport which is useful to the ILS system, ε_N is the total noise due to the emitters located inside the aircraft.

For the configuration defined by the N positions $(j_1, \ldots, j_N)$, the model of noise due to the PED emissions from inside the aircraft is defined by:

$$\varepsilon_N = \sum_{i=1}^{Nc} C_{j_i} E_i + \sqrt{\sum_{i=N_c+1}^{N} [C_{j_i} E_i]^2} \tag{5.2}$$

where $N = N_c + N_i$ is the total number of PEDs, with N_c the number of coherent sources (meaning that the phase is constant over time and equal to zero between two emitters at the level of the antenna) and N_i the number of incoherent sources (meaning that the phase is random over time between two emitters at the level of the antenna), C_{j_i} is the coupling function at the j_i-th position of interest, E_i is the emission of the i-th PED.

The first part of the noise is called the coherent addition of electromagnetic sources. The second part is called the incoherent addition of electromagnetic sources, also known as the root mean square addition (RMS signal).

Regarding susceptibility, for a given level of desired signal, it is possible to define the minimum level of undesired signal (the cumulative noise due to the N PEDs) above which the interpretation process of the desired useful signal is no longer feasible for the ILS system. This limit, in terms of the undesired signal, is called the level of susceptibility. It is noted $S_{susceptibility}$. It depends on the desired useful signal.

Finally, one can define the susceptibility safety margin, which makes it possible to separate the zone of interference from the zone of non-interference. It is defined in the following formula:

$$Z_N = S_{susceptibility} - \varepsilon_N \tag{5.3}$$

5.2.4 Regulatory and industrial stakes

- **Questions and problems from a regulatory point of view**

Current regulatory policy restricts the use of PEDs onboard aircraft on the basis that they could adversely affect the operation of aircraft systems. PEDs may be brought aboard by crewmembers or passengers, used 'stand alone' or connected to onboard aircraft systems, and might be found in the cabin or luggage compartments, in addition to other accessible compartments of the aircraft. In any case, these PEDs are subject to operational controls such as those contained in the standard JAR OPS 1.110, which states: 'An operator shall not permit any person to use, and no person shall use, on board an aeroplane a portable electronic device that can adversely affect the performance of the aeroplanes system and equipment.'

Therefore, if airlines or operators allow the use of PEDs by passengers on board their aircraft, procedures should be implemented to control their use and to ensure that the entire staff is aware of the safety issues and restrictions pertaining.

• Differing aircraft types within aircraft operators' fleets

In general, Aircraft Operators have a variety of aircraft types within their fleets. It is likely that, at any point in time, the level of transmitting PED use that is permissible will vary across the fleet, depending on the extent to which testing and analysis has been performed on the various aircraft types, on the results of such tests/analysis, and on the systems installed in the aircraft to communicate with and control onboard PEDs.

• Passenger expectations and industrial environment

Air travel today is increasingly becoming a normal part of life for a significant percentage of the world's population. This may lead passengers to expect that the use of portable electronic devices, upon which they have come to rely, should be acceptable everywhere, including onboard aircraft. Compounding this problem is the contrast between the dynamic, rapidly changing consumer electronics technology marketplace, and the slower development and certification cycles typical of the safety-conscious air transport environment. These factors suggest that restrictions on the use of certain devices must be clearly communicated and explained in order to ensure passengers' compliance and to maintain aircraft safety.

• Link between regulation and definition of a quantitative criterion

The safety constraints linked to the interference of PEDs with radio navigation systems have already been transformed into some regulatory standards. Following a classical safety analysis, because the ILS system is critical in some safety scenarios, the probability of the event 'loss of the ILS system' must be very low (lower than 10^{-9} per flight hour). This criterion has not yet been transformed from an aeronautical safety rule into a quantitative criterion. However, depending on the method of demonstration, two criteria have been selected for this paper:

- For deterministic methods: $Z_N > 0$
- For probabilistic methods: $P(Z_N \leq 0) < P_{Boundary}$, where $P_{Boundary}$ is a low probability (around 10^{-6}).

5.3 Underlying framework of the uncertainty study

5.3.1 Specification of the uncertainty study

The overall uncertainty framework is described by Figure 5.1. The deterministic model inputs $\underline{d}$ (i.e. design variables) are N, the total number of emitters, which is the sum of N_c, the number of coherent sources, and N_i, the number of incoherent sources; $S_{susceptibility}$, the level of susceptibility (the lowest value is taken into account as a 'worst-case' approach); and the frequency band of the receiver Δf (the ILS frequency band). The uncertain model inputs x are the emission level E_i of each emitter i, and the coupling function C_{k_i} at each position k_i of the $i - th$ emitter. The v.o.i. is the electromagnetic safety margin Z_N, defined by:

$$Z_N = S_{susceptibility} - \varepsilon_N \qquad (5.4)$$

In sum, the input/output variables are linked by the previous model:

$$\left(\begin{array}{l} x = (E_1, \ldots, E_N, C_1, \ldots, C_N, S_{susceptibility}) \\ d = (N, S_{susceptibility}, \Delta f) \end{array} \right) \xrightarrow{G} (Z_N) \qquad (5.5)$$

Various goals were assigned to this study:

- To propose a global methodology that would make it possible to monitor the risk better and thus define the R&T priorities better among all different industrial actors;

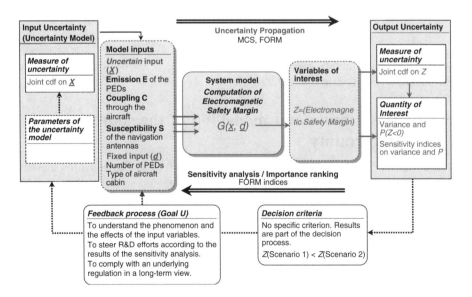

Figure 5.1 Overall uncertainty problem in the common framework.

- To update the uncertainty model (e.g. fix some unimportant model inputs);
- To adjust the design variables using an interpretation of the preliminary results;
- To take part in a long-term safety analysis.

This study is mainly driven by Goal U-UNDERSTAND, against the background of Goal C-COMPLY. The q.o.i. is the probability of interference. This is the probability that the Electromagnetic Safety Margin is negative, meaning physically that the cumulated level of noise received at the level of the ILS antenna is above the susceptibility threshold. This can be formalized by:

$$P_{Interference} = P(Z_N \leq 0) \tag{5.6}$$

For future studies, the computation of this probability of interference should be incorporated into a more global safety analysis. Thus, this computation should be linked to the probability of losing the critical function 'navigation capabilities by analogic means', which could induce a catastrophic event (crash of the aircraft during take-off or landing phases). Although very low values are required for the probability of interference, at this stage of the study no decision criterion is explicitly formalized.

Even if the main quantity of interest is the probability of exceeding a threshold, there is great interest in its associated sensitivity measures. The sensitivity results would enable the analyst to derive risk indicators and inform future R&T needs and data collection efforts. In this paper elasticity coefficients are used as importance measures. They are defined as:

$$e_\mu = \frac{\mu}{\beta} \cdot \left.\frac{\partial \beta}{\partial \mu}\right|_{u^*} \quad \text{and} \quad e_\sigma = \frac{\sigma}{\beta} \cdot \left.\frac{\partial \beta}{\partial \sigma}\right|_{u^*}$$

where β is the reliability index, μ is the mean value in the standard space, σ is the standard deviation in the standard space.

5.3.2 Description and modelling of the sources of uncertainty

The number of PEDs N is considered deterministic for each computation even if the variation of the probability of interference with regard to this variable might also be of interest. The level of emission E_i of each PED and the level of coupling C_{ji} attached to each PED are considered to be uncertain. On the other hand, the level of susceptibility $S_{susceptibility}$ is considered to be deterministic.

Electromagnetic emissions are managed by international standards to regulate the use of the frequency spectrum for the different end-users (radio, GSM, Blue Tooth, 'walkie-talkie', etc.). There is very little data collected on spurious emissions from PEDs. Indeed, even if PED manufacturers have to comply with

Table 5.1   Sources of uncertainty within emission data.

Source of uncertainty	Comment on the nature of uncertainty
Mode of use: time within the different working cycles of the emitter. Depends on the data being processed by the PED.	Partly reducible uncertainty if several basic modes in terms of emissions can be determined. Hard to be representative of all technologies. Partly irreducible as it is linked to passenger behaviour.
Ageing: effects due to thermal/mechanical/electrical over stresses.	Variability of performance. Reducible with better knowledge of PED technologies but few data/studies on this point.
Technology: effects linked to the electronics used for the design of the emitter (micro-controllers, chips, Printed Circuit Board).	Too many PEDs available worldwide. Partly reducible by gathering more data. Partly irreducible as it requires an extensive and costly survey policy.
Manufacturing process: variability of the printing of the circuits, assembling for one given type of PED.	Reducible if information from the manufacturer is available. Partly irreducible because of the numerous electronic manufacturers and the scarcity of studies on this topic.
Default modes: mix of ageing and mode of use	Irreducible
PED emission information beyond regulations (from manufacturers)	Reducible if PED manufacturers have the data.

regulatory standards of noise emission, they are mainly interested in the functional test, i.e. 'do they comply with the test or not?' For the purpose of this case study, emissions data have been obtained by measurement means for different protocols. Two main measurement set-ups are used: anechoic chambers and reverberating chambers.

Coupling data are also rare as they involve very expensive tests on aircraft (due to the immobilization of the aircraft). The rare data obtained are issued from single tests on one configuration of aircraft. Only one set of positions have been investigated inside the aircraft, based on perceived 'worst-case' locations. Yet, as proposed in this chapter, the exploration of all possible combinations facilitates a more objective and robust assessment.

The definition of the **susceptibility** of the ILS antenna is somehow uncertain, since this single value combines different signals and is a physical value which is

Table 5.2   Sources of uncertainty for coupling data.

Source of uncertainty	Comment on the nature of uncertainty
Position/orientation: the propagation of the electromagnetic fields is strongly linked to the position between the emitter and the ILS receiving antenna	Partly reducible if more data can be obtained on the coupling. Partly irreducible as it is linked to passenger behaviour.
Installation parameters: the same aircraft can be customized for different uses depending on the airline (low-cost vs first-class). This affects the propagation inside the cavity.	Aircraft manufacturers should characterize the propagation inside different configurations → very costly. Irreducible in practice.
Ageing: effects of the external environment on the aircraft (maintenance, reinstallation, damages, etc.).	Requires too many tests today. Irreducible in practice.
Manufacturing process: installation, tolerancing of the different parts and characteristics of the aircraft.	Requires too many tests today. Irreducible in practice.
Measurement protocol	Uncertainty of measurement.
Number of passengers inside the aircraft	The more passengers onboard, the more losses in any indoor propagation.

linked to a functional concept (the signal is not acceptable). The following table summarizes the various sources of uncertainty attached to the definition of the susceptibility threshold.

This study was undertaken within a probabilistic framework. For each uncertain model input, the most likely pdf was defined. Regarding emissions, the following probability density function translates the compilation of information obtained from several sources, from which a truncated normal law has been defined. The FCC and IEC standards make it possible to define the maximum level of emission. The minimum level of emission is related to the sensitivity of the measurements. The shape of the probability density function was defined by expert/engineering judgement and by using a database, obtained from public institutes and/or industrial actors (European Program EMHAZ, NASA reports, FCC & IEC standards).

$$E \;\rightarrow\; G_{Truncated}(\mu_E, \sigma_E, m_E, M_E)$$

$$\mu_E = 7,07.10^{-6} \text{ V/m } \|\sigma_E = 4.10^{-6} \text{ V/m} \| m_E = 1,26.10^{-6} \text{ V/m} \| M_E$$

$$= 5,01.10^{-4} \text{ V/m}$$

Table 5.3   Sources of uncertainty for susceptibility characterization.

Reason for uncertainty	Comment on the nature of uncertainty
Manufacturing process	Reducible if a good link with the antenna manufacturer can be established. Potentially reducible in cooperation with the antenna manufacturer.
Signal processing (capture effect) and possible data corruption	Irreducible.
Model uncertainty Functional/physical → details of the model	Model uncertainty. Irreducible thus far.

Regarding coupling, the following probability function is the result of a measurement campaign undertaken on a commercial aircraft for the D0-233 document. The coupling function was not obtained for all positions inside the aircraft. It is somehow representative of the most critical positions inside the cabin.

$$C \rightarrow G_{Truncated}(\mu_C, \sigma_C, m_C, M_C)$$

$$\mu_C = 6,68.10^{-5} \text{ V/V/m} \| \sigma_C = 5,5.10^{-6} \text{ V/V/m} \| m_C = 1.10^{-7} \text{ V/V/m} \| M_C$$

$$= 4,68.10^{-3} \text{ V/V/m}$$

The susceptibility threshold was obtained by a functional EMC test in a laboratory. The limit curve reveals the undesired signal level (due to the perturbation induced by the PEDs) at a given desired signal level (the useful signal coming from the airport). Preliminary studies showed the strong influence of the lower bound of the susceptibility threshold. This threshold is called the sensitivity threshold of the susceptibility level and is equal to $-104\,$dBm.

5.3.3   Uncertainty propagation and sensitivity analysis

In the typical scenario, the probability of interference is expected to be very low. So, even if the model is very simple (see Equation 5.2), the Monte Carlo method can be very expensive to run. To overcome this CPU time constraint, the FORM/SORM method (Madsen *et al.*, 1986) was chosen. This method gives acceptable results in comparison with Monte Carlo. Moreover, for some scenarios (e.g. with one PED emitting inside the aircraft), it is possible to demonstrate that the FORM hypothesis is conservative by curvature arguments. For the scenario with multiple emitters, the FORM method was used but no theoretical argument was found to justify the use of this method.

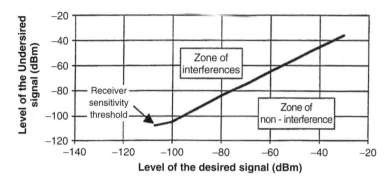

Figure 5.2 Limits of susceptibility.

5.3.4   Feedback process

The main goal of the research (partially already operationally achieved) is to identify the model inputs that are principally responsible for the uncertainty in the variables of interest. These results highlight what new emission data has to be collected in future initiatives.

5.4   Practical implementation and results

5.4.1   Limitations of the results of the study

Note that the following results have been obtained with a very simplified model and a small amount of data. Moreover, even if the statistical laws are realistic, they have been obtained by expert judgement. At this stage of the study, the results cannot be taken as guaranteed and are not applicable for regulation purposes.

5.4.2   Scenario no.1: effects of one emitter in the aircraft on ILS antenna (realistic data-set)

The 'worst-case' deterministic computation consists in taking the maximum level of emission E, the maximum level of coupling C and comparing this equivalent noise to the lowest level of susceptibility S: $E \rightarrow \max(E) = 5,01.10^{-4}V/m$, $C \rightarrow \max(C) = 4,68.10^{-3}V/V/m$ and $\min(S) \rightarrow 1,12.10^{-6}V$. Hence, the worst-case of the electromagnetic safety margin for one PED is equal to: $Z_1 = 1,12.10^{-6} - 5,01.10^{-4} \times 4,68.10^{-3} = -1,22.10^{-6}V$. This proves that some configurations give rise to potential interference.

The definition of 'worst case' is greatly facilitated by the monotonic behaviour of the model. If this feature could not have been inferred from a relatively simple mathematical formulation of the model, the definition of 'worst case' would have

Table 5.4   Main study characteristics.

Final goal of the uncertainty study	Goal U: To rank the importance of prospective R&T efforts. Goal C: To take part in a risk analysis process.
Variables of interest	Electromagnetic Safety Margin
Quantity of interest	Probability of exceeding a threshold.
Decision criterion	Decision criteria linked to a low probability.
Pre-existing model	Deterministic Analytic Model e.g. deterministic analytic computation of the electromagnetic safety margin for given emission and coupling levels.
Uncertainty setting	Standard probabilistic
Model inputs and uncertainty model developed	Uncertain model inputs: emission, coupling – pdf informed by expert judgement. Design variables: susceptibility, number of PEDs for each scenario.
Propagation method(s)	FORM validated by curvature arguments and Monte Carlo simulation.
Sensitivity analysis method(s)	FORM sensitivity analysis at the failure point.
Feedback process	Guidelines for future data collection. Insert new input variables into the coupling model.

been more difficult. Otherwise, monotonicity should be confirmed or the nominal values corresponding to the worst-case should be identified numerically.

In the probabilistic approach, the First-Order Reliability Method (FORM) was used in the calculation of the probability or interference and the results were validated by Monte Carlo simulation using the statistical laws previously presented: $E \rightarrow G_{Truncated}(\mu_E, \sigma_E, m_E, M_E)$, and $C \rightarrow G_{Truncated}(\mu_C, \sigma_C, m_C, M_C)$, both truncated Gaussian laws. S is taken at its minimum value, $S \rightarrow \min(S) = 1, 12.10^{-6}V$.

The results obtained with the FORM method (from the computation of the reliability index in the standard space) show that β is greater than 20 (the maximum distance in the standard space investigated by the research optimization algorithm), which leads to a probability of interference lower than 10^{-33}, which is the numerical accuracy of the software.

An issue arises when it is necessary to communicate around this value. Indeed, the classical 'worst-case' approach demonstrates that the data-set contains potential interference combinations, even while the probabilistic approach implies that these configurations are very rare.

5.4.3 Scenario no. 2: effects of one emitter in the aircraft on ILS antenna with penalized susceptibility

In this scenario the emission and coupling variables follow the same statistical laws, whereas a penalization is introduced by using lower values of the susceptibility threshold $S \in [10^{-8}V; 2.10^{-8}V]$.

For all the model inputs, the elasticity coefficient in Table 5.5 is negative. This means that an increase in the corresponding inputs (mean value or standard deviation) induces a decrease in the beta coefficient (reliability index) and thus an increase in the probability of interference. As expected, the lower the level of susceptibility, the higher the probability of interference. The values of the elasticity coefficients make it possible to rank the influence of the different model inputs:

- The mean values of C and E are the main drivers of the confidence index;

- The standard deviations of C and E are less influent to the variations of the confidence index.

The penalized susceptibility levels used in this scenario are very low: below the classical noise limit of an EM amplifier, lower than the 'real' susceptibility threshold by a factor of 100. Given the rapid decrease in the probability of interference when the level of susceptibility S_0 is increased, the EMC practitioner can be confident that the interference is very rare. In addition, the limit state in the standard space was drawn in order to validate the FORM hypothesis. The hyperbolic pattern observed reinforces the conservative aspect of this assessment.

Table 5.5 Reliability index vs susceptibility level for scenario no. 2.

S_0 (V)	S_0 (dBm)	Reliability Index β	Approximated $P_{Interference}$	Elasticity μ_C	Elasticity σ_C	Elasticity μ_E	Elasticity σ_E
$1,00.10^{-8}$	-180	$1,8863$	$0,0296$	$-3,23$	$-0,167$	$-0,868$	$-0,167$
$1,10.10^{-8}$	-179	$2,242$	$0,0125$	$-2,72$	$-0,144$	$-0,734$	$-1,035$
$1,50.10^{-8}$	-176	$3,63$	$1,43.10^{-4}$	$-1,69$	$-0,0989$	$-0,461$	$-0,999$
$1,70.10^{-8}$	-175	$4,30$	$8,47.10^{-6}$	$-1,43$	$-0,0870$	$-0,390$	$-0,993$
$2,00.10^{-8}$	-174	$5,29$	$6,03.10^{-8}$	$-1,17$	$-0,0757$	$-0,319$	$-0,987$

5.4.4 Scenario no. 3: 10 coherent emitters in the aircraft, ILS antenna with a realistic data set

The effect of the number of PEDs was investigated in this scenario. Ten coherent PEDs follow the same truncated Gaussian Laws E and C defined in the previous paragraph. With the data used in this study it was not possible to identify particularly bad locations for the sources. S is still equal to the sensitivity threshold $1,12.10^{-6}$ V/m.

As in scenario no.1, the FORM/SORM algorithms do not provide any 'design' point, as the optimization algorithms do not converge. This means that β is out of the domain of investigation of the optimization algorithms and therefore the probability of interference is greater than the resolution of the numerical implementation, that is to say, $\beta \geq 20$. Using a classical Monte Carlo method, only 10^7 samples could be computed and no interference events were found, which to a certain extent confirms the FORM results.

5.4.5 Scenario no. 4: new model considering the effect of one emitter in the aircraft on ILS antenna and safety factors

To increase the confidence obtained from the previous results, a penalized scenario was defined which is basically a change of model:

$$Z_1 = S_{susceptibility} - kC_1E_1 \qquad (5.7)$$

where k is a safety coefficient taking into account non-mastered effects. For the following numerical examples, E and C still follow the same truncated Gaussian law and S is taken at its minimum value. k varies between 600 and 1500.

For this model, the main driver of the reliability index is still the mean value of the coupling. The greater the value of k, the more influent the mean value of

Table 5.6   Reliability index vs susceptibility level for the scenario no.4.

Safety Factor k	Reliability Index β	Approximated $P_{Interference}$	Elasticity μ_C	Elasticity σ_C	Elasticity μ_E	Elasticity σ_E
600	4,85	$6,01.10^{-7}$	$-1,27$	$-0,08$	$-0,347$	$-0,989$
700	3,97	$3,65.10^{-5}$	$-1,55$	$-0,09$	$-0,423$	$-0,996$
900	2,75	$2,99.10^{-3}$	$-2,22$	$-0,122$	$-0,604$	$-1,02$
1200	1,65	$0,0498$	$-3,69$	$-0,187$	$-0,991$	$-1,08$
1500	0,97	$0,167$	$-6,28$	$-0,302$	$-1,66$	$-1,207$

the coupling. The mean value of the emission and the standard deviation of the coupling are the second most important model inputs, depending on the values of k.

5.5   Conclusions

A global methodology for tackling the effects of Portable Electronic Devices on radio navigation systems was presented in this chapter. The methodology is based on a common physical understanding of the global problem: the electromagnetic safety margin has to be determined from knowledge of the emissions of the PEDs, the coupling function through the aircraft and the susceptibility threshold of the ILS antenna. Then, since these model inputs remain uncertain for different reasons, a probabilistic framework is used to represent the uncertainty model in order to quantify the underlying risk. This risk is quantified by the probability of interference, which is intended to be low. The FORM method was applied to compute this probability and to derive elasticity coefficients that make it possible to rank the importance of the different model inputs.

This study was undertaken for R&T purposes. It has limitations due to the scarcity of data on emissions, the approximate quantification of the coupling function and the examination of only one failure mode of the antenna. Some questions remain open for future research, such as how to validate the FORM hypothesis for higher dimensions, or how to choose an efficient propagation algorithm.

The study met with a great deal of interest within the EMC regulatory committees and more data were collected after the first presentations. This should lead to a refinement of the study. Moreover, the sensitivity measures presented within this work should facilitate a technological survey and open a field of common understanding between safety analysis and EMC analysis.

References

EMHAZ-EADS (2003) *Design Guideline for PEDs Protection, EMHAZ-EADS CCR-REP 007, ref 2003 – 72737/3 – DCR/EX*, financially supported by the European Commission under G4RD-CT-1999-00093.

EUROCAE (2005) *ED 118 report*, EUROCAE association. Available online at: www.eurocae.org

Madsen, H.O., Krenk, S. and Lind, N.C. (1986) *Methods of Structural Safety*, New Jersey: Prentice Hall.

Mangeant, F., Vialardi, E., Christophe, J-J., Hoëppe, F. (2006) *Evaluation Par une Approche Probabiliste de la Menace Due Aux Appareils Électroniques Embarqués Dans des Avions Civils*, Congrès CEM '06 Saint-Mâlo.

RTCA (1996) *RTCA DO-233 report*, RTCA Inc., Washington. Available online at: www.rtca.org

RTCA (2006) *RTCA DO-294 report*, RTCA Inc., Washington. Available online at: www.rtca.org

6

Safety assessment of a radioactive high-level waste repository – comparison of dose and peak dose

6.1 Introduction and study context

Nowadays, many European countries, as well as several other countries with a similar level of industrial development, are attempting to demonstrate the feasibility of long-term disposal of spent nuclear fuel in engineered facilities hosted by deep geological formations. There are three potential host geological formations: salt, granite and clay. 'Performance assessment' is the tool used for demonstrating such feasibility. The performance assessment of a High-Level Waste (HLW) repository involves the modelling of the complete system, the representation of input uncertainties, their propagation through the system model and lastly the assessment of the resulting uncertainties in the variables of interest.

The system is considered to be divided into three parts: the near engineered facilities and the disturbed part of the geosphere (near field); the part of the geosphere that hosts the repository (far field); and the biosphere, the eventual sink of radioactive pollutants. Modelling such a system means modelling the inventory of radionuclides, the processes that cause the facility to deteriorate and which cause in the long term the release of radionuclides, and their transport through the geosphere and spread over the biosphere, which will ultimately produce doses in human beings. All those models will be integrated as components of the system model.

Three types of uncertainty arise in the performance assessment of an HLW repository: model uncertainty, scenario uncertainty and input uncertainty. Model

Uncertainty in Industrial Practice Edited by E. de Rocquigny, N. Devictor and S. Tarantola,
© 2008 John Wiley & Sons, Ltd

uncertainty arises because of a lack of scientific knowledge of some of the processes involved in an HLW repository. Of the various models proposed to explain the processes, some may be useful; with others, the range of validity is not clear. Scenario uncertainty is related to the long life of the radionuclides considered. The permanence of the pollutants (for periods of typically up to millions of years) makes it necessary to remember that the system could undergo dramatic changes as a result of many possible causes (seismicity, ice ages, human intrusion, etc.), which could produce worse conditions than normally expected. Finally, some model inputs (coefficients in model equations and boundary and initial conditions) are poorly known, and their uncertainties need to be properly assessed. In this piece of work the focus is on uncertain model inputs alone.

In order to assess the safety of the repository system, the uncertainty related to some variables of interest should be characterized. This requires the propagation of input uncertainties through the system model representing the HLW repository. Several variables of interest could be considered to assess the safety of the repository, but the most widely used is the 'dose over time'. In this paper the use of an additional variable of interest is proposed: the 'peak dose over time'. The study aims to shed some light on the relationship between the uncertain model inputs and the two variables of interest through uncertainty and sensitivity analysis. Another objective of the study is to compare the outcomes of the sensitivity analyses for both variables of interest.

6.2   Study model and methodology

The model under study reproduces the behaviour of a high-level radioactive waste repository and the contaminant disposed of. The repository is considered without any geometric complexity, as just a point. Engineered barriers are modelled through a containment time during which there is no release. After the containment period, the contaminant starts releasing at a fractional constant rate. Only one radionuclide is considered in this study, ^{129}I. This radionuclide was selected because of its relevance in many safety assessments already performed worldwide. The contaminant is carried by groundwater through two consecutive geosphere layers to the biosphere, where it makes its way into a water stream from which exposed population take drinking water. This model has 15 inputs, nine of which are affected by uncertainty. These model inputs are the initial inventory of ^{129}I (M_0), its decay rate (λ), the dose conversion factor (β), and all the other inputs that characterize the physical-chemical properties of the near field, both geosphere layers and the biosphere.

There are four components in this system model, which will be described in the following subsections: the source term model, geosphere models and the biosphere model.

6.2.1   Source term model

The source term model consists of a simple delay in the release, after which a fractional constant release begins (the release being proportional to the remaining quantity of contaminant). During the whole period, the inventory also decreases due to radioactive decay. Therefore, during the containment period, the inventory of ^{129}I decreases according to the following ordinary differential equation (ODE):

$$\frac{dM(t)}{dt} = -\lambda M(t) \qquad t \leq T, \tag{6.1}$$

while, after the containment period, the inventory of ^{129}I decreases according to the following ODE:

$$\frac{dM(t)}{dt} = -\lambda M(t) - kM(t) \qquad t > T. \tag{6.2}$$

The initial condition is $M(0) = M_0$. The flow of contaminant escaping from the facility to the first geosphere layer is given by

$$S(t) = kM(t). \tag{6.3}$$

Both containment time (T) and release rate (k) are considered to be uncertain.

6.2.2   Geosphere model

The transport of contaminant through both geosphere layers is simulated in one dimension $(1-D)$. Each geosphere layer is characterized by its length $(L^{(i)}$, where i stands for the layers; $i = 1$ corresponds to the first layer and $i = 2$ to the second). The transport equation is

$$R^{(i)}\frac{\partial C^{(i)}}{\partial t} + V^{(i)}\frac{\partial C^{(i)}}{\partial x} - d^{(i)}V^{(i)}\frac{\partial^2 C^{(i)}}{\partial x^2} = -\lambda R^{(i)}C^{(i)} \tag{6.4}$$

where $R^{(i)}$, $V^{(i)}$ and $d^{(i)}$ stand, respectively, for the retardation coefficient, the groundwater velocity and the dispersion length in the corresponding geosphere layer, which is indicated by i. $C^{(i)}$ stands for the concentration of contaminant in any position x at time t, so that formally it should be considered $C^{(i)}(x,t)$. Velocities of groundwater, retardation coefficients and lengths of both layers are considered to be uncertain model inputs, whereas dispersion lengths are known inputs. Null concentration of contaminant in both layers is considered for the initial condition. For the specification of the boundary conditions, it is assumed that the contaminant flow rate into the first layer is the flow coming from the facility. Moreover, the flow rate into the second layer is equal to the flow rate from the first layer; similarly, the flow rate into the biosphere is equal to the flow rate from the second geosphere layer.

6.2.3   The biosphere model

The biosphere model is very simple. It is assumed that the contaminant coming from the second geosphere layer gets into a stream used for drinking water. Therefore, the dose is a function of the proportion of water drunk by individuals. Mathematically it can be formulated as

$$D(t) = \beta \frac{w}{W} G^{(2)}(t) \tag{6.5}$$

where $G^{(2)}(t)$ is the flow rate coming from the second geosphere layer into the biosphere, β is the dose conversion factor, w is the individual's annual water consumption rate and W is the flow rate of the biosphere water stream.

6.3   Underlying framework of the uncertainty study

6.3.1   Specification of the uncertainty study

Two variables of interest are considered in this study: the dose over time, $D(t)$, and the peak dose. The dose over time is a dynamic variable whose values change over time in each computer run. The peak dose takes only one value per run, the maximum dose across time, yet the time point when the maximum dose is achieved (time to peak dose) may change from run to run. In principle, the dose over time is considered the main variable of interest; the peak dose is considered in this study as an alternative/complementary variable of interest. The appealing characteristic of the peak dose is that it is an upper bound for the doses obtained in each run.

The main goal of this study is Goal U-UNDERSTAND: to investigate the relative importance of each input for the two variables of interest. An additional goal is to simplify the model (Goal A-ACCREDIT), by identifying the model inputs that have no influence on the variables of interest and hence can be fixed to some value within their ranges of uncertainty. Both goals are useful to improve the design of the repository system.

The quantities of interest considered in this study are means, variances and specific quantiles (e.g. 50%, 95% and 99%) corresponding to the dose at different times, and the peak dose. The sample mean estimates the mean of the dose for each time point, and the mean of the peak dose. The theory of order statistics provides the framework to estimate any quantile of both variables of interest, including associated confidence intervals.

Certain sensitivity analysis techniques may provide inconsistent results. For example, if linear regression analysis were used, a low coefficient of determination (R^2) would disqualify the analysis. Therefore, additional approaches are tested (i.e. variance-decomposition based techniques and graphical methods). Moreover, using different techniques makes it possible to study the system model from different points of view, which is always desirable.

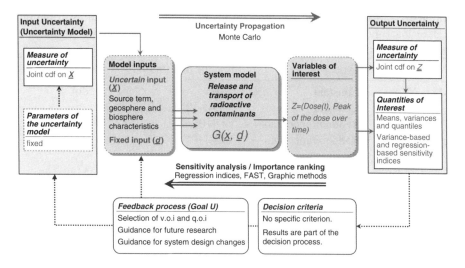

Figure 6.1 Overall uncertainty problem in the common framework.

6.3.2   Sources of uncertainty, model inputs and uncertainty model developed

As noted in the introduction, model input uncertainties alone are addressed in this study. Some model inputs are affected by random uncertainty, while for some others uncertainty is due to lack of scientific knowledge.

This model has 15 inputs: six of them are known (vector $\underline{d}$ of the model inputs), while the other nine are affected by uncertainty (vector $\underline{X}$ of the model inputs). Details of all the model inputs are given in Table 6.1. The uncertain inputs may be classified as:

- Uncertain inputs related to the source term: containment time (T) and contaminant fractional release rate (k);

- Uncertain inputs related to the transport of contaminant through the geosphere layers: water velocities $(V^{(i)})$, geosphere path lengths $(L^{(i)})$ and contaminant retardation coefficients $(R^{(i)})$. Index i indicates the layer;

- Uncertain input related to the biosphere: flow rate of the biosphere water stream (W).

The probability density functions for the nine uncertain inputs have been derived through a combination of experimental data and expert judgement. Complete independence between model inputs has been assumed. This strategy has proven useful in the first steps of any research project. The details of the model inputs are given in Table 6.1.

Table 6.1 Model inputs in the problem under study (U = uniform distribution, LU = logarithmic-uniform distribution; in both cases values shown in brackets denote lower and upper bounds).

Parameter	Type	Value/distribution	Units
M_0	Constant	10^2	Mol
λ	Constant	$4.41 \cdot 10^8$	y^{-1}
T	Random/uncertain	$U[10^2, 10^3]$	y
k	Random/uncertain	$LU[10^{-3}, 10^{-2}]$	y^{-1}
$V^{(1)}$	Random/uncertain	$LU[10^{-3}, 10^{-1}]$	$m \cdot y^{-1}$
$L^{(1)}$	Random/uncertain	$U[10^2, 5 \cdot 10^2]$	m
$d^{(1)}$	Constant	10	m
$R^{(1)}$	Random/uncertain	$U[1,5]$	Dimensionless
$V^{(2)}$	Random/uncertain	$LU[10^{-2}, 10^{-1}]$	$m \cdot y^{-1}$
$L^{(2)}$	Random/uncertain	$U[5 \cdot 10^1, 2 \cdot 10^2]$	m
$d^{(2)}$	Constant	5	m
$R^{(2)}$	Random/uncertain	$U[1,5]$	Dimensionless
w	constant	0.73	$m^3 \cdot y^{-1}$
W	Random/uncertain	$LU[10^5, 10^7]$	$m^3 \cdot y^{-1}$
β	constant	56	$Sv \cdot mol^{-1}$

6.3.3 Uncertainty propagation and sensitivity analysis

Monte Carlo simulation was used for the propagation of uncertainty. This is a well-known and well established method which makes it possible to map the uncertainty from the input space into the space of the variables of interest. The post-processing of the resulting sample is straightforward: many graphical techniques and statistical tools can be applied to gain information for the uncertainty analysis.

Three main sensitivity analysis techniques were applied: the Fourier Amplitude Sensitivity Test (FAST), Standardized Regression Coefficients (SRC) and Standardized Rank Regression Coefficients (SRRC). FAST (Cukier et al., 1978; Saltelli and Bolado, 1998; Saltelli et al., 1999) is a model-free technique that makes it possible to apportion the variance of the variable of interest to the different model inputs. The version of FAST used in this study provides estimates of main effects only (interactions are not addressed). Scatterplots and 'contribution to the sample mean' plots (Bolado et al., 2007) were also used.

Two different samples were taken. The first was a simple random sample of size 459. The selection of this sample size was based on the theory of tolerance intervals and order statistics (David and Nagaraja, 2003). The minimum sample size to obtain a one-sided 99%/99% tolerance interval from the sample is 459. Such a one-sided tolerance interval means that, a priori, the probability that at least the sample maximum is larger than the 99th percentile is 99%. Additionally, this sample size allows for the estimation of confidence intervals for interesting percentiles, as, for example, a 95% confidence interval for the 95th percentile of a variable of interest. This random sample was also used to do the regression analyses that allow the computation of the SRC and SRRC sensitivity indices. The second sample, of size 323, was obtained under the specific sampling scheme used by FAST (Saltelli and Bolado, 1998). While the random sample of size 459 can be used by a variety of sample-based sensitivity techniques (except FAST) in addition to uncertainty analysis techniques, the FAST sample can only be used to obtain FAST sensitivity indices.

Note that the quality of the samples has been checked before running the system model code. No pair of model inputs was found to show any statistically significant correlation and no sample input cumulative distribution showed any statistically significant departure from the theoretical cumulative distribution assumed.

6.3.4   Feedback process

The feedback process consists in fixing the model inputs that have been found non-influential through the sensitivity analysis. Such model inputs can be fixed at their nominal value in further steps of the safety assessment, reducing the effort and budget needed. The feedback process also affects inputs that have been identified as important in the analysis. Further research is planned to characterize better the uncertainty in these model inputs. This could also lead to an improvement in the system design.

Comparing both variables of interest is also a part of the feedback process. First they will be compared from the point of view of sensitivity. Influential model inputs will be identified for both variables of interest. Second, the study of their empirical distributions will be used to identify potential problems related to the estimation of confidence intervals for some quantities of interest. This information could be used to choose one or both to be used systematically as variables of interest in the safety analysis of this type of facility.

6.4   Practical implementation and results

6.4.1   Uncertainty analysis

The main target of the uncertainty analysis is to characterize, as accurately as possible, the variables of interest used to assess the safety of the repository. The characterization should be numeric and also graphic, in order to get as much information as possible from the samples used. A thorough analysis of the results may help to gain deeper insights into the characteristics of the variables of interest.

The figures below show some of the results of the uncertainty analysis. Figure 6.2 shows the evolution of uncertainty over time for the dose due to ^{129}I.

Table 6.2   Main study characteristics.

Final goal of the uncertainty study	To understand the model, identify the most influential inputs and their relative importance (Goal U). Model simplification is also a goal (Goal A).
Variables of interest	Temporal evolution of the dose. Peak dose over time.
Quantity of interest	Means, variances and different quantiles (50%, 95%, 99%) and associated confidence intervals.
Decision criterion	No formal decision criteria.
Pre-existing model	Model of release of radioactive contaminants from a high-level waste repository and their transport though the geosphere to the biosphere (system of coupled PDE's and ODE's).
Uncertainty setting	Standard probabilistic.
Model inputs and uncertainty model developed	Nine uncertain model inputs are considered, most of them affected by lack of knowledge uncertainty, only a few of them affected by random uncertainty.
Propagation method	Monte Carlo simulation.
Sensitivity analysis method	Variance-decomposition method (FAST). Regression techniques applied to raw values (SRC) and ranks (SRRC). Graphical methods (contribution to the sample mean plots and scatterplots).
Feedback process	To guide future research about the uncertain model inputs and to improve the design of the system. To support the choice of the variables and quantities of interest used to characterize the system.

The evolution of the 95th percentile of the dose, of its 95% confidence interval, of the mean dose and of the median are presented as the main quantities of interest.

An important conclusion obtained from the analysis of Figure 6.2 is the similarity between the trend in the curve showing the evolution of the mean and the trend followed by the 95% confidence intervals (the confidence band considering the evolution over time) for the 95th percentile. This implies that the mean is controlled by a few very large observations. The rest of the runs have almost no impact on the mean at any time. Additionally, at very early times the mean becomes even larger

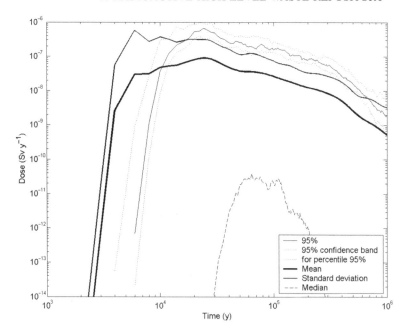

Figure 6.2 Evolution over time of the estimates of the mean, the median and the 95-th percentile (including its 95% confidence band) of the dose due to ^{129}I.

than the 95th percentile, which means that for that time period even fewer runs have a real impact. Another remarkable fact is the relation between the estimates of the mean and the estimates of the standard deviation over time. Standard deviation is approximately one order of magnitude higher than the mean at all times. The median is only significantly different from zero between $3 \cdot 10^4$ and $3 \cdot 10^5$ y. All these data show a picture of extremely (positively) skewed distributions, with a null lower bound and very long upper tails.

Figure 6.3 shows the evolution over time of the 459 runs considered in the study. Figure 6.4 and Figure 6.5 show the empirical distribution function of the dose for two times, 10^4 y and $5 \cdot 10^4$ y. These empirical distributions show the information corresponding to vertical cuts in Figure 6.2 and Figure 6.3 at the corresponding times. A relevant conclusion is that at most time steps only a fraction of the runs are non-null (approximately 23% at 10^4 y and 65% at $5 \cdot 10^4$ y). Moreover, non-null results are spread over ten orders of magnitude. This is in agreement with previous observations on the number of runs that effectively make a significant contribution to the mean.

In Figure 6.2, no confidence interval was given to the estimates of the mean, though they were provided for the 95th percentile. Exact confidence intervals may be given for any quantile estimate, provided that a minimum sample size is available (David and Nagaraja, 2003), independently of the distribution followed by the variable of interest (distribution-free confidence intervals). The case of the

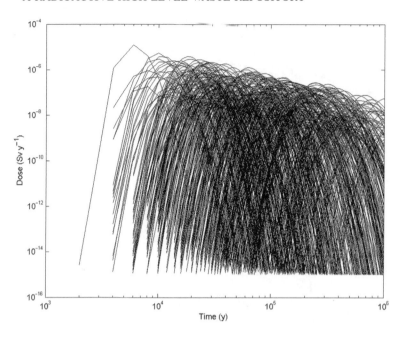

Figure 6.3 Evolution of the dose due to ^{129}I over time for 459 random runs.

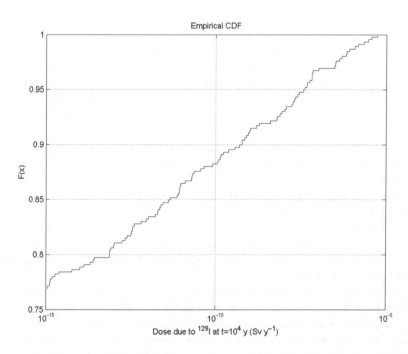

Figure 6.4 Empirical distribution function for the dose due to ^{129}I at t $=10^4$ y.

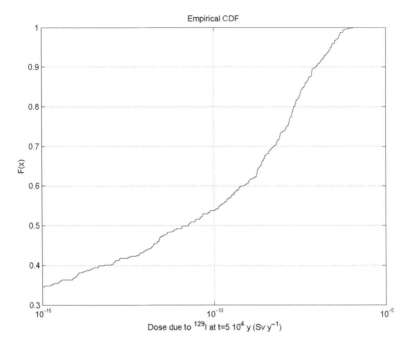

Figure 6.5 Empirical distribution function for the dose due to ^{129}I at $t = 5 \ 10^4$ y.

estimates of the mean is quite different. Exact confidence intervals may be given only for a few distributions (normal, log-normal and exponential, among others). If the distribution is not known, only approximate confidence intervals (assuming, for example, normality of the sample mean) may be assigned. Figure 6.4 and Figure 6.5 show the diversity of distributions that may be found for different time steps. As a result, when estimating confidence intervals for the mean dose, the reliability of confidence intervals is variable and unknown. Figure 6.6 shows the empirical distribution function for the peak dose. It is spread over only four orders of magnitude and fits quite well with a log-normal distribution (the p-value of the Kolmogorov test is 0.89). This means that an exact confidence interval may be assigned to the mean of the peak dose. Therefore, with available estimation techniques, quantiles and their confidence intervals may be equally well estimated for doses at different times and for peak doses; nevertheless, when estimating means, they may be correctly estimated only in the case of the peak dose; this cannot be guaranteed in the case of doses at different times.

6.4.2   Sensitivity analysis

This sensitivity analysis is based on two samples, the random sample of size 459 and the FAST sample of size 323. The former was used to estimate sensitivity indices based on regression models (SRCs) while the latter was used to estimate

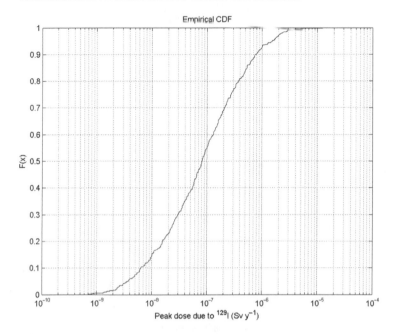

Figure 6.6 Empirical distribution function for the peak dose due to ^{129}I (log scale in x axis).

variance-based sensitivity indices (first-order effects' contributions to the variance of the variables of interest).

Figure 6.7 shows the results of the regressions (SRCs) of the dose over time for the nine uncertain inputs (raw values). It is difficult to draw conclusions from these results since R^2 never exceeds the value 0.2 and most of the time is around 0.1. The validity of this regression model for explaining the behaviour of the dose over time is very limited. Figure 6.8 shows the same type of results when the analysis is done on the ranks of the values. According to this analysis it is clear that $V^{(1)}$ is the most important model input. In fact, there is a clear correlation between the coefficient of determination of the regressions (R^{2*}) and the absolute value of the SRRC associated with $V^{(1)}$. The second most important inputs are $L^{(1)}$ and $V^{(2)}$ at early times and $R^{(1)}$ at late times. The rest of the inputs are practically non-influential.

Figure 6.9 shows the results of the FAST analysis. The results obtained are relatively different from the ones provided by the regression analysis. In this case, the two most relevant model inputs are W and $V^{(1)}$, usually in that order. The other inputs are definitely non-important. It is also essential to note that first-order effects are able to explain between 15% and 25% of the variability, depending on the time. This is not a great deal and shows the existence of important interactions. Figure 6.10 shows a contribution to the sample mean plot for the dose at 10^5 years. This plot shows that the impact on the mean W and $V^{(1)}$ is not the same in all their regions, clearly the lowest values of W contribute more to the mean of the dose

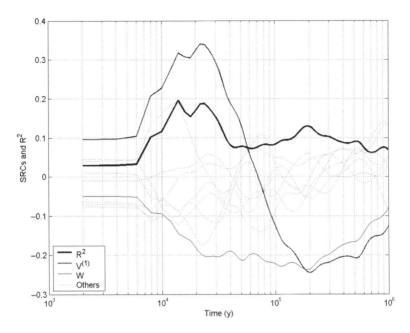

Figure 6.7 SRCs and R^2 versus time for the dose over time (raw values).

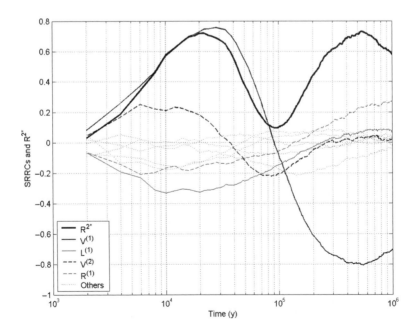

Figure 6.8 SRRCs and R^{2*} versus time for the dose over time (ranks).

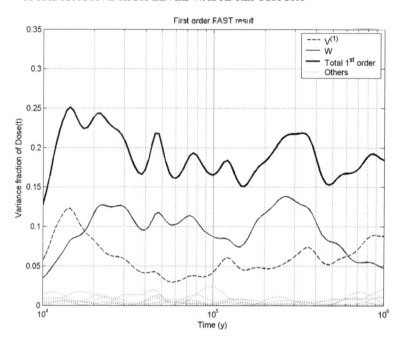

Figure 6.9 FAST sensitivity indices for dose over time.

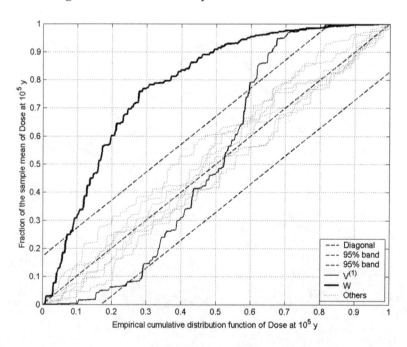

Figure 6.10 Contribution to the sample mean plot for the dose at 10^5 years.

Table 6.3 First-order results of FAST for peak dose and time to peak dose.

Uncertain model inputs	Fraction of the variance for time to peak dose	Fraction of the variance of the peak dose
T	0.00	0.00
k	0.00	0.00
$V^{(1)}$	0.50	0.17
$L^{(1)}$	0.10	0.01
$R^{(1)}$	0.08	0.02
$V^{(2)}$	0.00	0.02
$L^{(2)}$	0.00	0.00
$R^{(2)}$	0.00	0.01
W	0.00	0.24
First-order effects contribution to σ^2	0.68	0.47

than its largest values, while intermediate values of $V^{(1)}$ are the ones that really affect the mean of the dose at that time. This result is in agreement with the results provided by FAST.

Table 6.3 shows the results of the FAST analysis for the peak dose, and the time to peak dose. It can be seen that main effects are able to explain almost 50% of the variability of the peak dose. This means that interactions are responsible for only 50% of the variability of the peak dose, while they were responsible for up to 85% of the variability of the dose at some times. The most relevant inputs are W and $V^{(1)}$. The model explaining the behaviour of the peak dose is simpler than the model explaining the behaviour of doses at different times. In the case of the time to peak dose, up to almost 70% of the variability is due to main effects. In this case the most relevant inputs are related to the first geosphere layer, mainly $V^{(1)}$. Figure 6.11 shows the relation between peak dose and time to peak dose.

6.5 Conclusions

Uncertainty and sensitivity analyses were applied to the results of a computer code that describes the behaviour of a high-level waste repository. Two variables of interest were considered and compared: the dose over time and the peak dose. The objective of this study was to elicit information about the system model and to compare the outcomes obtained for both variables of interest. Regarding the analysis of the dose over time, the sensitivity techniques produce different results: while FAST identifies W and $V^{(1)}$ as the most important, regression techniques identify $V^{(1)}$ but not W. For the peak dose, FAST again identifies W and $V^{(1)}$ as the most important inputs, whose first-order effects together explain 40% of the variance of the peak dose. Some inputs, for example T, k, $L^{(2)}$ and $R^{(2)}$, show no influence on any

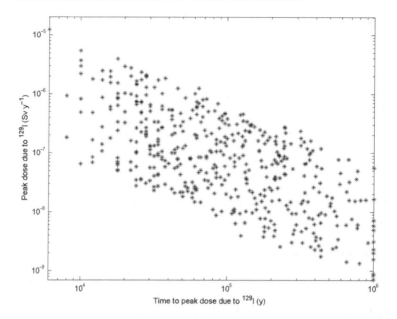

Figure 6.11 'Peak dose' vs. 'time to peak dose' (log scale on both axes).

variable of interest. They could be fixed to their nominal values with no real loss of information on the system. The analysis of the temporal evolution of the dose provides a comprehensive characterization of the model. It can be concluded that more accurate predictions of the dose over time and the peak dose could be obtained by focusing future R&D on improving knowledge of the inputs W and $V^{(1)}$.

Reliable confidence intervals can be estimated for any quantile of both variables of interest, provided that the sample size is large enough, and for the mean of the peak dose. The actual confidence associated with the confidence intervals for the mean of the dose over time is not well known.

References

Bolado, R., Castaings, W., Tarantola, S. *et al.* (2007) 'Estimation of the contribution to the sample mean and sample variance using random samples.' In *Proceedings of the Fifth International Symposium on Sensitivity Analysis of Model Output (SAMO) 2007.*

Cukier, R.I., Levine, H.B. and Shuler, K.E. (1978) Nonlinear sensitivity analysis of multiparameter model systems. *Journal of Computational Physics*, **26**, 1–42.

David, H.A. and Nagaraja, H.N. (2003) *Order Statistics*, Third edition, Chichester: John Wiley & Sons, Ltd.

Saltelli, A. and Bolado, R. (1998) An alternative way to compute Fourier amplitude sensitivity test (FAST), *Computational Statistics and Data Analysis*, **26**, 445–460.

Saltelli, A., Tarantola, S. and Chan, K.P.S. (1999) A quantitative model-independent method for global sensitivity analysis of model output. *Technometrics*, **41**(1), 39–56.

7

A cash flow statistical model for airframe accessory maintenance contracts

7.1 Introduction and study context

In transportation aeronautics maintenance costs are a driver of direct operating costs. Engines represent 45% of maintenance costs (TATEM). This case study was carried out in the field of engine maintenance contracts and focuses on line replacement units (LRU). Some contract conditions, namely guarantee periods and repair prices, have to be negotiated (Pajak, 2007). The buyer, airline and repair shop naturally all seek the most favourable contractual conditions. The seller or accessory manufacturer has concerns relating to cash flow, internal profitability rate and break-even time. Typically, such contracts cover a period of ten years. Therefore, a pre-existing statistical model forecasts an average cumulative cash flow curve, $Z(t)$, for each invoice calendar time, t, with percentile confidence interval $[Z_{min}(t), Z_{max}(t)]$ (Figure 7.1). The model was developed on Excel sheets (Massé, 2004). The curve changes according to contractual conditions.

7.2 The study model and methodology

7.2.1 Generalities

Considering a specific type of LRU, a representation of the cash flow $C(t)$ at the end of an invoice period t is

$$C(t) = N_o(t) \times (p_r - c) - N_u(t) \times c - c_s(t) \tag{7.1}$$

Uncertainty in Industrial Practice Edited by E. de Rocquigny, N. Devictor and S. Tarantola,
© 2008 John Wiley & Sons, Ltd

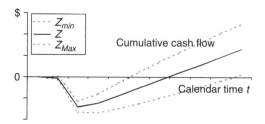

Figure 7.1 Pre-existing statistical model.

where $N_o(t)$ and $N_u(t)$ are the numbers of cumulated out-of-guarantee and under-guarantee LRU removals, p_r and c are the average sale price and the average cost of the repair of the LRU type under consideration, and $c_s(t)$, is the cumulative spare stock cost. The average cost c depends on the 'no fault found' (NFF) ratio and NFF cost. The cumulative spare stock cost, $c_s(t)$ at time t, of the considered invoice period is the sum, up to t, of the individual invoice periods' stock costs:

$$c_s(t) = n_s(t) \times p_r + \sum_{q \leq t} n_s(q) \times (s + m) \qquad (7.2)$$

where $n_s(q)$ is the stock size during period q, while s and m are the storage and stock management costs.

7.2.2   Level-1 uncertainty

At this stage the uncertainties are supposed to originate only in the times of occurrences of unscheduled removals of the LRU. Consequently, the cash flow $C(t)$, is a linear combination of two random variables $N_u(t)$ and $N_o(t)$ These are to be summed over all the LRU types included in the maintenance contract. From this fact, and from the central limit theorem, the distribution of the cash flow may be approached as normal, $(Z(t), \sigma^2(t))$:

$$Z(t) = E(N_o(t)) \times (p_r - c) - E(N_u(t)) \times c - c_s(t) \qquad (7.3)$$

$$\sigma^2(t) = \sigma^2(N_o(t)) \times (p_r - c) + \sigma^2(N_u(t)) \times c \qquad (7.4)$$

This approximate analytical approach has been validated by simulation.

7.2.3   Computation

To compute these statistical moments, the fleet entry-into-service schedule and average cumulative flight time per invoice period are considered (Figure 7.2).

From these data individual flight times under and out of guarantee for each set of a given LRU are cumulated for each invoice period. Let $u(t)$ and $o(t)$ be such

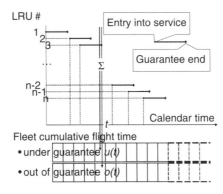

Figure 7.2 Cumulative flight times.

cumulative flight times under and out of guarantee at the end of a given invoice period t. Let F be the occurrence flight time cumulative distribution function (cdf) of the events leading to the LRU replacement under consideration.

If F is an exponential cdf (i.e. the LRU considered has a constant replacement rate), which is typical for electronic control units, the removal is said to be either 'on condition' or 'unscheduled'. According to practice, the distribution of the number $N_u(t)$ of replacements under guarantee until the end of the invoice period t is Poisson-distributed with parameter $E(N_u(t))$. Under such an hypothesis, and considering that the LRU removal times are independent, the variance $\sigma^2(N_u(t))$ of the number of replacements under guarantee is equal to $E(N_u(t))$. This simplifies to:

$$\sigma^2(N_u(t)) = E(N_u(t)) = \frac{u(t)}{\text{MTBUR}} \qquad (7.5)$$

where MTBUR is the mean time between unscheduled removals.

If F is not an exponential cdf (typical of a fuel pump or an actuator subject to ageing), there is usually a recommended preventative maintenance time between overhaul (TBO) with refurbishing as good as new. The mean time between removals is:

$$\text{MTBR}_{TBO} = \int\limits_0^{TBO} s \, . \, dF(s) + \text{TBO}(1 - F(\text{TBO})) = \int\limits_0^{TBO} (1 - F(s)) \cdot ds \qquad (7.6)$$

Considering that there is a relatively short TBO, the approximation below is acceptable:

$$\sigma^2(N_u(t)) \approx E(N_u(t)) \approx \frac{u(t)}{\text{MTBR}_{TBO}} \qquad (7.7)$$

The average removal cost is

$$c = \frac{c_f \times F(\text{TBO}) + c_o \times (1 - F(\text{TBO}))}{\text{MTBR}_{TBO}}$$

where c_f and c_o are the average replacement costs between and at overhauls, respectively. Equations (7.5) and (7.7) also apply to the number of removals out of guarantee $N_o(t)$ where $u(t)$ is replaced by $o(t)$.

7.2.4   Stock size

The sizes of the bases are continually adjusted to the traffic as the entries into service occur. The criterion for size adjustment is the minimization of cost under acceptable flight cancellation due to the stock rupture occurrence rate. The optimal stock size n_s for a given LRU type in a given base during an invoice period depends on:

- the cumulative flight time during the invoice period devoted to the storage base under consideration and, consequently, the number of bases;
- the turnaround calendar time (TAT) for the LRU and base under consideration.

7.3   Underlying framework of the uncertainty study

7.3.1   Specification of the uncertainty study

- **Pre-existing model and variables of interest**

The variables of interest are the cumulative curve of the seller's average cash flow forecast during the contract period $Z(t)$, and the confidence percentiles $Z_{min}(t)$ and $Z_{max}(t)$. As shown above, the pre-existing model is an analytical approximation. The only uncertainties considered in the model are the times of unscheduled removals.

- **Final goals of the uncertainty study**

The final goal of the uncertainty study is to determine which uncertain inputs are the most influential on the uncertainty in the variables of interest, thereby providing an understanding of the relative contribution of uncertainty sources (Goal U-UNDERSTAND). An additional objective of the study is to quantify the order of magnitude of the quantities of interest.

- **Quantities of interest and decision criteria**

The quantities of interest are the variances of the bounds $Z_{min}(t)$ and $Z_{max}(t)$, which constitute an aid to decisions related to the proposal of contract conditions (when the contract is negotiated) and logistics and supply chain management (when the contract has been signed).

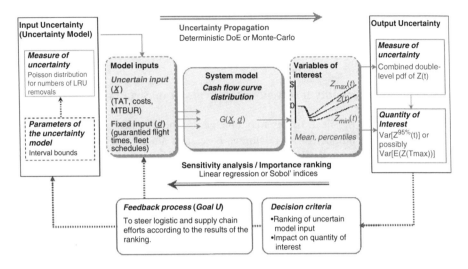

Figure 7.3 Overall uncertainty problem and common framework.

7.3.2 Description and modelling of the sources of uncertainty

- **Model inputs**

The pre-existing model takes into account the fleet entry-into-service schedule and average cumulative flight time per aircraft per year. The contract buyer implicitly covers the risks due to improper claim on these quantities. The pre-existing model also takes into account the conditions of the contract listed in the introduction; namely:

1. guaranteed flight times before unscheduled removal for each type of LRU; and

2. removal prices, in dollars, once the guarantee period has expired.

These values are subject to negotiation before the contract is signed. They are known with certainty once the contract has been signed. Finally the uncertain inputs that will be considered in this study are:

1. the number of spare-part bases and the traffic ratio devoted to each base;

2. the times for logistical turnaround (TAT);

3. the costs of removal, 'no fault found' (NFF), storage and stock management;

4. the dollar-euro exchange rate;

5. the ratio of removals with NFF; and

6. the LRU time-to-removal distribution parameters.

The LRU distribution parameters are summarised by MTBUR or MTBR (these in turn depend on the NFF ratio). The contract seller implicitly covers the risks due to improper evaluation of these parameters (consequently, this study focuses on these parameters).

• Sources of uncertainty involved: nature/types of uncertainty

During the life of the fleet, routes may change and new routes may be launched. Consequently, the number of spare storage bases and attached relative traffic may fluctuate. This is not under the control of the seller. It has an impact on the global stock size and consequently on the stock costs.

Logistics and supply chain events or neglected controls may occur for materials, sub-parts or sources of the accessories. This may have an impact on the logistical turnaround times (TAT) of the repair, i.e. the times between in-service LRU removal and repaired LRU return to storage bases. The number of bases has an impact on the necessary stock sizes. The evolution of the dollar-euro exchange rate is also unpredictable. Contracts are usually denominated in dollars, while European companies usually also have some costs in euros. Some supplies and subcontracting in dollar zones may dampen the fluctuations. Certain financial insurance methods may stabilize the fluctuations for a period. However, the model considers prices in their specific currencies at a given exchange rate.

Finally, the only uncertain inputs that are covered by the pre-existing model are the times to removal of the LRUs. These are handled through renewal processes involving Weibull's or, more simply, exponential distributions. However, distribution parameters, namely of size and shape (but especially of size), may evolve during fleet life, as well as knowledge of these parameters. Before entry into service of a new model of equipment, the reliability parameters are known only through field experience, similarity with other equipment, reliability test trends and sub-component reliability models. During servicing some unpredicted malfunctions may occur. Consequently, there are uncertainties in the time-to-removal distribution parameters. It can be seen that this last kind of uncertainty is of level 2: as the pre-existing model is already intrinsically probabilistic in its equipment reliabilities, the uncertainty setting may be considered double-level probabilistic.

• Building of the uncertainty model

In order to take these sources of uncertainty into account, two methods are used: an adjusted linear model by means of design of experiments (DoE) and estimation of Sobol' sensitivity indices (Sobol', 1993).

The DoE method is the most classic in the aircraft industry. It is part of 'six-sigma' quality (Deming, 1966; Schonberger, 2007) and robust design (Phadke, 1989). The linear model assumption is

$$Z = G(X^1, \ldots, X^n) + \varepsilon = \mu + a_1 X^1 + a_2 X^2 + \cdots + a_n X^n + a_{1,2} X^1 X^2$$
$$+ a_{1,3} X^1 X^3 + \cdots + a_{n-1,n} X^{n-1} X^n + \varepsilon \qquad (7.8)$$

where X_i are the normalized $(-1, +1)$ model inputs, Z is a variable of interest and ε is a standard, normally distributed error. The variance of Z is expressed by

$$\sigma^2(Z) = \sum_i \left(\frac{\partial G}{\partial x^i} (E(X^1, \ldots, X^n)) \right)^2 \cdot \sigma^2(X^i) + \sigma^2(\varepsilon) = \sum_i a_i^2 \cdot \sigma^2(X^i) + \sigma^2(\varepsilon)$$

(7.9)

Unlike the DoE approach, the Sobol' method offers the advantage of not imposing the quadratic hypothesis on the pre-existing model. The deterministic pre-existing model G is executed a certain number of times on an input sample generated with a quasi-random design, and a corresponding sample from the distribution of the variable of interest $Z = G(\underline{X})$ is obtained. The method of Sobol' provides a decomposition of the variance of the empirical distribution of the variable of interest and, correspondingly, sensitivity indices of different types can be estimated for all the model inputs. The 'first-order index' (or main effect), S_i, explains the sensitivity of Z due to X^i; the 'second-order index', S_{ij}, explains the sensitivity of Z due to the pure interaction between X^i and X^j, without considering the main effects. A 'total effect index', S_{Ti}, gives the overall effect of X^i (including its interactions with any other input) on Z. These sensitivity indices are appropriate in the case of independent inputs.

The DoE approach requires the upper and lower bounds of each uncertain model input. These are evaluated as follows:

- Discussions with the buyer will probably give some insight into how fleet routes are expected to evolve and into the relative traffic split. This includes the number of storage bases. The upper and lower bounds typically differ by one or two units.

- Dollar-euro exchange rate hypotheses are decided at company level.

- Turnaround times are a complex matter of logistics and supply chain management. Upper and lower bounds represent back-up and targeted values.

- Repair, 'no fault found' and storage costs combine euro and dollar values. Back-up and targeted values are considered for dollar/euro splits and unitary costs.

- Reliability parameters and 'no fault found' ratio bounds are a matter of field experience, reliability growth testing, reliability prediction, in-service event management and health management.

The first two items are out of the seller's control, in contrast to the latter three. Furthermore, hypotheses on these bounds may evolve during the contract period.

7.3.3 Uncertainty propagation and sensitivity analysis

Uncertainty propagation, with both DoE and Sobol' approaches, requires knowledge of the distribution of each uncertain model input. As there are no grounds for modal distributions, the distributions are assumed uniform between the lower and upper bounds. Under this assumption the variances are $b^2/12$, b being the bound amplitude under consideration.

Under the DoE linear model assumption of Equation 7.1, X^i being $[-1, +1]$ normalized and supposed uniformly distributed, Equation 7.9 simplifies to

$$\sigma^2(Z) = \frac{1}{3} \cdot \sum_i a_i^2 + \sigma^2(\varepsilon) \tag{7.10}$$

Consequently, an estimation of the variance of Z is given by

$$\hat{\sigma}^2(Z) = \frac{1}{3} \cdot \sum_i \hat{a}_i^2 + \hat{\sigma}^2(\varepsilon) \tag{7.11}$$

According to the final goal, the importance ranking of each input X^i may be viewed in terms of the mean effect of X^i on Z:

$$Effect(X^i) = \overline{Z}^{x_i^+} - \overline{Z}^{x_i^-}. \tag{7.12}$$

while the effect of the interaction between X^i and X^j can be defined as:

$$Effect(X^i, X^j) = \left(\frac{\overline{Z}^{x_i^+ x_j^+} + \overline{Z}^{x_i^- x_j^-}}{2} \right) - \left(\frac{\overline{Z}^{x_i^+ x_j^-} + \overline{Z}^{x_i^- x_j^+}}{2} \right) \tag{7.13}$$

being $\overline{Z}^{x_i^+}$ and $\overline{Z}^{x_i^-}$ the variable of interest Z calculated at $X_i = 1$ and $X_i = -1$, respectively, and averaged over the values of the other inputs.

7.3.4 Feedback process

Logistics and supply chain management are informed by the importance ranking of the model inputs and their impact on the variance in the variable of interest. If the variance is unacceptable in terms of financial risks for the company, the inputs with the highest sensitivity are further investigated in order to reduce their uncertainty and to obtain, at a second stage, a better prediction of the variable of interest.

Some upper and lower bounds for the model inputs are defined, respectively, as back-up and targeted values. This is the case for TAT and costs. For instance, the costs tend to become dependent on the dollar-euro conversion. A speed-up and enhancement of action towards targets may make it possible to move the back-up value towards the targeted value. This decreases the sensitivity index and the standard deviation of the variable of interest.

As in-service experience is capitalized upon, the reliability parameters are known with better accuracy. By contrast, the dollar-euro upper bound may increase.

7.4 Practical implementation and results

The approach described in this chapter is demonstrated on a typical maintenance contract for a fleet of engine electronic control units.

Table 7.1   Main study characteristics.

Final goal of the uncertainty study	Importance ranking of the input uncertainty sources (Goal U). Order of magnitude of the quantity of interest.
Variables of interest	Cumulative curve of seller's cash flow forecast during the contract period in mean, $Z(t)$, and with confidence percentiles, $Z_{max}(t)$ and $Z_{min}(t)$.
Quantity of interest	Variance of $Z_{max}(t)$ or possibly variance of $Z(t)$ at the end of contract period.
Decision criterion	No formal decision criteria: the decision process involves the consideration of model input rankings and their impact on quantity of interest.
Pre-existing model	Analytical approximation of the distribution of $Z(t)$ for each invoice at date t.
Uncertainty setting	Double probabilistic setting
Model inputs and uncertainty model developed	Uncertain: number of spare-part bases and the traffic ratio devoted to each base; the repair logistical turnaround times (TAT); repair and storage costs; the future \$/€ exchange rate (variability model); accessory reliabilities summarized by the time to removal distribution parameters (epistemic). Fixed: guaranteed flight times before unscheduled removal for each accessory; fleet entry-into-service schedule and average cumulative flight time per aircraft per year.
Propagation method	Monte Carlo and quadratic calculation based on ANoVA linear approximation.
Sensitivity analysis method	ANoVA linear model verified with Sobol' indices.
Feedback process	Logistic and supply chain management in accordance with sensitivity rankings and impact on quantity of interest.

7.4.1   Design of experiments results

The DoE approach is first used on $2^9 = 512$ experiments. The classical Fischer-Snedecor test, applied to the effects calculated by Equation 7.12, based on the ratio *sum squares of model/sum of squares of residuals*, is accepted for all inputs (the p-values are negligible). This makes it possible to reject the hypothesis that all

Table 7.2 Significant model inputs.

Model Input	p-value
MTBUR	0.00
TAT	0.00
Number of bases	0.00
$/€ rate	0.00
NFF ratio	0.00
Storage cost	0.00
Removal cost	0.00
Stock management cost	ns
NFF cost	ns

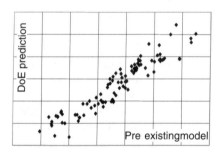

Figure 7.4 Match between the existing and the surrogate model.

model inputs are non-influential. The Student test applied to the a_i coefficients of the model (see Equation 7.8), shows that two model inputs are non-significant (*ns*), while for the others the p-value (probability of being null) is close to zero (Table 7.2).

The estimation of the quantity of interest is conducted on the reduced model, applied only to the significant inputs. In order to validate the reduced model, a sample is created of size 100 for each significant input, with the bounds considered in DoE. The predicted values show a good match with the actual values from the pre-existing model (Figure 7.4).

The effects of each significant input and the effects of the interaction between two inputs are plotted in Figure 7.5 and Figure 7.6.

Figure 7.5 shows that the most important input is MTBUR, which has a positive effect on cash flow. The second most important input is TAT, which, on the contrary, has a negative effect on cash flow. Figure 7.6 shows that MTBUR interacts mostly with TAT. In particular, looking at interaction patterns (not presented here), for a given increase of MTBUR, high TAT levels impact more on cash flow than low TAT values. Figure 7.6 also shows that MTBUR interacts with the number of bases

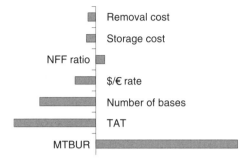

Figure 7.5 Importance ranking with DoE.

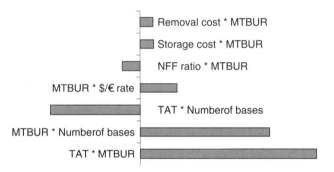

Figure 7.6 Interactions with DoE.

and their product has positive effects on cash flow. On the other hand, the product of TAT and the number of bases has a negative effect on cash flow.

7.4.2   Sobol's sensitivity indices

The Sobol' method was coded in Visual Basic in Excel. The distributions of the model inputs are the same as in DoE. Two samples, each of size 10000, were generated for the model inputs (see Equation 7.14) and the model was evaluated for each sample point. The sets of model realizations were used to estimate the first-order indices and the total effect indices (see results in Figure 7.7).

The first-order Sobol' indices for six model inputs (i.e. dollar-euro exchange rate, stock management cost, 'no fault found' cost, 'no fault found' ratio, removal cost and storage cost), are roughly null. These inputs have a negligible impact on the variability of the variable of interest. The results confirm those obtained from the linear model.

The largest first-order Sobol' indices are scored for MTBUR, the number of bases and the TAT. The MTBUR explains roughly 60% of the variability of the cash flow, while the TAT and the number of bases have an impact of 11%.

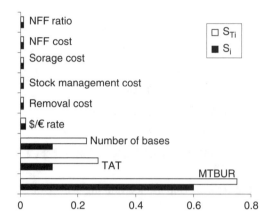

Figure 7.7 First-order, S_i, and total, S_{Ti}, Sobol' indices.

By comparing first-order Sobol' indices to total Sobol' indices for the three most important inputs, it is possible to discover whether the input is influential by itself or in conjunction with other inputs. The fact that the total index is somehow larger than the corresponding first-order index means that the three inputs contribute to the uncertainty of the cash flow, with some interactions as well.

The sum of first-order indices equals 0.89, very close to 1. This means that 11% of the variance of the cash flow is explained by such interactions.

The accuracy of the sensitivity method depends on the sampling size used. The estimates of the Sobol' index were replicated with another sample and a new set of simulations, in order to test for the robustness of the sensitivity indices. The new set of indices was found to be very similar to the previous, confirming the adequacy of the sample size used.

7.4.3   Comparison between DoE and Sobol' methods

The Design of Experiments approach makes it possible to identify influential inputs quickly. However, the results are based on a linear model with interactions, which involves a quite restrictive hypothesis. On the other hand, the Sobol' method does not require any hypothesis on the pre-existing model. Consequently, this method has more general validity. However it may be computationally expensive when the pre-existing model is expensive to run: in this study the analysis took 1.5 hours.

7.5   Conclusions

The pre-existing model takes into account the fleet entry-into-service schedule and cumulative flight times at invoice calendar times. The contract buyer implicitly covers the risks due to improper claim on these quantities. The pre-existing model also takes into account the guaranteed flight times before unscheduled removal for each LRU covered by the maintenance contract. These values are known without uncertainty once the contract has been signed. Finally, the uncertain model inputs, for which a sensitivity analysis was carried out, are the number of spare part bases and the traffic ratio devoted to each base, the repair logistic turnaround times (TAT), the repair and storage costs, and, finally, the accessories' reliabilities summarized by the mean time between unscheduled removals (MTBUR). This last kind of uncertainty is of level 2: the pre-existing model already being intrinsically probabilistic in its reliability component, the uncertainty setting can be seen as a double-level probabilistic.

The final goal is to rank the importance of the uncertain model inputs. This is required for the feedback decision criterion for logistics and supply chain management. Another objective is to show the order of magnitude of the quantity of interest and the relative contribution of uncertainty model sources. This can be done analytically through the linear model, which can therefore be seen as a surrogate model. Two kinds of sensitivity approaches were considered: the DoE and the Sobol' method. In this study, the DoE linear model with interactions yields the same results as the method of Sobol'.

The TAT, the MTBUR and the number of sites are the most influential inputs on the variance of the cash flow. Even if the reduced model is not really validated for forecasting, the DoE approach is acceptable for the detection of influential model inputs. It also helps to avoid a large number of expensive simulations.

References

AIR and COSMOS (2000) *Les Enjeux de la Maintenance*, N° 1774 Décembre 2000.

Deming, W. E. (1966) *Some Theory of Sampling*, New York: Dover Publications.

Massé J. R. (2004) '*Air fleet in-service guaranties inference.*' ESReDA 26th seminar and IMS '04, May 2004.

Pajak, F. (2007) *Hispano-Suiza Customer support centre*, available online at: www.hispano-suiza-sa.com

Phadke, M. S. (1989) *Quality Engineering Using Robust Design*. New Jersey: Prentice Hall.

Schonberger, R. J. (2007) *Best Practices in Lean Six Sigma Process Improvement*. New York: John Wiley & Sons, Ltd/Global Publishing Industry

Sobol', I. (1993) Sensitivity analysis for non-linear mathematical models, *Mathematical Modelling and Computational Experiment*, **1**, 407–414.

TATEM (Technologies and techniques for new maintenance concepts), available online at: www.tatemproject.com

8

Uncertainty and reliability study of a creep law to assess the fuel cladding behaviour of PWR spent fuel assemblies during interim dry storage

8.1 Introduction and study context

In the 1990s the *Commissariat à l'Energie Atomique* (CEA) undertook to study the long-term evolution of nuclear spent fuel in storage and disposal conditions in order to help decision-makers design a policy for long-term management of spent fuel. For this reason, the CEA and *Electricité de France* (EDF) launched a common research project in 1998 called PRECCI (French acronym for 'Research Programme on the long-term Evolution of Irradiated Fuel Assemblies'). The purpose of PRECCI was to predict the long-term evolution of spent fuel in storage and disposal conditions and the potential release of radionuclides in nominal and incidental scenarios. During dry storage, fuel cladding undergoes a deformation by creep under the effect of internal pressure and temperature, which can lead to a cladding rupture and consequently a release of radionuclides in the cask. The amount of information available on the subject is relatively substantial but certain physical mechanisms are still poorly understood.

The aim of this paper is to explain the interest in uncertainty and sensitivity analysis in the context of the research project, and especially to identify the main sources of uncertainty which further R&D efforts should address. The paper focuses on methodology – consequently, some *a priori* influential physical parameters and assumptions are not taken into account in the system model used here. A

Uncertainty in Industrial Practice   Edited by E. de Rocquigny, N. Devictor and S. Tarantola,
© 2008 John Wiley & Sons, Ltd

specific fuel with cladding of Zircaloy 4 was considered in the study. The approach developed in the paper is structured into three parts: statistical analysis of model responses based on Monte Carlo simulation, sensitivity analysis of the variables of interest according to uncertain model inputs, and reliability analysis, in order to determine which uncertainty sources need to be better understood or controlled. This approach is currently applied to more up-to-date and complete models.

8.2   The study model and methodology

The creep model for fuel cladding used in PRECCI is described in this section. The model is based on tests of creep behaviour of short duration (up to 60 days) performed on sections of cladding taken from low parts of the four-cycle-old fuel rods at the temperature range of (380 °C; 420 °C) and at the stress range of (150 MPa; 250 MPa) (CEA, 2005).

The temporal evolution of the circumferential strain ε depends on the cladding temperature T and the stress σ:

$$\frac{d\varepsilon(t)}{dt} = V_0 \times \sinh\left(\frac{\sigma(t)}{\sigma_c}\right) \times \exp\left(-\frac{T_a}{T(t)}\right) \times (\varepsilon(t) + \varepsilon_0)^{-p}$$

with $T_a = 32000\,K$, $\varepsilon_0 = 0.000045$, $\sigma_c = 34\,MPa$, $V_0 = 3.47 \times 10^7\,s^{-1}$ and $p = 0.9$.

The circumferential strain ε, without dimension, results from the integration of the differential equation given above: $\varepsilon(t + dt) = \varepsilon(t) + \dfrac{d\varepsilon(t)}{dt} \cdot dt$. The stress σ, which applies to the cladding, is given in the following expression:

$$\sigma(t + dt) \approx P_i(t + dt) \times \frac{D_{m0}}{2e_0} \times (1 + \varepsilon_c(t))^2$$

$$\approx P_i(t + dt) \times \frac{D_{m0}}{2e_0} \times (1 + 2\varepsilon_c(t)) \tag{8.1}$$

where $P_i(t)$ is the internal pressure of the rod (MPa) at the time t, D_{m0} is the average internal diameter of the cladding at the end of the irradiation (m), e_0 is the width of the cladding at the end of the irradiation (m) and ε_c is the conventional strain of the cladding related to the true strain ε by: $\varepsilon(t) = \ln(1 + \varepsilon_c(t)) \approx \varepsilon_c(t)$.

The evolution of the cladding temperature T depends on the temperature scenario (see Section 8.2.2), and more specifically on the following two uncertain inputs during the transport phase: the duration of the transport phase (Nj2, see Figure 8.1) and the maximal value of the fuel cladding temperature during this phase (Temp).

- The internal pressure $P_i(t)$ of the rod at time t is established from the burn-up TC (i.e. the total energy released per unit of mass in a nuclear fuel), the only operational uncertain input available to date, from the evolution of the cladding temperature T, the evolution of the free volume

V_v and the inventory of fission gases located within grain-boundaries at the end of irradiation FGB. The models of their evolutions have been built from databases developed at the CEA's Department of Fuel Study: $P_i(t) = G(TC, T(t), FGB, FGB\_A, V_v(t), Rpii)$, $V_v(t) = F(TC, Rvvi, d\varepsilon(t))$ and $FGB = H(Rfgb)$, where FGB_A is the duration of the release into the free volume of the fission gases located within grain-boundaries at the end of irradiation, and $Rpii$, $Rvvi$ and $Rfgb$ are three random variables that model the uncertainty around the experimental values for, respectively, the initial internal pressure, the initial free volume and the quantity of fission gases initially located at the grain boundaries in the fuel rod.

8.2.1   Failure limit strain and margin

The analysis of the creep tests results performed by the CEA for the Zircaloy-4 cladding material coupled with a theoretical study made it possible to establish a conservative criterion of failure adapted to the storage, which is expressed as an acceptable maximum strain according to the stress and to the amount of irradiation. The higher the irradiation, the more severe the criterion (Limon and Lehmann, 2004):

$$\varepsilon_r(t) = \frac{1}{2m(t)} \exp\left(-a\phi^b\right) \tag{8.2}$$

where

$$m(t) = \frac{\sigma(t)}{\sigma_c} / \tanh\left(\frac{\sigma(t)}{\sigma_c}\right) \tag{8.3}$$

where ϕ is the neutron fluence, and a and b are constants.

The margin, the main object of this study, is the minimum of the difference between the failure limit deformation and the true circumferential strain:

$$M_r = \frac{\min}{t \in [0; T]}[\varepsilon_r(t) - \varepsilon(t)] \tag{8.4}$$

where T is the duration of the scenario.

8.2.2   The temperature scenario

The deformation of the fuel cladding by creep is strongly dependent on the evolution of the temperature in the rod. It is assumed that this temperature follows a pattern of three phases (see Figure 8.1): during the *first phase*, the fuel is extracted from the cooling pool where it was stored at a temperature of 25°C, its temperature increases quickly up to a maximal temperature of between 400 and 420°C; the *second phase* corresponds to the conditioning and the transport of the fuel to the storage site over a period of a few weeks, during which the fuel temperature remains at its maximum; during storage, the *third phase*, the fuel temperature decreases slowly to reach a temperature of 150°C after a period of approximately five years.

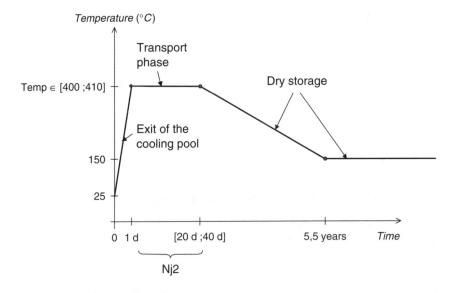

Figure 8.1 Temperature scenario for the fuel.

8.3   Underlying framework of the uncertainty study

8.3.1   Specification of the uncertainty study

Three variables of interest are considered in the study: the strain ε, the stress σ applied to the cladding, and the minimal margin M_r. The system model was defined in Section 8.2. As shown by the differential equations in Section 8.2, the variables of interest ε and σ are dependent and are a function of time. The minimal margin value M_r is a function of the strain ε and the acceptable maximum strain ε_r. The final goal of this study is Goal U-UNDERSTAND, i.e. understanding the influence of uncertainty sources, and prioritizing research efforts. In order to fulfil this goal, several quantities of interest are studied:

- the variance of the minimal margin, but also of the values of the strain ε and the stress σ at the time t_m of reaching of the minimal margin;

- an exceedance probability, defined by the minimal margin lower than 0. Such an exceedance probability is also called a 'failure probability' as 'M_r *lower than 0*' means the loss of the integrity of the cladding.

As this study is a part of a research programme not governed by an existing regulation, there are no formal criteria for the decision-making. The work was

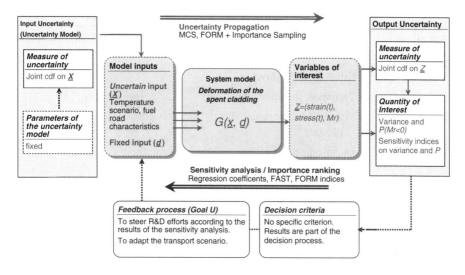

Figure 8.2 Case study in the common framework.

done in the standard probabilistic uncertainty setting. The uncertainty modelling is described in Section 8.3.2, below.

8.3.2   Description and modelling of the sources of uncertainty

The uncertain model inputs considered in the study are:

- two inputs linked to the transport scenario: the transport duration ($Nj2$) and the temperature during the transport phase ($Temp$);

- one input due to the burn-up range of the fuel inside the container: the burn-up (TC);

- two inputs to characterize the fuel after irradiation: the initial internal pressure ($Rpii$) and the initial free volume ($Rvvi$) in the fuel rod;

- two inputs to characterize the spent fuel: the quantity of fission gases (FG) ($Rfgb$) at the grain-boundaries and the duration of the release of a part of these gases into the free volume of the spent fuel rod (FGB_A).

The last three items refer to uncertainties of fuel properties. All these uncertain model inputs are assumed to be uniformly distributed and statistically independent. Uniform distributions were adopted because the only available information is that those inputs are bounded. Table 8.1 gives the minimal and maximal values, obtained principally through expert judgement.

Table 8.1   Description of the uncertain model inputs.

Uncertain model inputs	Description	Range interval
TC	Burn-up (GW.day.ton^{-1})	[30;60]
Nj2	Transport duration (d)	[20;40]
Temp	Maximal temperature during the transport phase ($^{\circ}$C)	[400;410]
FGB_A	Duration of the release of the FG initially located at the grain-boundaries (years)	[5;50]
Rfgb	Random variable characterizing the uncertainty in the quantity of FG initially located at the grain boundaries	[0;1]
Rpii	Random variable characterizing the uncertainty in the initial internal pressure in the fuel rod	[0;1]
Rvvi	Random variable characterizing the uncertainty in the initial free volume in the fuel rod	[0;1]

8.3.3   Uncertainty propagation and sensitivity analysis

Simple Random Sampling is used for the Monte Carlo method (Gentle, 2003; Hammersley and Handscomb, 1979; Rubinstein, 1981). For the minimal margin M_r the results of the uncertainty analysis are summarized by a scatterplot and a histogram. For the three variables of interest, the sensitivity analysis is based on Pearson correlation coefficients, supplemented, when the assumption of linearity cannot be proven, by Spearman correlation coefficients (Saltelli *et al.*, 2000) and Sobol' variance decomposition (Sobol', 1993). The exceedance probability that M_r is lower than 0 is computed with FORM-SORM methods (Melchers, 1999) and validated by Importance Sampling (Ripley, 1987). From the coordinates of the design point obtained by FORM some sensitivity results are computed: the importance measures that describe the contribution of the uncertain inputs to the probability, and the elasticities, in order to provide (if possible) some recommendations.

8.3.4   Feedback process

The aim of this study is to identify the main sources of uncertainty to which further R&D efforts should be devoted, with secondary aims of gaining a more precise description of the influent uncertainty sources or of modifying the transport scenario in order to make it safer. Table 8.2 summarizes the scope of the case study.

Table 8.2   Main study characteristics.

Final goal of the uncertainty study	Goal U: to understand the influence of the uncertainty sources and prioritize R&D efforts.
Variables of interest	Strain, stress applied to the cladding and minimal margin.
Quantity of interest	Variance analysis and associated sensitivity indices. For the minimal margin: histogram, and exceedance probability that the margin is below than 0 and associated sensitivity indices.
Decision criterion	No formal criterion. Results are part of the decision process.
Pre-existing model	One system model representing the evolution of the variables of interest with time (duration of the transport phase).
Uncertainty setting	Standard probabilistic framework.
Model inputs and uncertainty model developed	Several uncertainties linked to the temperature scenario and the characteristics of fuel rods. Uniform distribution modelled principally through expert judgement.
Propagation method	Monte Carlo, FORM and Importance Sampling.
Sensitivity analysis method	Pearson and Spearman coefficients, FAST, and FORM indices.
Feedback process	To steer R&D efforts according to the sensitivity analysis results. To adapt the transport scenario.

8.4   Practical implementation and results

The system model was implemented as a Matlab® procedure. Phimeca Software® was used for the exceedance probability calculations, and a coupling with Matlab was done. For the other calculations, functionalities available in some Matlab toolboxes were used.

8.4.1   Dispersion of the minimal margin

In this section, a short analysis of the dispersion of the minimal margin M_r is presented. Figure 8.3 shows the results of margin minima calculated by 1500 simulations according to time. It is observed that no negative margin, thus no

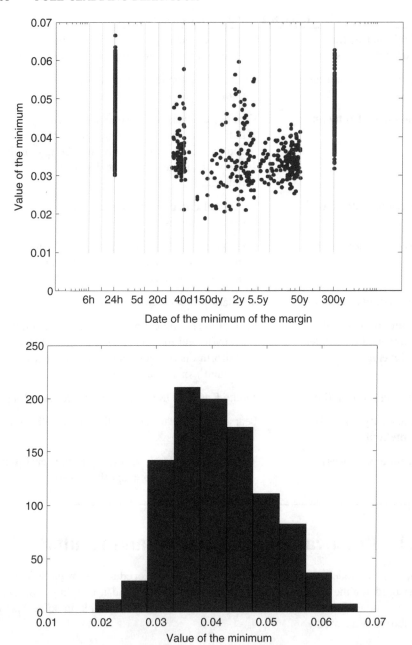

Figure 8.3 Position of the minimum of the margin (1500 simulations) and histogram of the minimal margins over 300 years (1000 simulations).

failure, was obtained, but especially that there are many simulations in which the minimal margin was obtained at the exit of cooling pool (24 h) and at 300 years. The histogram of the values of the minimal margin over 300 years for 1000 simulations is rather regular and mono-modal.

8.4.2   Sensitivity analysis

• First sensitivity analysis

The quantities of interest are the variance of the minimal margin M_r, but also of the values of the strain ε and the stress σ at the time t_m of reaching of the minimal margin, denoted respectively as $\varepsilon(t_m)$ and $\sigma(t_m)$. This analysis is based on the calculation of correlation coefficients (Pearson and Spearman), supported or not by the calculation of the coefficient of determination R^2 and the coefficient of determination based on ranks R^{2*}. The Pearson coefficients investigate the existence of linear relationships between uncertain model inputs and variables of interest. On the other hand, the Spearman coefficients are calculated on ranks and investigate the existence of a monotonic relationship between uncertain inputs and variables of interest. The determination coefficient R^2 is generally one of the best criteria to judge goodness of fit, in particular for the linear regression. The closer it is to 1, the more the adjustment can be regarded as 'good'. The rank determination coefficient R^{2*} corresponds to the same calculation as R^2 but on the rank of the data and is used to characterize the existence of a monotonic relation between the uncertain inputs and variables of interest.

Table 8.3 reports the results where only coefficients higher than 0.5 (in bold) are regarded as significant. The values of R^2 close to 1 for σ and M_r show that there is a close-to-linear relation between the uncertain model inputs and these variables of interest, which is not the case for the $\varepsilon(t_m)$. Consequently, it can be concluded that the information provided by the Pearson coefficient should be of interest for the variables of interest $\sigma(t_m)$ and M_r. The values of R^{2*} close to 1 for $\sigma(t_m)$ and M_r show that there is a monotonic relation between the input and these variables of interest, which is not the case for the $\varepsilon(t_m)$. Consequently, it can be

Table 8.3   Correlation coefficient of Pearson and Spearman.

Pearson coef.	R^2	FGB_A	TC	Temp	Nj2	Rfgb	Rpii	Rvvi
$\varepsilon(t_m)$	0.19	−0.2	0.34	0.17	−0.20	0.13	−0.13	0.12
$\sigma(t_m)$	0.79	−0.13	**0.67**	0.003	−0.13	0.11	**0.62**	−0.04
M_r	0.9	0.11	**−0.87**	0.004	0.11	−0.09	−0.42	−0.002
Spearman coef.	R^{2*}	FGB_A	TC	Temp	Nj2	Rfgb	Rpii	Rvvi
$\varepsilon(t_m)$	0.25	−0.20	0.36	0.31	0.05	0.16	−0.07	0.12
$\sigma(t_m)$	0.83	−0.10	**0.68**	−0.014	−0.04	0.08	**0.64**	−0.036
M_r	0.90	0.10	**−0.87**	0.002	0.02	−0.08	−0.4	−0.007

stated that the relation with $\sigma(t_m)$ and Mr is linear and monotonic, and the burn-up TC is the most influential uncertain model input for both responses. The effect of variations in $Rpii$ is comparable for $\sigma(t_m)$.

- **Sensitivity analysis in non-monotonic contexts**

In the previous analysis, the indicators R^2 and R^{2*} could not highlight the existence of a linear relation between ε and the uncertain inputs for the creep law. Therefore, a more computationally intensive, variance-based sensitivity analysis is carried out using the FAST method, implemented with the software *FAMAT* (FAST on MATLAB developed at CEA). For 14000 runs per uncertain input, the time required to estimate the total sensitivity indices is approximately five days (for a PC Pentium, 2 GHz, 512 Mb of RAM).

Table 8.4 highlights that the variability of the model outputs is influenced primarily by the burn-up. Very high total sensitivity indices of $\varepsilon(t_m)$ are observed for all the uncertain inputs. Indeed, the calculation of the strain is expressed using a formula where all the terms are products; it implies that the terms of interactions are important and it explains why all uncertain inputs contribute equally to the variance of $\varepsilon(t_m)$. This application illustrates the ability of the FAST method to capture interactions, in contrast to the sensitivity indices used previously.

8.4.3 Exceedance probability analysis

The purpose of this analysis is to determine the rate of rod failures against time. A Monte Carlo method seems unsuitable as a very low exceedance probability is expected due to the results of Section 8.4.1. Since there are some difficulties of convergence for the application of FORM, it is decided to apply a penalizing transformation[1] to the definition of the margin,

$$M_r = \varepsilon_r - \varepsilon \quad \Rightarrow \quad M' = \frac{\varepsilon_r}{\varepsilon \times k} - 1$$

Table 8.4   Total sensitivity indices by FAST.

Total indices of sensitivity	TC	Temp	Nj2	FGB_A	Rfgb	Rpii	Rvvi
$\varepsilon(t_m)$	0.95	0.84	0.82	0.84	0.82	0.86	0.84
$\sigma(t_m)$	0.61	0.07	0.08	0.08	0.07	0.46	0.08
M_r	0.84	0.06	0.05	0.05	0.05	0.22	0.05

---

[1] Another choice for the transformation is to penalize an uncertain or a fixed model input. Such an approach is useful if the penalized input is a possible controlled variable, such as a design variable or an environmental variable. Among the set of uncertain inputs in this application, only the variables defining the transport phase are of this kind. As no argument has been found for justifying a change in the bounds of the uncertainty range, the transformation used above has been preferred.

Table 8.5   Results relative to the coefficient k of the margin M'.

k	FORM (P_f)	SORM (P_f)	Importance Sampling (P_f)	Variation coefficient	FORM index (β_{HL})	SORM index (β_S)	IS index (β_{IS})
7.34	1.16 10^{-4}	6.16 10^{-6}	9.7 10^{-6}	31%	3.68	4.37	4.3
7.5	2.51 10^{-4}	3.51 10^{-5}	8.5 10^{-6}	23 %	3.48	3.97	4.3

Table 8.6   Importance factors relative to the coefficient k of the margin M'.

k	FGB_A	Nj2	Rfgb	Rpii	Rvvi	TC	Temp
7.34	0.036	0	0.126	0.17	0.025	0.533	0.11
7.5	0.027	0	0.134	0.18	0.022	0.519	0.12

with k a positive real greater than 1, and to carry out new FORM-SORM calculations with the obtained margin M'. Importance Sampling is then applied using 100 sampling points, from a Gaussian distribution centred on the design point obtained by FORM.

Table 8.5 presents the results for the exceedance probability (P_f) calculations carried out for various values of the coefficient k of the margin M'. The first results of convergence are obtained starting from $k = 7.34$. The Importance Sampling result shows that the SORM result, by the Breitung formula, is a good approximation of the exceedance probability; the SORM result is inside the approximated 95% confidence interval of the Importance Sampling result. This probability is very small, and, in combination with the additional margin defined by k, it can be considered that the probability of M_r being lower than 0 is negligible over 300 years.

Table 8.6 shows that the exceedance probability comes mainly from the burn-up (TC), and a group of three variables of *a priori* equal importance (*Rpii, Temp, Rfgb*). If it is necessary to reduce the exceedance probability, the elasticities for the reliability index and the exceedance probability could be used; these elasticities are defined as a normalized sensitivity to the standard deviation and the average for each uncertain input. The analysis of the elasticities shows that a decrease in the exceedance probability could be obtained by a decrease in the standard deviation of the burn-up (*TC*).

8.5   Conclusions

This paper illustrates the relevance of uncertainty and sensitivity analysis in the context of a particular research project, through a description of an industrial application. The Pearson and Spearman coefficients highlighted the strong influence of the burn-up *(TC)* on the calculation of the margin. The variance-based sensitivity analysis confirmed the dominating influence of this uncertain model input on all the variables of interest and the sensitivity of the strain to all the uncertain inputs. The exceedance probability that M_r is lower than 0 appears negligible over 300 years. However, this analysis highlighted the fact that only the burn-up *(TC)* influences the exceedance probability in the application presented. Regarding the feedback process, knowledge of the burn-up of the fuel rods and the fission gas release from the grain boundaries scenario should be improved. Moreover, some recommendations for the improvement of the transport process should be produced. As stated in the introduction, the system model and the data used are not state of the art, and some influential physical parameters and assumptions have not been taken into account. Consequently, the most influential parameters really could be otherwise. The system model is currently being updated through the integration of new information and data, and the application of the approach developed in this paper is ongoing.

References

CEA (2005) *Synthesis on the Spent Fuel Long Term Evolution*. CEA Report CEA-R-6084.

Gentle, J. E. (2003) *Random Number Generation and Monte Carlo Methods*, New York: Springer.

Hammersley, J. and Handscomb, D. (1979) *Monte-Carlo Methods*, London: Chapman & Hall.

Limon, R. and Lehmann, S. (2004) A creep rupture criterion for Zircaloy-4 fuel cladding under internal pressure. *Journal of Nuclear Materials*, **335**, 322–334.

Melchers, R. E. (1999) *Structural Reliability Analysis and Prediction* (2nd edition), Chichester : John Wiley & Sons, Ltd.

Ripley, B. D. (1987) *Stochastic Simulation*, New York: John Wiley & Sons, Ltd.

Rubinstein, R. Y. (1981) *Simulation and the Monte-Carlo Method*, New York: John Wiley & Sons, Ltd.

Saltelli, A., Chan, K. and Scott, E. M. (2000) (Eds) *Sensitivity Analysis: Gauging the worth of scientific models*, Chichester: John Wiley & Sons, Ltd.

Sobol', I. (1993) Sensitivity analysis for non-linear mathematical models, *Mathematical Modelling and Computational Experiment*, **1**, 407–414.

9

Radiological protection and maintenance

9.1 Introduction and study context

Maintenance actions in nuclear power plants may expose operators to ionizing radiations. Therefore, protections such as lead shields are installed to limit the global radiation dose received during maintenance interventions. But the choice of the optimal radioprotection solution – which makes the dose 'As Low As Reasonably Achievable' (the origin of the title of the ALARA approach) – is generally not obvious. Several radiological sources with various impacts on the different working locations have to be taken into account, and installing protections also implies exposure to radiation. A robust estimate of the total dose for each possible radioprotection scenario would obviously be helpful in making an appropriate decision (IAEA, 2002; ICRP, 1989).

As is shown in this chapter, *Electricité de France* (EDF) uses 3-D numerical simulation to estimate the expected doses, and numerous sources of uncertainty are identified in the process. In order to check the robustness of the final decision, uncertainty studies are now considered necessary. The final goals of the study are:

1. control of the order of magnitude of the dose of the chosen scenario (Goal C-COMPLY);

2. robust choice of the optimal radioprotection scenario under uncertainties (Goal S-SELECT);

Uncertainty in Industrial Practice   Edited by E. de Rocquigny, N. Devictor and S. Tarantola,
© 2008 John Wiley & Sons, Ltd

3. identification of main uncertainty contributors for future research efforts (Goal U-UNDERSTAND).

9.2   The study model and methodology

The total dose received by operators during the execution of a maintenance program depends on:

- the maintenance operators' behaviour, i.e. the path followed in the maintenance area and the durations of exposure at the different maintenance spots;

- the radiological context, i.e. the dose absorbed per unit time at these spots;

- the efficiency of the radiological protections used by the operators (a new radiological context).

For the determination of the radiological context (before and after the installation of protections), a software called PANTHERE has been developed. The inputs for this software consist in a 3-D description of the geometrical context (position, size, shape and material type of objects in the maintenance area – see Figure 9.1, below, for an illustration), and in a 3-D description of the radiological sources (position, size, shape, type of radionuclides, and overall activity).

All these model inputs are determined via expert knowledge and feedback of experience. The geometrical context is built using plans of the power plants, with some geometrical simplification (where justified) for certain complex objects in order to boost the speed of the numerical model. A database of radiological sources has been built over the years, to provide an image of the 'average' sources (in terms of radionuclide composition) that are encountered in the field; again, geometrical simplifications of the sources are made for computational reasons.

Figure 9.1 3-D representation of the geometrical context.

To describe the outputs of PANTHERE, the following notation will be used:

$$\underline{y} = H(\underline{x}_0, \underline{x}_1) = \begin{bmatrix} H^1(\underline{x}_0, \underline{x}_1) \\ \vdots \\ H^k(\underline{x}_0, \underline{x}_1) \end{bmatrix} \tag{9.1}$$

where $\underline{y} = \{y^1, \ldots, y^k\}$ is the vector of computed dose rates (the superscript i in y^i refers to the spatial location), $\underline{x}_0$ denotes the vector of main inputs (geometrical context – without protections – and radiological sources), and $\underline{x}_1$ denotes the optional inputs concerning radiological protections (position and characteristics of the protections). By convention, if no protection is taken into account, the dose is given by $\underline{y} = H(\underline{x}_0)$.

The final aim of a radioprotection study is to identify the optimal radioprotection strategy among a list of possible scenarios $s_1, \ldots, s_N$ ($N << 10$ in a typical study). The absorbed dose z_i associated with s_i, is estimated as follows:

$$z_i = G(\underline{x}_0, \underline{x}_{1,i}, \underline{t}_i, \underline{r}) = \underbrace{\left[\sum_{j \in P_i} t_i^j \times H^j(\underline{x}_0) \right]}_{\substack{\text{dose received during} \\ \text{installation of protections}}} + \underbrace{\left[\sum_{j \in M} r^j \times H^j(\underline{x}_0, \underline{x}_{1,i}) \right]}_{\substack{\text{dose received during} \\ \text{maintenance tasks}}}$$

$$\tag{9.2}$$

where P_i is the set of working locations for the installation of the radiological protection planned in scenario s_i, t_i^j is the exposure time to install the protection planned in scenario s_i at location j, $\underline{x}_0$ are the inputs related to the geometrical context and the radiological sources, M is the set of working locations where maintenance tasks are planned, r^j is the exposure time due to a maintenance task carried out at location j and $\underline{x}_{1,i}$ are the inputs related to radiological protections used in scenario s_i.

Once the dose z_i has been computed for each s_i, the decision-maker is interested in identifying the scenario that minimizes the total dose. However, uncertainties affecting the dose can make the choice of the optimal scenario difficult. First, it is impossible to forecast perfectly the timing and movements of the maintenance team and thus the durations of exposure $\underline{r}$ and $\underline{t}_i$. Secondly, PANTHERE does not give a perfect picture of radiological reality, for the following reasons:

- **modelling uncertainties**: geometrical simplifications have been considered in the code to make simulations faster;

- **input uncertainties**: the numerical values of PANTHERE inputs are affected by a lack of knowledge concerning the exact context of a study: objects' dimensions (e.g. the thickness of a pipe) are not known perfectly because of construction tolerances; furthermore, the average composition and activity of the radiological sources simulated in PANTHERE do not match those of the real sources;

- **numerical uncertainties**: the solution of the core equations of PANTHERE implies numerical integrations that may also contribute to uncertainty.

Until recently, these uncertainties were assessed in two ways. First, a preliminary adjustment of the solving algorithms was carried out in order to control for numerical uncertainties (this would be good practice for every numerical code). Secondly, some inputs of PANTHERE were estimated on the basis of a relatively small set of dose rate measures $\underline{y}_{mes}$ carried out on the field. The estimates were obtained by minimizing the cumulative quadratic distance between the measures $\underline{y}_{mes}$ and the model results $G(x_0)$,

$$\tilde{\underline{x}}_0 = \arg\min_{\underline{x}_0} \left\{ \sum_{i=1}^{k} (H^i(\underline{x}_0) - y^i_{mes})^2 \right\}, \tag{9.3}$$

and then the doses were calculated via the calibrated model. However, this procedure is not efficient because of the numerous uncertain inputs involved. Obtaining reliable estimates of the model inputs when there is no indication as to which input is most likely to be estimated (information that could be obtained through sensitivity analysis) can be very demanding. This is why a more comprehensive uncertainty treatment is now envisaged in radioprotection studies, the emphasis being on input uncertainties (Bouchet, 2006).

In this case study it is assumed that the maintenance programme refers to a system located in an area that includes eight radiological sources. Maintenance actions have to be carried out in two different spots. On the whole, the model G has approximately forty model inputs: three durations of exposure, and the others of geometrical and radiological context. Eight simplified radioprotection scenarios are considered: the i-th scenario consists in the installation of protection that removes the activity of source i.

A probabilistic framework is used: the input $\{x_0, x_{1,i}, \underline{t}_i, \underline{r}\}$ of Equation 9.2 becomes a random vector $\{\underline{X}_0, \underline{X}_{1,i}, \underline{T}_i, \underline{R}\}$. The probability distributions of all these random variables are chosen mainly by expert/engineering judgement. For some components of $\underline{X}_0$ – more precisely, the activity of the radiological sources – expert knowledge is not accurate enough; a second phase of calibration is therefore carried out by comparing measures and PANTHERE results to adjust their probability distributions (this can be seen as an extension of Equation 9.3 to this probabilistic context, and will be described in Section 9.3.2). In any case, the random variables are supposed independent mainly for the sake of simplicity, even if the introduction of dependencies is an important future prospect.

In this probabilistic framework, the estimate z_i of the dose for the i-th scenario computed via Equation 9.2 becomes a random variable Z_i:

$$Z_i = G(\underline{X}_0, \underline{X}_{1,i}, \underline{T}_i, \underline{R}) \tag{9.4}$$

It is then useful to compute for each scenario the mean expected dose μ_i, which quantifies the 'average efficiency' of the scenario, and the 95% confidence interval $[z_i(\inf), z_i(\sup)]$, which quantifies the uncertainty range around this mean value:

$$\mu_i = E[Z_i], \quad \Pr[z_i(\inf) \leq Z_i \leq z_i(\sup)] = 95\% \tag{9.5}$$

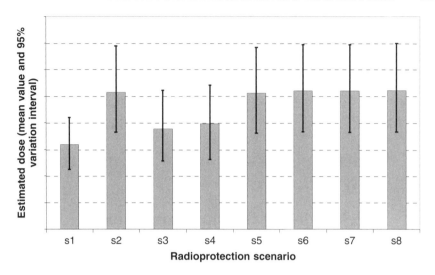

Figure 9.2 Estimation of the expected doses and of their uncertainties (95% variation intervals).

The mean dose values provide a preliminary ranking of the radioprotection scenarios in order of radiological dose. However, these rankings should be treated with caution, because the confidence bounds among the different scenarios can have different widths and (partially or totally) overlap with one another. For example, Figure 9.2 shows that scenario s_1 seems to be the most favourable, but its confidence bound overlaps with those of scenarios s_1 and s_3 (see Section 9.4).

The presence of such uncertainty does not mean that it is impossible to make a robust decision. Let us consider two scenarios s_i and s_j and the associated dose estimates $Z_i = G(\underline{X}_0, \underline{X}_{1,i}, \underline{T}_i, \underline{R})$ and $Z_j = G(\underline{X}_0, \underline{X}_{1,j}, \underline{T}_j, \underline{R})$. Moreover, assume that $\mu_i < \mu_j$ (so that scenario s_i seems to be the better choice), and that the uncertainty intervals $[z_i(\text{inf}), z_i(\text{sup})]$ and $[z_j(\text{inf}), z_j(\text{sup})]$ overlap (this is the case, for instance, in scenarios s_1, s_3 and s_4 in Figure 9.2). As seen in the definition of Z_i and Z_j, the sources of uncertainty $\underline{X}_0$ and $\underline{R}$ are common to both scenarios, while $\underline{X}_{1,i}$ and $\underline{T}_i$ are scenario-specific. It is then quite interesting to compute:

$$\Pr[Z_j < Z_i] = \Pr[G(\underline{X}_0, \underline{X}_{1,j}, \underline{T}_j, \underline{R}) < G(\underline{X}_0, \underline{X}_{1,i}, \underline{T}_i, \underline{R})] \qquad (9.6)$$

If this probability is close to zero, it means that the importance of the specific sources $\underline{X}_{1,i}$ and $\underline{T}_i$ is negligible compared to that of the common sources: choosing scenario s_i is then a robust decision, even if the 95% confidence interval is large. On the contrary, if $\Pr[Z_j < Z_i]$ is large, it means that the specific sources are important enough to put the final decision seriously into question. As will be seen in Section 9.4, the probabilities computed for the pairs (s_1, s_3) and (s_1, s_4) in the case study turned out to be reasonably low, confirming that the choice of s_1 is robust despite uncertainties. In any case, a large uncertainty interval for the dose is never satisfactory. In such cases the uncertain input sources that influence the dose most should be identified via sensitivity analysis.

In a typical investigation, one run of PANTHERE lasts a few minutes. Given this time constraint and the importance that these results can have for the planning of plant maintenance, it seems reasonable to spend a few hundreds of Monte Carlo simulations on the uncertainty and sensitivity analysis (1000 simulations were run in this case study).

9.3 Underlying framework of the uncertainty study

9.3.1 Specification of the uncertainty study

Figure 9.3 synthesizes the framework of the uncertainty study. The different steps of the process described are detailed in the subsequent paragraphs.

In order to achieve the three main goals of the study, the uncertainty study involves the computation of several quantities of interest capable of 'measuring' the uncertainty of the variables of interest. These q.i.s are:

- the mean expected dose μ_i which provides the average dose associated with scenario s_i;

- the standard deviation σ_i and the 95% confidence interval [z_i(inf), z_i(sup)], which provide the uncertainty of the estimated dose;

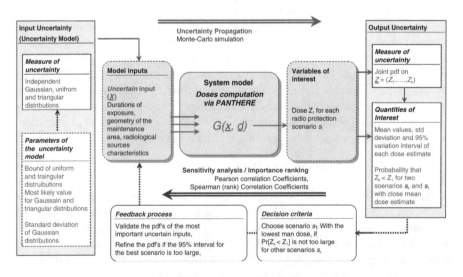

Figure 9.3 Overall uncertainty problem in the common framework.

- the probability $P[Z_j < Z_i]$ for two scenarios with close mean values $\mu_i < \mu_j$. This probability indicates whether or not the choice of scenario s_i is a robust decision.

9.3.2 Description and modelling of the sources of uncertainty

Recall the equation used to assess the dose Z_i for scenario $s_i : Z_i = G(\underline{X}_0, \underline{X}_{1,i}, \underline{T}_i, \underline{R}) = G(\underline{U})$. The inputs of the pre-existing model G can be classified into three categories:

Geometrical description of the location (in $\underline{X}_0$ and $\underline{X}_{1,i}$)

This category of inputs includes the geometry of the objects located in the maintenance area (e.g. pipes, or lead shields included in the radioprotection scenario): coordinates, dimensions (e.g. length, diameter and thickness of a pipe), and material composition.

Description of the radiological sources (in $\underline{X}_0$)

The second category of inputs refers to radiological sources. Each source is described by its coordinates, dimensions and composition (type of radionuclides and their activity).

Human behaviour (in $\underline{T}_i$ and $\underline{R}$)

The last category accounts for the path followed by maintenance operators in the maintenance area and the time spent at the various working locations.

The geometry of the location is not perfectly known: the information available comes mainly from conception plans, and the position of objects may vary slightly between two plants. Dimensions and material composition may also vary among similar objects since the specifications imposed on installers admit a certain tolerance. Moreover, some characteristics of the radiological sources (composition in terms of radionuclides, overall activity) may be imperfectly known. Another type of uncertainty comes from modelling simplifications carried out in the study. The shape of some objects and radiological sources is simplified for computational reasons. The third type of uncertainty is related to human factors. The behaviour of maintenance operators cannot be forecast perfectly in advance, and this has an impact on both the durations of exposure (a maintenance task can be carried out slightly slower or faster than expected) and the maintenance locations (the operator may be positioned slightly differently around the expected spot). Let us now examine how the uncertainty model has been built for these different types of uncertainties.

As far as geometry and material composition are concerned, plans and specifications provide 'most-likely values' (at least) and tolerance intervals (quite often). For each input concerned, this information is translated into terms of probability density function using simple rules. If the tolerance interval is judged 'absolute' by the experts (i.e. a value outside this interval is impossible), then a triangular distribution is used. Sometimes the tolerance interval is corrected if experts believe it necessary. The uncertainty of the thickness of pipes is, for instance, thought to be non-symmetric – unlike the theoretical tolerance interval – simply because the cost of raw materials makes the largest values of the interval less likely. If the tolerance interval is thought to be 'non-absolute' (i.e. values outside this interval are possible, even if unlikely) but symmetric, Gaussian distributions are used (e.g. the bounds of the tolerance interval correspond to 2.5% and 97.5% quantiles). 'Non-absolute and non-symmetric' cases have already been encountered. The inputs describing radiological sources are assumed to be normally distributed and background experience was used to choose the mean and standard deviation values.

For the inputs characterizing model uncertainty, uniform distributions are used to include all possible bounds provided by experts who have been studying the impact of geometrical simplifications in deterministic studies. Gaussian distributions are also used to model the uncertainty of the coordinates of the maintenance locations. The durations of exposure are the most difficult inputs to characterize because of the multiplicity of factors influencing the time spent at each location. Uniform distributions are used in this study. Note that, ideally, dependencies between the various model inputs should be acknowledged. However, at present, the choice of an appropriate model for such dependencies has been judged too difficult to make. Therefore, in this case study the model inputs are considered independent.

In most radioprotection studies a few measures of dose rates are available. They are used to update the uncertainty model by adjusting some of the model inputs' pdfs described above. A simplified solution to this inverse problem is proposed here. The aim is to find a diagonal matrix $\underline{\alpha}$ so that the dose measurements fit with the probabilistic model:

$$\underline{Y} = H(\underline{\alpha} \cdot \underline{X}_0) + \varepsilon_{mes} \qquad (9.7)$$

where $\underline{X}_0$ denotes the set of model inputs with pdfs initially chosen by experts, ε_{mes} denotes the measurement uncertainty, and $\underline{\alpha}$ is a diagonal matrix, which is used to calibrate the model (each source of uncertainty X_0^i is multiplied by a coefficient α^i). Only a few α^i's are allowed to be different from 1, and they concern the activity of some of the radiological sources. The matrix $\underline{\alpha}$ is estimated by the maximum likelihood (ML) method. This is greatly simplified by a property of PANTHERE equations: they are linear with respect to the activities of the radiological sources. Therefore, the previous equation can be re-written as follows:

$$\underline{Y} = \varphi(\underline{\alpha}) \cdot H(\underline{X}_0) + \varepsilon_{mes} \qquad (9.8)$$

where φ is a simple analytical function. Then, the optimization algorithm of the ML method requires only a preliminary assessment of the pdf of $H(\underline{X}_0)$ (obtained by

Monte Carlo simulation – 1000 runs of H); the CPU cost remains acceptable. Note that without the property of linearity, Monte Carlo simulations would have been required to assess the pdfs of $H(\underline{\alpha} \cdot \underline{X}_0)$ for numerous values of $\underline{\alpha}$: the problem would simply have been too CPU-consuming.

9.3.3   Uncertainty propagation and sensitivity analysis

Now that the probability distribution functions of the uncertain inputs have been chosen, the problem is to use this information in order to assess the quantities of interest defined in Section 9.3.1. Monte Carlo simulation is used for this purpose. Since all the q.i.s depend on the computation of the doses (Equation 9.4),

$$Z_i = G(\underline{X}_0, \underline{X}_{1,i}, \underline{T}_i, \underline{R}) = G(\underline{U}),$$

the steps are the following. First, a sample $\{\underline{u}_1, \ldots, \underline{u}_K\}$ is drawn randomly according to the distribution of $\underline{U}$. Then K runs of the pre-existing model G are carried out to obtain a sample $\{z_{i,1}, \ldots, z_{i,K}\}$ for all scenarios s_i. The estimates of the quantities of interest are then derived from a simple post-treatment. For instance:

$$\mu_i = E[Z_i] \approx \frac{1}{K} \sum_{k=1}^{K} z_{i,k}, \; \Pr[Z_j < Z_i] \approx \frac{1}{K} \sum_{k=1}^{K} 1_{\{z_{j,k} < z_{i,k}\}} \tag{9.9}$$

where 1 denotes the indicator function.

Monte Carlo simulation also provides confidence intervals for these quantities of interest to measure the degree of precision of such estimates. Confidence intervals are examined to determine whether the number K runs of the pre-existing model (mainly imposed by CPU time constraints) is sufficient. Monte Carlo simulation is not efficient if $\Pr[Z_j < Z_i]$ is close to 0. However, this is not a real problem, since the point is to compare s_i and s_j only if $\mu_i - \mu_j$ is close to zero (i.e. if the two scenarios have similar average dose). In such cases (see Section 9.4) a few hundred simulations are generally adequate to support a decision.

Some measures of importance can indicate which uncertain components of $\underline{U}$ are the major contributors of uncertainty. In this study the sensitivity analysis is carried out by estimating Pearson and Spearman correlation coefficients (Helton, 1993, Saltelli et al., 2000) using K equal to a few hundreds. However, these measures have some limitations. For instance Spearman correlation coefficients work correctly only if G is monotonic, as is the case here. The application of more rigorous variance-decomposition methods is envisaged for the future.

9.3.4   Feedback process

The uncertainty study does not end after this uncertainty/sensitivity step. Indeed, the sensitivity step helps to identify the inputs which are the major 'uncertainty drivers'. For these inputs, a refinement of their pdfs will automatically lead to more precise results (smaller confidence interval $[z_i(\text{inf}), z_i(\text{sup})]$) and more robust decisions.

Table 9.1   Main study characteristics.

Final goal of the uncertainty study	Robust choice of the optimal radioprotection scenario under uncertainties (Goal S). Control order of magnitude of the dose in the chosen scenario (Goal C). Identify main uncertainty contributors prior to future research efforts (Goal U).
Variables of interest	Vector of doses associated with each radioprotection scenario.
Quantity of interest	Average dose in each scenario and 95% confidence interval for the dose in each scenario. Probability that the dose in a scenario is smaller than the dose in other scenarios.
Decision criterion	Determine the scenario in which the dose is minimal.
Pre-existing model	PANTHERE 3-D simulation code.
Uncertainty setting	Standard probabilistic setting.
Model inputs and uncertainty model developed	3-D description of the maintenance area: probability distribution functions chosen either by expert judgement or via an inverse (estimation) method. Durations of exposure: probability distribution functions chosen by expert judgement.
Propagation method	Monte Carlo
Sensitivity analysis method	Rank correlations
Feedback process	Validation and potential refinement of the probability distribution functions of major uncertainty drivers.

Time constraints make it impossible to devote the necessary effort to characterizing uncertainty for all inputs. Therefore a preliminary study uses 'conservative' probability distribution functions which are not fully realistic, but which are believed by experts to encompass the true uncertainty. The uncertainty model is refined in a second step after the most important inputs have been highlighted through sensitivity analysis. Therefore these tools are extremely relevant, especially in this study where the number of uncertain inputs is relatively large (dozens of inputs). The following table summarizes the characteristics of the case-study.

9.4   Practical implementation and results

The parts of the uncertainty study consuming CPU-time are all based on the Monte Carlo method. This situation is quite easy to handle in practice, since each run of the pre-existing PANTHERE model is totally independent of all the others. Parallel operations of several computers can therefore be used without any difficulty in order to achieve satisfactory precision in the results. Today the time constraint seems to be a limit only to applying more thorough sensitivity analysis methods, such as the Sobol' variance decomposition (even if this possibility might be investigated in the future). Under these conditions, the implementation of the inverse method for the uncertainty model would certainly have been a problem, but, fortunately, the linearity of PANTHERE with respect to some of its inputs can be used to circumvent it.

The results of the case study were obtained via 1000 Monte Carlo simulations. The estimates of the mean dose and of the 95% variation interval can be found in Figure 9.2, which answers to Goal C of the study (controlling the order of magnitude of the dose in a scenario). For every radioprotection scenario, the variation intervals match approximately $[\mu_i\text{-}30\%, \mu_i+30\%]$, and the coefficient of variation α_i/μ_i is close to 15%. This rather large uncertainty is a consequence of the pessimistic probability distributions chosen for the model inputs. Thus, the first outcome of the study is that further refinements of the uncertainty model and better assessment of the uncertainty of the model inputs are required.

The second outcome of the study, which concerns Goal S (robust choice of the optimal radioprotection scenario under uncertainties), is far more positive. Figure 9.2 shows that s_1 seems to be the best scenario (according to its minimum mean dose), but s_3 and s_4 are also possible candidates, given the overall uncertainty. Yet the Monte Carlo simulations provide the following results:

$$\Pr[Z_3 < Z_1] \approx 17\% (95\% - \text{confidence interval} = [16.6\%, 17.4\%])$$

$$\Pr[Z_4 < Z_1] \approx 9\% (95\% - \text{confidence interval} = [8.7\%, 9.3\%])$$

This shows that scenario s_1 can already be selected with satisfactory confidence. Note that the confidence intervals given above are obtained assuming that each Z_i is distributed normally (as shown by a statistical test applied to the 1000 simulations).

The last outcome of the study concerns Goal U (identification of main uncertainty contributors for future research efforts). Among the dozens of uncertain inputs in the case study, Pearson and Spearman correlation coefficients show that the main contributors of uncertainty to the dose are the durations of exposure $\underline{R}$ (two inputs). The absolute values of the correlation coefficients between the dose and each duration of exposure are rather high (around 0.5), while correlations with the other uncertain inputs have absolute values of below 0.2. Note that for the durations of exposure, Spearman and Pearson coefficients are quite close, which is quite natural given that the relationship between the dose and these inputs is linear (see Equation 9.2). These results suggest that the modelling of the link between the dose and the durations of exposure, at present quite simplified and conservative, will have to be further investigated.

9.5   Conclusions

An uncertainty study was carried out to analyse the robustness of radioprotection decisions. A probabilistic framework was used in a simplified example to model uncertainties in the inputs of EdF numerical simulation code PANTHERE. The various probability distributions were chosen mainly through expert judgement. An inverse method was also implemented in order to exploit dose rate measurements and to adjust the most poorly known aspects of the uncertainty model.

Monte Carlo simulation was then used to obtain an estimate of the uncertainty of the PANTHERE outputs, with a significant but acceptable CPU cost (a few hundred runs of the model). This uncertainty is summarized by a set of quantities of interest. Each quantity contributes to achieving one or several goals of the study: making a maintenance decision that is robust despite uncertainties, assessing the uncertainty of the radiation dose associated with the optimal scenario, and understanding the main inputs that explain this uncertainty.

This piece of work is a first approach to uncertainty in the radioprotection field for EdF. Two objectives may be pursued in the future: modelling the dependencies between sources of uncertainty, and refining the uncertainty model of uncertain inputs related to human factors.

References

Bouchet, J.L. (2006) *Intégrer la gestion des incertitudes dans la démarche métier. Radioprotection: un point clé pour valoriser un patrimoine ALARA. In: Séminaire de la Société Française de Radioprotection, 28–29th November 2006, Saclay.*

Helton, J.C. (1993) Uncertainty and sensitivity analysis techniques for use in performance assessment for radioactive waste disposal, *Reliability Engineering & System Safety,* **42**(2–3), 327–367.

IAEA, Agence Internationale de l'Energie Atomique (2002) Optimization of Radiation protection in the control of occupational exposure. *AIEA Safety Series,* **21**.

ICRP (1989) Optimization and decision-making in Radiological Protection. *Annals of the ICRP,* **55**.

Saltelli, A., Chan, K. and Scott, E.M. (2000) (Eds) *Sensitivity Analysis,* Chichester: John Wiley & Sons, Ltd.

10

Partial safety factors to deal with uncertainties in slope stability of river dykes

10.1 Introduction and study context

Approximately half of the Netherlands is situated below or at around sea level. Dykes have been constructed throughout the centuries to keep the sea out. Nonetheless, the dyke system has failed from time to time. Several failure mechanisms are known to cause dyke failure. The most important mechanism is failure due to overflowing or wave overtopping which causes the inner slope of a dyke to erode. This was the dominating mechanism during the 1953 flood in the south-western part of the Netherlands. Other important failure mechanisms are slope instability, piping, revetment erosion and the failure of a hydraulic structure.

Since the 1950s, probabilistic techniques have been developed to allow for an economically optimal design, and an optimal level of protection was determined in the 1960s. All known failure mechanisms have to be taken into account in any risk analysis to ensure reliability and to observe the established optimal safety criteria. This study focuses on the mechanism of slope instability of the inner slope, also called sliding or slope failure. The slope fails when it is unable to support its own weight (see Figure 10.1). Slope stability becomes a problem especially during enduring flood waves (TAW, 2001), when water pressures can develop inside the dyke's body, causing its weight to increase. Meanwhile, the resistance against sliding decreases because the water pressures lower the soil's shear strength. A number of design guidelines and regulations have been developed to avoid slope instability, thereby to ensure sufficient reliability. The guidelines are based on a

Uncertainty in Industrial Practice Edited by E. de Rocquigny, N. Devictor and S. Tarantola,
© 2008 John Wiley & Sons, Ltd

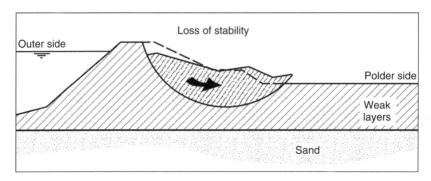

Figure 10.1 Slope instability of the inner slope (TAW, 2001). Reproduced with permission from ENW.

safety format using partial safety factors (PSFs) to guarantee a level of safety; however, an integrated scientific approach towards these partial safety factors is unavailable. In this study a scientific basis for partial safety factors is sought.

10.2   The study model and methodology

10.2.1   Slope stability models

The strength of a dyke with respect to slope stability is expressed in terms of a slope instability margin. In geotechnical literature, this margin is usually called the safety factor, but to avoid confusion the term slope instability margin (SIM) is used in this study. The SIM is defined by the driving force (Md) and by the maximal resisting force (Mr). Mr is a function of the shear strength; Md is a function of the weight of the soils. In the case that all quantities are perfectly known, a SIM higher than 1.0 means that the slope is stable and a SIM lower than 1.0 means the slope will fail. The driving force is the result of the weight of the dyke soils. The longer a flood wave lasts, the further the water penetrates into the dyke embankment and hence, the heavier the soils. The resisting force is determined by the shear strength of the soil. The shear strength may be expressed by two quantities: cohesion (pressure independent component) and friction angle (pressure dependent component). The determination of the SIM can be quite complex, since many possible failure planes could occur. Two kinds of models have been developed to calculate the SIM, based on the methods of slices and on Finite Element Methods (FEM) (see Figure 10.2). Models based on the methods of slices assume circular failure planes and may therefore not always give accurate results. With FEM models, the shear strength parameters are usually lowered until the structure looses its stability. The SIM can then be defined as the initial strength parameters divided by the parameters at failure. The failure planes are not restricted to being circular; they may take any shape. Although FEM is assumed to be more accurate than the slices methods, calculation times are usually longer. Note that the soil density acts both on the driving forces and on the resisting forces. This study focuses on the FEM model.

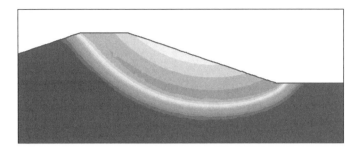

Figure 10.2 Calculating the Slope Instability Margin with a FEM model. The contours indicate the shear stresses.

10.2.2 Incorporating slope stability in dyke design

When designing or assessing the reliability of a dyke, the slope stability is usually calculated conditional to a normative water level. The normative water level depends on the location of the dyke (see Section 10.3.2). The dyke is usually higher than the normative high water. As the water is standing against the dyke, it will gradually infiltrate. The water level inside the dyke is called the phreatic line and is shown in Figure 10.3 with 'general'. This development of the phreatic line is a time dependent process and can be calculated with ground water flow models. The water infiltration influences the slope stability through the increase of the weight of the soil and the decrease of the shear strength.

In brief, the design of a river dyke would involve the following steps:

- Determine the dyke and subsoil profile; determine the normative water level and the corresponding water pressures (based on TAW, 2001).

- Take soil samples and determine characteristic (95% quantile) soil properties.

- Calculate design values by dividing characteristic values by the partial safety factors (PSF) (see Equation 10.1).

- Calculate the slope SIM with the design values – a SIM equal or larger than 1.0 means the dyke is stable. To ensure sufficient reliability, the SIM has to

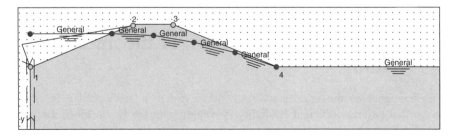

Figure 10.3 Slope stability check conditional to water level.

fulfil a minimal requirement, varying between 1.1 and 1.3, depending on the area's required level of protection and the dyke ring length.

$$c_d = \frac{c_{char}}{\gamma_c} \quad \varphi_d = \frac{\varphi_{char}}{\gamma_\varphi} \tag{10.1}$$

where γ_c is the partial safety factor of the cohesion, c_{char} is the characteristic value of the cohesion and c_d is the design value of the cohesion. γ_φ is the partial safety factor of the friction angle, φ_{char} is the characteristic value of the friction angle and φ_d is the design value of the friction angle.

The engineer has a number of options for adjusting the dyke design, should an insufficient SIM be found. Adding a stability term is one possibility; creating a less steep slope is another. Both measures usually result in a higher SIM. Safety is incorporated into this framework in several steps: the calculation is conditional to a normative water level, conservatively determined soil profiles and water pressures, the 95% quantile characteristic values and the partial safety factors.

10.2.3   Uncertainties in design process

In the design framework, as presented in Section 10.2.2, uncertainties are introduced in all four steps:

- schematization of the soil profile (dyke height, slopes, subsoil layers);
- schematization of the water pressures (water level, phreatic line, water pressures in subsoil layers);
- schematization of the soil properties (cohesion, friction angle, soil density);
- the choice of a calculation model (Bishop's method of slices, finite element method).

The calculation of the partial safety factor of the cohesion and of the friction angle is the objective of this chapter. Uncertainties in the soil profile and in the determination of water pressure are not taken into account, nor are model uncertainties. As will be explained in Section 10.3, this kind of design methodology implies a form of probabilistic treatment of uncertainty that fits within the generic framework of the book, although the domain-specific terminology, which comes from structural safety, may sound different.

10.3   Underlying framework of the uncertainty study

The final goal of the uncertainty study is to realize a methodology for dyke design that ensures sufficient reliability to comply with the Dutch Flood Defence Act. Traditionally, conservativeness is introduced into the calculation methodology (see Section 10.2.2) in several steps. One is by means of PSFs on two of the model inputs. One of the constraints of this chapter is to keep the PSF format intact, according to standard practice in this kind of engineering application in the Netherlands. In other words, the result of uncertainty treatment has to appear in deterministic terms, through factors covering the uncertainty of model inputs. However, the values of the factors should be the outcome of a probabilistic calculation, or probabilistic treatment of uncertainty. Safety is thereby incorporated in the 95% characteristic values, the PSF and the normative water level with corresponding phreatic line (see Section 10.2.2).

10.3.1   Specification of the uncertainty study

The approach developed by Ciria (1977) is used to determine the optimal PSFs. According to this approach, all uncertainties have first to be identified and covered in a safety format, before safety factors can be calculated. The safety format describes how to deal with the uncertainties and which (groups of) uncertainties are covered by which partial safety factors. The determination of a safety format is an iterative process. Some exploratory calculations have to be made and the practical implementation of the PSFs has to be considered. After the choice of a safety format, the PSFs are calculated. The whole framework is shown in Figure 10.4. Specific aspects of the framework are elaborated in the following sections.

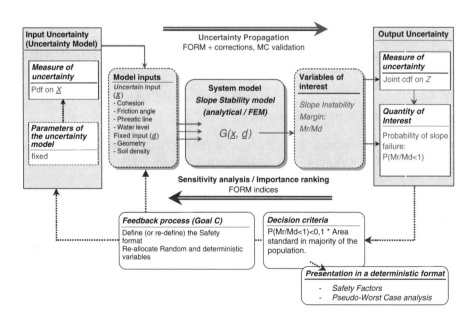

Figure 10.4 Overall uncertainty problem in the common framework.

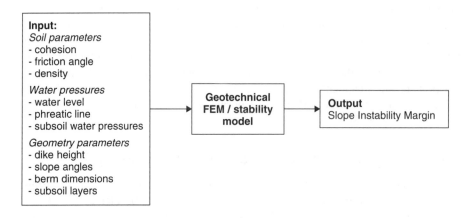

Figure 10.5 Input and output of the pre-existing model.

- **Pre-existing model and variables of interest**

In this study, the pre-existing model is a geotechnical FEM model as discussed in Section 10.2 (an analytical stability model could also be used). The model inputs and the variable of interest are summarized in Figure 10.5. The variable of interest is the SIM. The PSFs have to ensure sufficient reliability in case a dyke has been designed with a SIM larger than 1.0.

Different problems come to light when reliability analysis is carried out via FEM-based models; the most important is the CPU demand. It was therefore decided in this study to replace the FEM model with a simplified analytical limit state function as shown in Section 10.3.3.

- **Final goals of the uncertainty study**

The final goal of the uncertainty study is to comply with an absolute threshold (Goal C-COMPLY). A set of PSFs has to result in a probability of slope failure that is less than 10% of an area's safety standard. The Netherlands is divided into 53 dyke ring areas. Every dyke ring has a certain level of protection, the so-called safety standard. The safety standard per dyke ring is listed in the Dutch Flood Defence Act and varies from 1/1250 per year to 1/10000 per year. This means that a Normative water level with a probability of exceedance of, for example, 1/1250 per year has to be withstood. This general requirement has been elaborated by TAW (1989) into requirements for the different failure mechanisms of dykes. The maximum probability of slope stability somewhere along the dyke ring was determined to be 10% of the safety standard:

$$P_{slope,ring} = 1/10 \cdot P(WL \geq NWL) \tag{10.2}$$

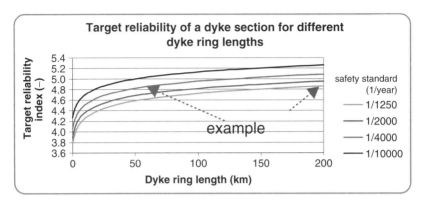

Figure 10.6 Target reliability as a function of the dyke ring length and safety standard.

where $P_{slope,ring}$ is the probability of slope failure somewhere along the dyke ring, WL is the water level, NWL is the normative water level and $P(WL \geq NWL)$ is the safety standard of the dyke ring. The maximum allowable probability of slope failure of a dyke section can be derived, taking into account the length of the dyke ring, the length of a section of dyke that fails, the correlation between dyke sections and the probability of flooding given slope instability:

$$P_{section} = \frac{1/10 \cdot P(WL \geq NWL)}{\left(1 + \alpha_{ring} \cdot \frac{L}{l}\right) \cdot P_{f|instability}}$$ (10.3)

where $P_{section}$ (equal to $P(Md{>}Mr)$) is the probability of slope failure of a dyke section; α_{ring} takes into account both correlation and differentiated contribution of dyke sections to total probability of failure. If $\alpha_{ring} = 1$, the dyke sections are dependent. If $\alpha_{ring} = 0$, the sections are independent. Values between 0 and 1 indicate a certain degree of correlation. TAW (1989) suggests an α_{ring} value of 1/30. L is the length of the dyke ring, l is the representative length of the slope failure and is an indication of the correlation length. TAW (1989) suggests a value of 50 m. $P_{f|instability}$ is the probability of flooding due to slope failure. (TAW, 1989, suggests a value of 1.0 in case slope failure is correlated with MHW.)

The result is the target reliability (derived from $P_{section}$) of a dyke section as a function of the dyke ring length and of the safety standard of the area (see Figure 10.6), keeping the representative length of failure and α_{ring} constant. No codes are available for the allowable deviations of the target reliability. There are many possible dyke configurations and it is impossible to take all into account. A reasonable choice has to be made to cover the whole dyke population.

10.3.2   Description and modelling of the sources of uncertainty

• Model inputs

The model inputs shown in Figure 10.5 are described below (details are given for uncertain inputs only). The *water levels* are best described by a frequency exceedance plot of the river discharges, in this case the river Rhine. Both the annual maxima and the peak over threshold method could be used, giving slightly different results. A Gumbel distribution is fit through 150 years of annual data to produce the frequency exceedance plot. This water flow distribution is then transformed into a cumulative distribution function (cdf) of the water level at a given location on the riverside. This is done through a hydraulic model not specified here. The final cdf of the water level depends on the location of the dyke. An example of the cdf of a water level is shown in Figure 10.11.

The phreatic line (PL, see Figure 10.3) influences the stability both by lowering effective stresses, and thus the shear strength, and by increasing the weight of the soils. Groundwater flow models are usually adopted to determine the position of the PL, but there is still uncertainty about the final position. A standard deviation of 0.5m for the PL is assumed in this study.

The partial safety factors should be applicable for all types of *soil* in the Netherlands. There are three levels of soil variability. First, although there are many classes of soil, usually only three are considered in dyke design: clay, peat and sand. Second, within each soil class, the relative contribution of c and φ can differ within the different types. Third, each property of a soil type has spatially distributed uncertainty. A two-step strategy is applied to incorporate soil uncertainties. First, the type and class is assumed to be known and only the intrinsic (spatial) uncertainty is propagated. Secondly, the different classes and types are accounted for in the downstream derivation of the PSFs. For a given type of soil, soil properties fluctuate in horizontal and in vertical directions (see Figure 10.7). Variations in soil properties can be characterized by the coefficients of variation (CV) and the fluctuation scales.

In relation to modelling those uncertainties with normal (or sometimes lognormal) distributions, international literature shows a wide range of CV for friction angle and cohesion (Kanning, 2005). A combination of these values supplemented with engineering judgement is used to determine the values for the Dutch situation (see Table 10.2).

Table 10.1   Coefficients of variation on strength parameters.

Shear strength parameter	CV before averaging	Effective CV
Cohesion	0.275	0.151
Friction angle	0.150	0.083

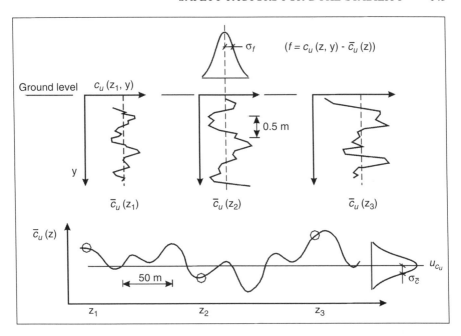

Figure 10.7 Fluctuation shear strength (Cu) in vertical and horizontal directions (TAW, 2001). Reproduced with permission from ENW.

In fact, the pre-existing model (0-D) takes only a point value for the soil properties as model inputs, whereas in reality this uncertainty is spatially distributed (2-D or even 3-D). A certain averaging of the spatial uncertainty reflected in those coefficients of variation has to be done to produce the CV of the point model input. Considerable geostatistical literature discusses this point. The horizontal fluctuation scales now determine only how many possible failure planes can occur in one dyke ring (see Section 10.3.1). A review of available sources shows that horizontal fluctuation scales vary between 25 and 100m, with an average of 50m (Kanning, 2005). The vertical fluctuation scales are relatively small compared to the height of a failure plane. Hence, the uncertainty in vertical direction partly averages out, resulting in lower coefficients of variation of the failure plane. A review shows that vertical fluctuation scales vary between 0.1 and 6m (Kanning, 2005), while failure planes can be up to 20m deep. The result of this averaging is that the CV over a failure plane, or effective CV, is approximately half of the point CV.

- **Sources of uncertainty involved: nature/types of uncertainty**

Soil uncertainties are usually regarded as epistemic uncertainties, since there is only one realization of the subsoil. However, due to the limited number of measurements, a very dense and unrealistically costly measurement density would be necessary to determine exactly the subsoil composition. Besides, measurement errors and non-reproducibility of the measurement results in significant uncertainties. There is still discussion on the classification of uncertainty in soil properties. For instance, in van Gelder (1999) these are regarded as inherent uncertainties in space because it will never be feasible to remove the uncertainty in soil parameters. The water level, on the other hand, is mainly an aleatory uncertainty. There is however an epistemic aspect to be found in the uncertainty of the distribution type and parameters. The position of the phreatic line is mainly a model uncertainty. Since the position of the line has to be predicted during water levels that have a very low probability, hardly any measurements are available and so models have to be used for the prediction. A component of the uncertainty in the phreatic line is epistemic: the input in the phreatic line prediction performs better if more data is available. The position of the phreatic line is determined conditional to the water level, hence incorporating the aleatory uncertainty of the water level.

10.3.3   Uncertainty propagation and sensitivity analysis

- **Stepwise simplification of the model before propagation**

A FEM analysis is usually time-consuming. For this reason, the FEM model is represented by an analytical limit state function (Z), based on the failure criterion that is defined in the FEM model.

$$Z = SIM - 1 \tag{10.4}$$

The previous section shows the many uncertain inputs in the FEM model. Many realizations need to be calculated to obtain the probability of failure, resulting in extensive calculation times. Simplifying the Z function to Equation 10.5 and applying linearizations requires only a few FEM calculations. This simplification is further elaborated in Section 10.3.3 and in Bakker (2005):

$$Z = SIM - 1 = G(X, d) = G(\varphi, c, NWL, PL, d) \tag{10.5}$$

The uncertain model inputs (X) include, as described in Section 10.3.2, four variables: φ, c, WL and PL. The deterministic inputs (d) are all other non-modelled uncertainties (geometry, density, etc.). For a given set of WL and PL, G can be approximated by an analytical function of φ and c around the result of a single FEM calculation. By applying FORM inspired corrections (see Section 10.3.3), $G(X,d)$ can be approximated using the steps described in the following paragraph.

- **Calculation procedure**

Instead of propagating the four uncertain model inputs together to determine the failure probability, which would need many FEM calculations, the general strategy is the following:

- Calculate SIM_1 with one FEM calculation, based on mean values of c and φ, and compute (analytically) the failure probability conditioned on both NWL and PL (i.e. $P(G < 0|PL, NWL)$) through FORM calculations (Madsen et al., 1986; Baecher and Christian, 2003) of the analytical model (very quick to do).

- Calculate SIM_2, the SIM with one standard deviation of the phreatic line, based on one FEM calculation. Compute the conditional failure probability including PL uncertainty (i.e. $P(G < 0|NWL)$) through a first-order FORM-inspired correction of the failure probability conditioned to PL and the variance of PL.

- Calculate SIM_3, the SIM with one decimation height deviation and all other variables with the mean values (Decimation height is the water level difference corresponding to exceedance probability difference of a factor 10). Based on SIM_3, build an (exponential) interpolation of the conditional failure probability as a function of WL ($P(G(X, d) < 0|WL)$).

- Calculate the total probability of failure by integrating the result of step 3 over the probability density of the water level.

Hence, only three FEM calculations are needed (one in each of steps 1, 2 and 3). This approach is based on a model developed by Bakker (2005).

- **Step 1: The conditional SIM model**

The limit state function of the conditional SIM model is:

$$Z = SIM - 1 \approx \frac{Cu - Cu_c}{Cu_c} \tag{10.6}$$

where Cu is undrained shear strength according to Equation 10.7, Cu_c is the undrained shear strength at failure according to Equation 10.8 and SIM is the slope instability margin.

$$Cu = c\cos(\phi) + P\sin(\phi) \tag{10.7}$$

where c is the cohesion, approximated by a normal distribution with μ_c and σ_c; φ is the friction angle with μ_φ and σ_φ and P is the average soil pressure, mainly determined by the dyke height and by the soil density. Equation 10.8 shows the relatively strong contribution of the friction angle to high values of P, which is mainly the case for relatively high river dykes. Cohesion is more dominant in the lower, more cohesive flood defences.

$$Cu_c = c_c\cos(\phi_c) + P\sin(\phi_c) \tag{10.8}$$

where c_c is the cohesion at failure according to Equation 10.9 and ϕ_c is the friction angle at failure according to Equation 10.10.

$$c_c = c_{char}/SIM_1 \tag{10.9}$$

$$\phi_c = \arctan(\tan(\varphi_{char})/SIM_1) \tag{10.10}$$

where c_{char} is the characteristic 95% quantile used for calculating SIM_1 in the FEM model. φ_{char} is the characteristic 95% quantile of the friction angle used for calculating SIM_1 in the FEM model. Actually, any value could be input into the FEM model (i.e. means), affecting the outcome of SIM_1, though the characteristic value is used to comply with the example of Section 10.3.3. SIM_1 is the slope instability margin corresponding to step 1, for which the values c_{char}, φ_{char}, mean PL (μ_{PL}) and the normative WL (NWL) are used. Equation 10.6 is evaluated with a FORM algorithm – the result is the reliability index (another way of expressing the probability of failure) $\beta_{c,\varphi|PL,NWL}$ conditional to PL and NWL.

• **Step 2: Interpolation of the phreatic line**

The influence of PL is accounted for by constructing a simple, linearized response surface (see also Bakker, 2005). One additional FEM analysis is needed (using a higher phreatic line, PL_2, instead of PL_1, see Figure 10.8; the result is SIM_2) to calculate the new reliability index according to Equation 10.11. In order to use Equation 10.11, PL must be assumed to be normally distributed and to have relatively minor variations. For the calculation of SIM_2, c_d, φ_d, $\mu_{PL} + \sigma_{PL}$ and NWL are used.

$$\beta_{c,\varphi,PL|NWL} = \frac{\beta_{c,\varphi|PL,NWL}}{\sqrt{1 + \left(\frac{\partial \beta_{c,\varphi,PL|NWL}}{\partial PL} \right)^2 \cdot \sigma^2(PL)}} \tag{10.11}$$

$$\frac{\partial \beta_{c,\varphi,PL|NWL}}{\partial PL} = \frac{\partial \beta_{c,\varphi|PL,NWL}}{\partial (SIM_1)} \cdot \left(\frac{SIM_1 - SIM_2}{PL_1 - PL_2} \right) \tag{10.12}$$

where $\beta_{c,\varphi,PL|NWL}$ is the reliability index including uncertainties in c, φ and PL, conditional to NWL. $\beta_{c,\varphi|PL,NWL}$ is considered in step 1; $\sigma^2(PL)$ is the standard deviation of PL and $\partial \beta_{c,\varphi|NWL}/\partial PL$ is calculated by finite differences using SIM_1 and SIM_2. $\partial \beta_{c,\varphi|PL,NWL}/\partial SIM_1$ follows from step 1 (not elaborated here).

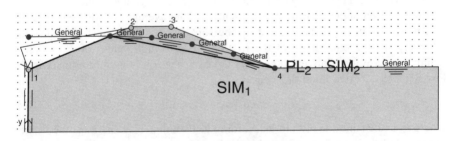

Figure 10.8 Calculating SIM1 and SIM 2.

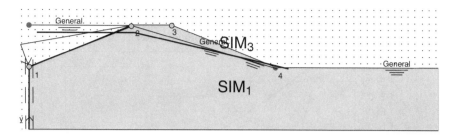

Figure 10.9 Incorporating WL.

• **Step 3: Failure probability as a function of the WL**

The determination of the failure probability as a function of WL is done in a similar way to step 2. One additional FEM calculation is needed, using WL_3 (NWL + additional height) instead of WL_1 (i.e. NWL, resulting in SIM_3 (see Figure 10.9). For the calculation of SIM_3, c_d, φ_d, μ_{PL} and $NWL + 0.55\,m$ (=decimation height) are used. The probability of failure as a function of WL is given by Equation 10.13 and 10.14 assuming a linearization around the normative water level NWL. The correction applied in step 2 may not be used because the WL behaves in a non-Gaussian manner.

$$Pf_{c,\varphi,PL}(WL) = \phi(-\beta_{c,\varphi,PL}(WL)) \tag{10.13}$$

$$\beta_{c,\varphi,PL}(WL) = \beta_{c,\varphi,PL|\mathrm{NWL}} + \frac{\Delta\beta_{c,\varphi,PL|\mathrm{NWL}}}{\Delta\mathrm{HWL}}(WL - NWL) \tag{10.14}$$

in which $\beta_{c,\varphi,PL|NWL}$ follows from step 2 and NWL is the normative WL, typically with an exceedance frequency between 1/1250 and 1/10000 per year. $\Delta\beta_{c,\varphi,PL|NWL}/\Delta WL$ is calculated in a similar way to step 2. The result is the probability of failure as a function of WL (see Figure 10.10).

• **Step 4: Calculating the probability of failure**

Using a Gumbel distribution of the water level (see Figure 10.11), integration results in the annual probability of failure including water level uncertainties according to Equation 10.1

$$Pf_{c,\varphi,\,\mathrm{PL,WL}} = \int\limits_{WL=0}^{WL=\infty} f_{Gumbel}(WL) \cdot Pf_{c,\varphi,PL}(WL)dWL \tag{10.15}$$

in which $f_{Gumbel}(WL)$ is the Gumbel distribution of the water level and $Pf_{c,\varphi,Pl}(WL)$ follows from step 3.

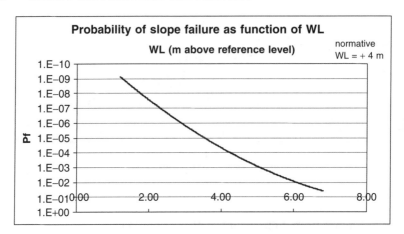

Figure 10.10 Failure probability (function of WL).

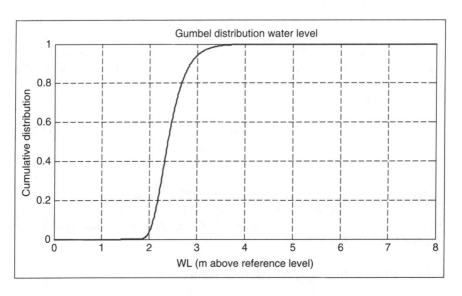

Figure 10.11 Gumbel distribution water level.

- **Propagation example of the conditional model with FORM and Monte-Carlo**

An example of a clay dyke with height = 5m and soil density = 18 kN/m³ is considered. The corresponding P is 32 kN/m². c is assumed to have a Gaussian distribution with $\mu_c = 6.6$ kN/m², $\sigma_c = 1.0$ kN/m², $\mu_\varphi = 28.3°$ and $\sigma_\varphi = 2.6°$. The *NWL* of the dyke is 4m, corresponding to an exceedance frequency of 1/1250

Table 10.2   Calculation example.

	c_d (kN/m^2)	φ_d($^{\circ}$)	PL(m)	WL(m)	SIM(-)	β(-)	Pf
Step 1	5	25	0	4	1.252	$\beta_{c,\varphi|PL,NWL} = 4.3$	8 10^{-6}
Step 2	5	25	+1	4	1.123	$\beta_{c,\varphi,PL|NWL} = 3.9$	5 10^{-5}
Step 3	5	25	0	5	1.132		
Step 4						$\beta_{c,\varphi,PL,NWL} = 4.9$	5 10^{-7}

per year with a decimation height of 0.55m. Furthermore, $\sigma_{PL} = 0.5$m. A FEM model is generated with the following inputs: $c_{char} = 5$ kN/m2, $\varphi_{char} = 25^{\circ}$, $WL_1 = NWL = +5$ m and PL_1 according to Figure 10.8. SIM_1 is calculated using these inputs. SIM_2 is calculated using $PL_2 = PL_1 + 1m$ according to Figure 10.8. SIM_3 is calculated using PL_1, and $WL_3 = WL_1 + 1m$. The results are presented in Table 10.2.

The calculated reliability index of 4.9 corresponds to a dyke ring with a length of over 200 km in the case of a safety standard of 1/1250 per year and a dyke ring of 60 km in the case of a dyke ring with a safety standard of 1/4000 per year (see Figure 10.6). The FORM sensitivity coefficients for c, φ, PL and WL are 0.56, 0.66, 0.39 and 0.31, respectively. The FORM approximation of step 1 allows a quick computation of failure probabilities with usually good results for linear failure limit state functions. In the case of very non-linear limit state functions, multiple design points may be encountered. The FORM approximation is validated with a crude Monte Carlo comparison. Ten million MC runs resulted in a failure probability of $5.5 \cdot 10^{-6}$, with a residual standard deviation of 13%. This is assumed to be accurate enough to validate the FORM results.

10.3.4   Feedback process

The feedback process initially includes the configuration of the safety format and finally the validation of the FORM calculations with Crude Monte Carlo (see Section 10.3.3). Initial calculations, taking into account all model inputs, show that the most important contributions to the probability of failure are cohesion, friction angle, water level and deviations in phreatic line. Other (geometrical) inputs usually do not vary significantly and are therefore deemed deterministic. The soil density influences both driving and resisting forces and does not influence the reliability very greatly. It is therefore also deemed deterministic. The safety format used ultimately includes two partial safety factors, one related to cohesion and one to the friction angle, according to current design practice. The uncertainties covered by the PSF include uncertainties in the soil properties and uncertainty due to variations in the phreatic line (which are not explicitly covered with a specific PSF but are included in the two PSFs on cohesion and friction angle – see Figure 10.4).

Table 10.3   Table of study characteristics.

Final goal of the uncertainty study	Goal C – compliance with a threshold. Safety factors to ensure that dykes comply with an area dependent threshold.
Variables of interest	Slope stability margin
Quantity of interest	Probability of slope failure $P(Mr/Md<1)$
Decision criterion	$P(Mr/Md<1)<0.1 *$ Area dependent standard
Pre-existing model	Geotechnical slope stability software
Uncertainty setting	Standard probabilistic (+ deterministic format)
Model inputs and uncertainty model developed	Soil (strength) parameters, geometry, water pressures. Gaussian pdfs modelled by statistical estimates and expert judgment.
Propagation method	FORM with Monte Carlo validation
Sensitivity analysis method	FORM sensitivity indices
Feedback process	To determine safety format (which variables deterministic and which random) based on FORM indices/initial calculations.

10.4   Practical implementation and results

The results of one full computation of the failure probability were given in Section 10.3.3. The final result is given below, which is the relation between the reliability index (which is another way to express a probability of threshold exceedance) and the PSF (which is the presentation of the results in a deterministic format). Due to historical reasons, dykes are not designed on the basis of a full probabilistic calculation, but on the basis of partial safety factors (PSFs), see Section 10.2. The PSFs can now be calculated according to Equation 10.16.

$$\gamma_{c,\varphi} = \frac{1 - 1.64 \cdot V_{c,\varphi}}{1 - \alpha_{c,\varphi} \cdot \beta \cdot V_{c,\varphi}} \tag{10.16}$$

In the most conservative approach, one set of PSFs has to guarantee sufficient safety for all possible dyke configurations. A better specified set of PSFs could lead to a more economical design, but scenarios of dyke configurations would have to be determined and the safety format would become more complicated. An analysis of a dyke database shows that the average height is approximately 5m. This height is used to represent all dykes.

Table 10.4   Soil scenarios for different materials.

Range	Clay		Peat		Sand
	Cohesion (kN/m²)	Friction angle(°)	Cohesion (kN/m²)	Friction angle(°)	Friction angle(°)
Low representative value	1	17.5	2.5	10	27.5
Average representative value	10	22.5	10	17.5	32.5
High representative value	25	30	17.5	22.5	37.5

The PSFs are derived for three materials (clay, peat and sand) according to current design practice. There are many different configurations possible; these are assumed to be represented by the configurations as shown in Table 10.4.

Incorporating the different soil scenarios results in sets of PSFs as a function of the reliability index (see Section 10.2). This relation is not yet the final communication to the designers, since it lacks a decision criterion and remains somehow too detailed in relation to soil properties. The first step towards a more practical format is to derive one relation between β and *PSF*, covering the soil types with a sort of weighted average. The black line in Figure 10.12 represents the relation between β and *PSF* in the majority of the dyke population (regarding their soil types). If the example of Section 10.3.3 were to be used, the *PSF* would be 1.3 for *c* and 1.2 for φ. The second step is to relate β to the target reliability of Figure 10.6, i.e. the decision criterion. The result of this is the relation between PSF, dyke ring length and safety standard (see Figure 10.13). The designer has only to determine the dyke ring length and the area's safety standard to be able to determine the PSF. The partial safety factors are derived for the cohesion and for the friction angle; all safety standards are considered. Distinction is made between clay, sand and peat, but the results for clay only are presented in this chapter. A small uncertainty in the *PL* is incorporated into the derivation of the *PSFs*, since in practice a conservative *PL* would be used in the calculation. Theoretically, the design point should be used for *PL* and *WL*, but this is practically not desirable.

The purpose of this case study is not to provide the real final values for application in the Netherlands, since this would, of course, involve other considerations, more detailed studies, reviews, and decision-making which is well beyond the scope of this chapter. However, Figure 10.13 shows the type of format in which the results of the implicit probabilistic uncertainty assessment are typically given to the design processes. The numerical values are given for a virtual 'standard case' but should not in any way be compared to existing or future official codes.

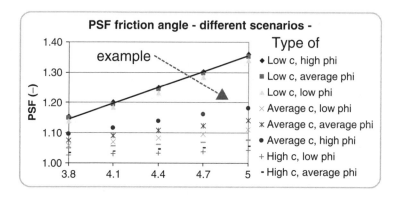

Figure 10.12 Relation between PSF and reliability index.

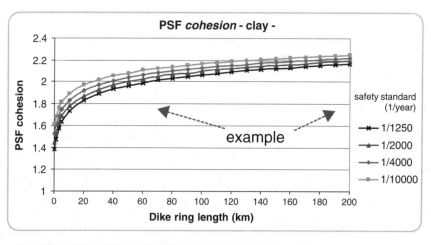

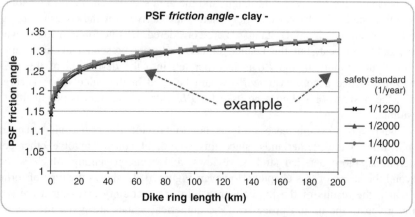

Figure 10.13 PSFs cohesion and friction angle.

10.5   Conclusions

This study has resulted in a format for deriving partial safety factors for the design of river dykes. The required safety with respect to slope stability of the inner side of river dykes, as interpreted from the law, is covered for the majority of dykes. In the first step, all uncertainties are identified and the safety format is chosen. Two partial safety factors are used to cover the soil uncertainties; one related to cohesion and one to the friction angle, according to standard design practice. The second step is to calculate the actual values of the PSFs. FORM is used to propagate soil uncertainties, in which simplified FORM corrections are used to incorporate uncertainties in the phreatic line and outside water levels. Based on the FORM (correction) results, the relation between the *PSFs* and the reliability index is established. The FORM method was checked for several cases with a crude Monte Carlo simulation and the deviations found were considered to be satisfactorily small. Since the pre-existing model is FEM-based, the limit state function had to be simplified in order to perform a fast reliability analysis. The error made is unknown and should be determined. The variations in subsoil layer and model uncertainties are not incorporated into the safety format. Additional research is necessary to address their influence on the reliability. The main contributions to the *PSFs* are uncertainties in soil properties, phreatic line and soil scenarios. These are all epistemic (hence reducible) uncertainties, partly based on engineering judgement. Additional research on soil databases might result in a reduction in the uncertainties and a decrease in the *PSFs*. The relation between *PSF* and reliability index is not workable in practice and not unique because of other variables that need to be specified. For the implementation of the *PSFs*, these are converted into the relation between dyke ring length and an area's safety standard. This is more practical for engineers designing river dykes. However, many sets of *PSF* can lead to the required reliability. One (conservative) set covers all scenarios. More specific sets, applicable only to certain scenarios, allow for better designs. An optimal solution has yet to be defined, but this is outside the scope of the study.

References

Baecher, G. B. and Christian, J. T. (2003) *Reliability and Statistics in Geotechnical Engineering*, Chichester: John Wiley & Sons, Ltd.

Bakker, H.L. (2005) *Failure probability of river dykes strengthened with structural elements.* In *Proceedings of the International Conference on Soil Mechanics and Geotechnical Engineering (ICSMGE) 2005*, Osaka, Japan.

Ciria (1977) *Ciria report 63: Rationalisation of safety factors in structural codes*. Available online at www.ciria.org.uk.

Kanning, W. (2005) *Safety format and calculation methodology slope stability of dykes – a probabilistic analysis on geotechnics to determine partial safety factors,* MSc thesis Delft University of Technology, Delft. Available online at www.hydraulicengineering.tudelft.nl.

Madsen, H. O., Krenk, S. and Lind, N. C. (1986) *Methods of Structural Safety*, New Jersey: Prentice Hall.

TAW, Technical Advisory Committee on Water Defences (1989) *Guideline for the design of river dykes. Vol. 2 Lower River Area – appendices*, pp.13, 87–124. Delft Uitgeverij, Waltman.

TAW, Technical Advisory Committee on Water Defences (2001) *Technical report 'Water Retaining Earth Structures'*. Expertise Network on Flood Protection (ENW), formerly known as Technical Advisory committee for Flood Defence. Available online at www.tawinfo.nl

Van Gelder, P.H.A.J.M. (1999) 'Risks and safety of flood protection structures in the Netherlands'. In *Proceedings of the Participation of Young Scientists in the Forum Engelberg 1999 on Risk and Safety of Technical Systems – in View of Profound Changes*, pp.55–60, 93.

11

Probabilistic assessment of fatigue life

11.1 Introduction and study context

Designing against fatigue is of primary importance for aeronautical structures. Current industrial practice is based on extensive testing of the structures to be conceived or certified. The classic procedure for the assessment of the fatigue resistance of a given object is largely based on experimental data, from which a mean curve is extracted, and from which in turn design curves are derived using safety factors. The procedure is described in Section 11.3.1.

At present there is great interest in using more simulations in order to reduce the amount of testing involved and to gain better control over the margin used. This entails the use of advanced fatigue criteria coupled with uncertainty analysis. In this study a numerical approach is taken to simulate the fatigue data and thus to obtain a safe curve corresponding to a desired quantile for a structural test. The design exercise is carried out without considering a load safety factor (see Section 11.3.1) and the results are compared to those obtained using the classic procedure. The basic principle in the simulation-based approach is to propagate information about the mean fatigue life and the scatter, from elementary coupons (slice of test material) to other coupons or (ultimately) to structural details.

11.2 The study model and methodology

11.2.1 Fatigue criteria

Fatigue-life assessments are customarily based on the use of stress criteria. For a specimen subjected to N cycles at load level $S(t)$, giving rise to a local stress $\sigma(t)$, the function $f_N(\sigma(S, \ldots), \beta)$ denotes a fatigue criterion, where β represents a set

Uncertainty in Industrial Practice   Edited by E. de Rocquigny, N. Devictor and S. Tarantola,
© 2008 John Wiley & Sons, Ltd

of material properties (typically the so-called 'fatigue limits') which are determined on simple coupons at a given number of cycles N. The general idea underlying fatigue criteria is to split the stress-lifetime space into two parts: a safe region where $f_N < 0$, and an unsafe region where $f_N > 0$.

The stress criterion described in this study is based on the spatial average of a classic (macroscopic) criterion over a damaged area defined by a microscopic criterion. Its formulation is given briefly below (for additional details see Schwob et al., 2006). The safe region of the stress-lifetime space is defined by:

$$criterion_N(\sigma(S, \ldots), \beta) = \frac{1}{vol(V)} \iiint_V \left(\sqrt{\langle T_a^2 \rangle} + \alpha(N) P_{\max} \right) dV - \beta(N) \leq 0$$

(11.1)

where α and β are deterministic quantities for the material, determined by experiments in mean fatigue life, T_a and P_{max} are tensor fields depending on σ; and V stands for the damaged volume encompassing the point under consideration. This volume is defined by another ('mesoscopic') criterion, the formulation of which may be debated. The following formulation was used in this study:

$$V = \left\{ M, T_a(M) \frac{\|\nabla T_a(M)\|}{Max_P \|\nabla T_a(P)\|} > G(N) \right\}$$

(11.2)

where G is determined from experimental mean fatigue-life data. The three parameters α, β, and G are determined from three fatigue limits (tensile fatigue tests on plain specimen, with stress ratio $R(R = S_{min}/S_{max})$ $R_{0.1}$, and on open-hole specimen $R_{0.1}$ and R_{-1}). The experimental points, from which these fatigue limits are determined, are represented in Figure 11.5.

11.2.2   System model

The objective of designing against fatigue is to identify the limit between the safe and the unsafe regions. This limit curve, usually named the SN curve (or Wöhler curve), is determined by solving the following equation in (S,N):

$$f(\sigma(S, geometry, \ldots), N) = 0$$

(11.3)

The solution of this non-linear equation may require a large number of finite element calculations. Under some mathematical hypothesis on the f function, it provides a function, fatigue, of the number of cycles N such that:

$$\forall N > 0, ves(\sigma(fatigue(N, \alpha, \beta, G, geometry, \ldots), geometry, \ldots), N) = 0 \quad (11.4)$$

That is, $S = fatigue(N)$ represents the equation of the Wöhler curve and ves denotes a volume-averaged implicit equation (see Schwob et al., 2006, for details). Note that the variable of interest in the study is neither S nor N but the function $S = fatigue(N)$. In other words, the analysis is carried out on the Wöhler curve itself, which is considered a random variable for which different quantiles must be estimated. In this system model the stress is typically obtained through a detailed 3-D finite element model, and a specific programme is used to solve the various non-linear equations using a Newton-Raphson scheme.

11.3 Underlying framework of the uncertainty study

11.3.1 Outline of current practice in fatigue design

Before presenting the proposed probabilistic method, a brief outline is given below of how fatigue design is typically handled. The procedure may vary depending on the company and the product, but the general principles may be summarized as follows:

- Collection of experimental data on coupon (material level) and representative specimens (structural detail level, more or less problem-specific). This data is a collection of observations (the load applied vs observed number of cycles to failure). The fatigue load is typically periodic and is generally summarized by two scalar values. Greater availability of more and better data generally allows for more robust analyses. However, tests are quite expensive and, as a rule of thumb, less than 20 specimens are normally used. A typical example of a fatigue database is shown in Figure 11.1.

- Determination of the mean curve fitting these data. This mean curve is usually determined by least-square fitting. From this mean curve a 'safe curve' can be deduced which should be able to take the experimental scatter into account. This curve can be obtained simply by translation of the mean curve, using a

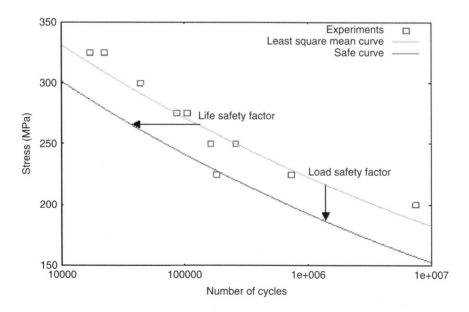

Figure 11.1 'Safe curve' from experiment.

life safety factor as defined in design manuals (the use of such a factor – see Figure 11.1 – is customary in fatigue analysis).

- Although not always recognized as such, this 'safe' curve is an approximation of an $\alpha\%$ iso-probability curve of failure (α is linked to the rules specified in the design handbook used to construct the safe curve, which are usually derived from statistical principles and empirical observations). The usual way of determining the corresponding α is to postulate a lognormal distribution for the number of cycles to failure along with a coefficient of variation.

- Computation/determination of the fatigue load being applied to the material coupon/structural detail. This load can be of constant or variable amplitude. In the following, only constant amplitude loadings are considered; some adjustments would therefore need to be made for variable amplitude loadings (typically the use of Miner's rule in conjunction with Rainflow analysis of the load). This load is then multiplied by deterministic safety factors in order to take into account various effects, such as surface finish, environmental effects, difference between laboratory testing and in-service behaviour, etc. C will denote the global safety factor corresponding to these effects. Instead of being applied on the load, the safety factors may also be applied to mechanical properties or to the safe curve.

- The material coupon/structural detail is deemed safe if the point of coordinates (N,S) (where N is the minimum desired number of cycles that the detail has to undertake and S is the load corrected by the safety factors) is below the safe curve. In other words, the 'decision criterion' is:

$$P_{failure}^{N}(C^*load) < \alpha\% \tag{11.5}$$

This approach involves many approximations. The safety margins associated with these approximations are not well controlled, in particular those linked to the determination of the 'safe curve' and to the contribution of life and load safety factors. Since a set of 20 experiments has to be performed in order for most of the details to be analysed, improving current knowledge would require many experiments.

11.3.2   Specification of the uncertainty study

Figure 11.2 provides the overall framework of the uncertainty study, while an overview of the system model is given in Figure 11.3. In this figure RFL stands for random fatigue limit and denotes a particular regression model used for the statistical analysis of the experimental data (see Pascual and Meeker, 1999).

The variable of interest considered in the uncertainty study is the SN curve, i.e. a function of the number of cycles.

The formal use of a probabilistic method to assess fatigue life of structures is relatively new in the aeronautical industry, though not always feasible, given

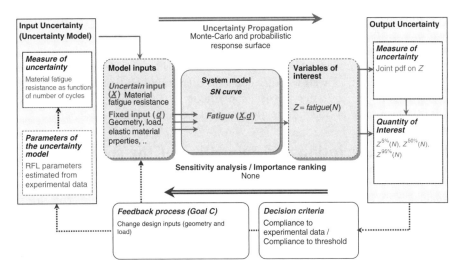

Figure 11.2 Overall uncertainty problem in the common framework.

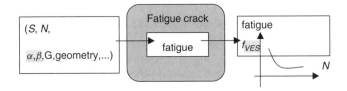

Figure 11.3 The deterministic simulation model.

the high number of uncertainty sources involved and the number of assumptions currently made to build the simulation model. The main goal of the uncertainty study in the long term could be to show compliance with a threshold, i.e. to check that the loading applied to a given structure (the global safety factor, for a given C) for at least N cycles causes less than $\alpha\%$ of the specimen to fail:

$$P^N_{failure}(C^*load) < \alpha\% \tag{11.6}$$

The current study aims to provide an assessment of the safety margins in statistical terms and to rationalize the use of safety factors, but not to reject established practices. In order to provide designers with a convenient tool, the study will focus on an intermediate goal, namely the generation of pSN curves (i.e. SN curves for a given quantile, $\alpha\%$, of the fatigue-life distribution, $\alpha\%$ being the 50% and 95% failure probabilities with 95% confidence, for example). These will give a fair estimate of the expected scatter in structures early in the design process. In addition, they allow the designer to evaluate quickly and graphically whether Equation 11.6 has been verified for a given pair (design, load). These *pSN* curves may actually be defined as quantiles of the variable of interest, *fatigue*.

The quantity of interest is the 95% quantile (with 95% confidence) of the variable of interest, as well as other quantiles, depending on the degree of safety required. The value of α for the decision criteria in (6) typically depends on the risk of catastrophic accident due to failure of the object under consideration.

11.3.3   Description and modelling of the sources of uncertainty

A number of uncertain inputs are considered in the system model: geometrical inputs (diameter of the hole, thickness of the sample, etc.) and inputs related to material properties (fatigue inputs $\alpha(N)$, $\beta(N)$ and $G(N)$, Young's modulus, etc.). In addition, the manufacturing process of each sample (surface finish, etc.) induces uncertainties which are not explicitly considered here. Other inputs can play a role in fatigue life, such as environmental effects and uncertainties in the cyclic state of stress in the sample. To take these effects into account, one should either model their effect explicitly (which may not be an easy task) or use additional safety factors.

Most of the uncertainty sources listed above are aleatory, or a mixture of aleatory and epistemic, and are treated as purely aleatory uncertainties. Purely epistemic uncertainties arising in this study are essentially due to the lack of fit of the fatigue model and to the estimation of the random variables (after a statistical treatment of the experimental data involving a regression model). Epistemic uncertainties will not be considered in this study.

For the material properties, only one uncertain model input will be considered: the macro fatigue resistance $\beta(N)$. All other inputs are considered deterministic. It is thus implicitly assumed that:

- Manufacturing, geometry and loads are consistent from test to test. This is debatable but achievable to the first order if care is taken in the preparation of material coupons.

- The uncertainty of the cyclic state of stress is negligible. A thorough study of the cyclic evolution of the stress should be conducted in order to investigate the validity of this hypothesis, but preliminary tests support the assumption. The same hypothesis is adopted for the uncertainty of the elastic inputs.

- The loading S applied to the specimen has been very well controlled in the laboratory experiments considered in this study, i.e. no appreciable error in the load has been induced by the instrumental controller.

These hypotheses are very restrictive, especially for complex structures, but they allow for a simplification of the model which is acceptable for simple material coupons (i.e. validated through sensitivity analysis). The uncertainty model will be enriched with more uncertain inputs in future studies. For instance, for specimens representative of assemblies, the following inputs should be accounted for: hole size, fastener size, fastener tension and friction coefficients. Young's moduli for the fasteners and plates should also be added. For specimen with a residual dent, one should consider the residual dent depth as an additional input.

The uncertainty of the macro fatigue resistance $\beta(N)$, in terms of the probability density function, is estimated through an inverse model from experiments on simple coupons (the procedure is described in Section 11.4). If additional uncertain inputs were to be considered, they should be evaluated by expert judgment and/or measurements (for example, the diameter of the hole of an open-hole specimen would be considered to be uniformly distributed between boundaries determined by the precision of the manufacturing process).

11.3.4   Uncertainty propagation and sensitivity analysis

The most comprehensive way to obtain the whole family of pSN curves is to propagate uncertainties using Monte Carlo simulation (Gentle, 2003; Hammersley *et al.*, 1979; Rubinstein, 1981). A Monte Carlo sample of size 3000 is generated from the pdf of the $\beta(N)$. This size is restricted by the computational time required by the model. In fact, to solve a non-linear equation in N for a sufficiently refined set of $(S_j)_{1<j<n}$, finite element calculations are required. In order to reduce the computational time related to the generation of the sample, a 'degraded' system model is used: Equation 11.4 is solved for one particular stress level for each realization of β.

This consistent simplification of the model enables more Monte Carlo simulations to be performed for a given CPU cost. However, the outcome of the degraded system model is no longer a function of the number of cycles representing the fatigue limits, but only of one particular number of cycles to failure, corresponding to a particular stress level and a particular realization of the random variable β. By changing the stress level considered in the degraded system model, it is possible to generate a scatterplot (see Figure 11.6, from which the pSN curves can be reconstructed (using, for example, the model of Pascual And Meeker, 1999).

11.3.5   Feedback process

If the decision criterion is not fulfilled, the probabilistic model suggests that the expected scatter is not well covered by the handbook values. The source of discrepancy should be investigated, for instance with sensitivity analyses or by reviewing the hypothesis of the probabilistic model. Ultimately, if the comparison suggests that the handbook values are not restrictive enough for a particular case, the designer may have to change the design inputs (these are the geometrical parameters of the detail being analysed) or choice of material (to increase the fatigue resistance, for example), or make some overall structural change in order to reduce the load.

Table 11.1   Table of study characteristics.

Final goal of the uncertainty study	Goal C – compliance with a threshold
Variables of interest	Fatigue curve (*SN* curve) delimiting safe/unsafe regions. It is a function of the number of cycles.
Quantity of interest	Main quantiles of the variable of interest (5%–50%–95%)
Decision criterion	Compliance with experimental data and with threshold.
Pre-existing model	Deterministic model requiring numerous calls to finite element analysis.
Model inputs and uncertainty model developed	Fatigue resistance (uncertain) + design input (fixed): Cumulative distribution function of fatigue resistance.
Propagation method	Monte Carlo and use of probabilistic response surface.
Sensitivity analysis method	None
Feedback process	Changes in design inputs

11.4   Practical implementation and results

11.4.1   Identification of the macro fatigue resistance $\beta(N)$

The estimate of the probability density function of the macro fatigue resistance $\beta(N)$ is obtained through an inverse model from experiments on simple coupons. The procedure for the estimation of $\beta(N_0)$ for a fixed N_0 is presented here. It has to be repeated for a sufficiently refined set of N_0's. First, some basic properties of the fatigue model are described.

• **Qualitative properties of the model**

The stress fatigue criterion used in this study can be formulated as:

$$g(S, N) \leq \beta(N) \qquad (11.7)$$

where $g(S,N)$ is some equivalent stress function (supposed deterministic for the material coupons) and β represents the fatigue threshold not to be exceeded. For a fixed stress S_0 the fatigue model should possess certain properties:

- • $f : N \mapsto g(S_0, N) - \beta(N)$ should be strictly monotonically increasing, which means simply that, because the criterion estimates the severity of the loading, it should be larger when N is larger (since the larger N, the more severe the loading). Similarly, $h : S \mapsto g(S, N_0) - \beta(N_0)$ should be strictly monotonically increasing.

- $f(N)$ takes negative values when N is close to zero (if there is no loading the criterion should predict no failure) and positive values when N increases indefinitely (eventually the sample will break, even for a very low applied stress).

- Denoting two realizations of the model input β with $\beta_i : N \mapsto \beta_i(N)$ and $\beta_j : N \mapsto \beta_j(N)$, the following equivalence holds: $\exists N, \beta_i(N) \leq \beta_j(N) \Leftrightarrow \forall N, \beta_i(N) \leq \beta_j(N)$.

Note that these properties do not need to be fulfilled for every kind of test and specimen geometry, but only for the test on which the estimation of the distribution of β is based. Assuming that these properties hold for a fixed S_0, one has: i) $\exists! N_R(\beta)$, $f(N_R) = 0$, where N_R is the number of cycles to failure (i.e. the observed quantity), $\beta_i < \beta_j \Leftrightarrow N_R(\beta_i) < N_R(\beta_j)$ and $N_R(\beta) \leq N_0 \Leftrightarrow \beta(N_0) \leq g(S_0, N_0)$

- **Analysis of fatigue experiments**

A typical fatigue experiment database is shown in Figure 11.4. The stress range is denoted by $[S_{min}; S_{max}]$. An observation is a pair (number of cycles to failure, stress level applied) denoted by (N_R, S). For a given value of stress, S (the control variable of the experiment), the value N_R is observed. From a set of such experiments conducted on different specimens to the point of failure, the cumulative distribution

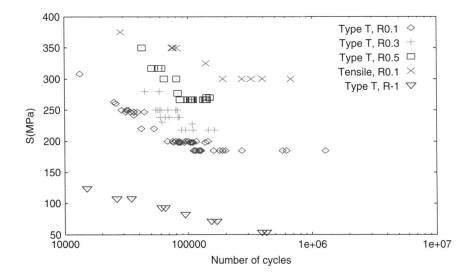

Figure 11.4 Fatigue database for the alloy 2XXXT3.

function of the number of cycles to failure for a given stress can be obtained, using, for example, the model described in Pascual and Meeker (1999).

In order to build a relationship between the cumulative distribution function of N_R and that of $\beta(N_0)$ for a given N_0, a stress level S_0 is first chosen. The probability of failure before N_0 is denoted by P_{S_0} $(N_R < N_0)$, giving:

$$P_{S_0}(N_R < N_0) = P(\beta(N_0) \leq g(S_0, N_0)) = \int_{\omega < g(S_0, N_0)} P_\beta(\omega, N_0)\, d\omega \qquad (11.8)$$

where $\omega \mapsto P_\beta(\omega, N_0)$ is the probability density function of $\beta(N_0)$. The left–hand side of this equation can be determined experimentally if S_0 is one of the stress levels tested, and provided that there are enough specimens tested at this stress level (i.e. the proportion of samples which have broken before N_0 at S_0). If not, the value of this probability has to be estimated from available data by using statistics on a regression model (Pascual and Meeker, 1999). Allowing S_0 to vary over $[S_{min}; S_{max}]$, the value of the cumulative density function F_R of $\beta(N_0)$ is obtained for the interval $[g(S_{min}, N_0); g(S_{max} N_0)]$ as:

$$F_R(g(S, N_0)) = P_S(N_R < N_0) \qquad (11.9)$$

- **The identification procedure**

The entire estimation procedure was executed for two light alloys (2XXXT3, 6XXXT78). The results for the 2XXXT3 alone (for which a dedicated fatigue database has been realized) are described here for the material coupons. The results of the procedure are similar for the other material. This dedicated test campaign was conducted at EADS Innovation Works. Two different configurations were tested, namely simple tensile and open-hole specimen. Several loading ratios R (S_{min}/S_{max}) were investigated for the open-hole specimen. The database is represented in Figure 11.4.

Using this database it is possible to estimate the cdf of $\beta(N)$ for three specimens out of the five available (i.e. tensile R0.1, open hole R0.1 and open hole R0.5). The application of the statistical evaluation procedure to the open-hole R0.1 database for different numbers of cycles N_i, $(1 < i < n)$ yields the cumulative distribution functions of $\beta(N_i)$, $(1 < i < n)$.

11.4.2   Uncertainty analysis

The uncertainty in $\beta(N)$ was propagated through the fatigue model (Equation 11.4), and the resulting fatigue curve (the pSN curve) delimiting safe/unsafe regions was examined. The case of the open-hole specimen tested with a loading ratio $R = 0.3$ was studied using both the Monte Carlo procedure and a simpler propagation method consisting of a sampling of the response followed by statistical analysis of the results. The simpler method was then used to analyse the structural case considered in this study.

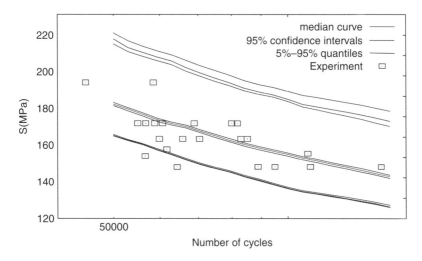

Figure 11.5 pSN curves.

The post-processing of Monte Carlo results is straightforward. The estimation of the 5%–50% and 95% curves for the open-hole specimen tested with a load ratio $R = 0.3$ is presented in Figure 11.5, together with the experimental data. Confidence intervals for such percentile curves are obtained using order statistics (David and Nagaraja, 2003).

It can be seen that the predictions are in reasonable agreement with the experimental data. However, due to the small number of experimental points, a quantitative testing of the fit between predictions and experiments would be pointless. In order to validate the model, additional testing should be performed on some particular case studies.

The sampling of the response, yields the following cloud of points, depicted in the next figure together with experimental points.

A statistical analysis of these simulated points yields the desired set of *pSN* curves. Confidence intervals can be attached to this evaluation procedure, based on the statistical analysis, but unlike the Monte Carlo approach, order statistics may no longer be used. Figure 11.7 shows the 5%–50% and 95% quantiles. These predictions are also compared to those obtained by Monte Carlo.

The two procedures for uncertainty analysis produce similar results, although the simpler sampling method is slightly more conservative (which gives some confidence in the use of such methods for this kind of study).

Concerning the structural test (specimen with a residual dent subjected to a cyclic tensile loading), the computational time needed for one simulation run is about one hour (using one 2 GHz processor). The response sampling method was used in this case and the results are given in Figure 11.8.

The predictions are in reasonable agreement with the experiments, including for the scatter. The estimated pdfs of the *SN* curves, and in particular the mean values

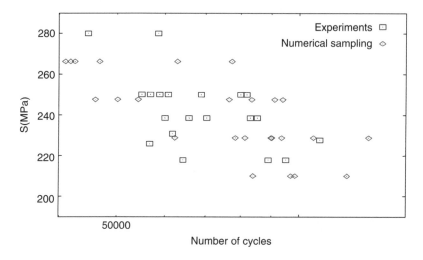

Figure 11.6 Response sample versus experimental results.

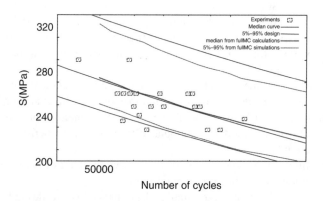

Figure 11.7 Uncertainty analysis of the SN curves. Comparison of Monte Carlo simulations and response surface results.

of the curves, may be affected by assumptions made in the simulation model (the accuracy of the predictions of the fatigue criterion has not yet been extensively checked). Owing to specific stress distributions in structural specimens, the qualitative response of the model may indeed be affected by the formulation of the mesoscopic criterion. Its validity should be further analysed before the quantitative accuracy of the probabilistic model is assessed. For comparison the 'handbook approach' described in the introduction was also applied to this particular test using a life safety factor of 2 (see Figure 11.8). It is customary in this approach to estimate the quantile curve based on an assumed distribution of the fatigue life for a fixed

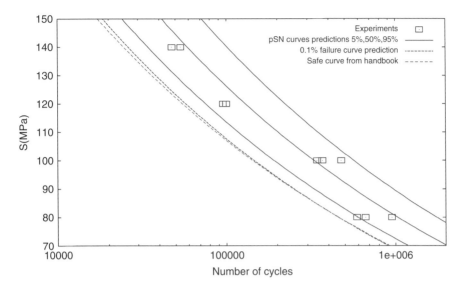

Figure 11.8 6XXXT78 fatigue design curves for a specimen with residual dent.

stress. Assuming a lognormal distribution of the number of cycles to failure, and selecting a reasonable value for the coefficient of variation (i.e. 2%), this approach equates the safe curve with a safety factor of 2 to the 0.02% failure probability curve. However, in the case under consideration, the result was found to be nearly equivalent to the 0.1% iso-probability curve through our probabilistic approach.

11.5   Conclusions

A probabilistic framework for fatigue analysis of small coupons was described in this chapter and comparisons of results from structural experiments were presented. The procedure is based on a stress-based criterion which is currently the basic tool for the assessment of fatigue. In particular, a random sampling of the response followed by a statistical treatment proved to be quite efficient for practical applications, and is in this case a viable alternative to more costly Monte Carlo simulations. Note that the intermediate results obtained through this procedure (the production of *pSN* curves) exhibit similar statistical quality to the experimental data currently used in the design/certification process. With further research and improvements the proposed approach could be integrated into a classic fatigue assessment scheme, replacing more costly experimental data.

References

David, H.A. and Nagaraja, H.N. (2003) *Order Statistic* (3rd edition), Chichester: John Wiley & Sons, Ltd.

Gentle, J. E. (2003) *Random Number Generation and Monte Carlo Methods*, London: Springer.

Hammersley, J. and Handscomb, D. (1979) *Monte-Carlo Methods*, London: Chapman and Hall.

Schwob, C., Chambon, L. and Ronde-Oustau, F. (2006) 'Fatigue crack initiation in stress concentration areas'. In *Proceedings of the 16th European Conference on Fracture*, Alexandropoulis, Greece, Gdoutos, E., Springer.

Pascual, F.G. and Meeker, W.Q. (1999) Estimating fatigue curves with the random fatigue-limit model. *Technometrics*, **41**(4), 277–302.

Rubinstein, R.Y. (1981) *Simulation and the Monte-Carlo Method*, New York: John Wiley & Sons, Ltd.

12

Reliability modelling in early design stages using the Dempster-Shafer Theory of Evidence

12.1 Introduction and study context

In the recent past demand for uncertainty modelling in the fields of automotive reliability and functional safety has grown. Since the release of the German industry guideline VDI 2206, proposing the V-Model as a process model to cover the entire development cycle VDI (2004), the challenge has been to carry out reliability assessment mainly in the integration phase (Figure 12.1). The growing importance of the IEC 61508 guideline (IEC, 2001), including in the area of mechatronic design, has led to a further growth in interest in quantitative prediction.

The research focus has shifted from qualitative models without uncertainty treatment to prediction models analysing and controlling the uncertainties (Jäger and Bertsche, 2004). In automotive systems, Safety Integrity Levels (SIL 1-4) according to IEC 61508 are used to define system safety requirements. Uncertainty analyses of the predicted system safety provide both a robust way to demonstrate that the system complies with the target failure measure and an indicator for possible violation of these targets.

Quantitative reliability and safety prediction in an early design stage needs to deal with uncertainties from various sources. Information on the failure probabilities for the system components is obtained from expert estimates, tests and experience from past projects. Models have a low level of detail and may not be an accurate description of the real failure behaviour.

In this study a fault tree analysis of an automatic transmission from the ZF-AS Tronic product line is considered and an investigation carried out into whether the system can comply with the target failure measure of SIL 2. However, the proposed

Uncertainty in Industrial Practice   Edited by E. de Rocquigny, N. Devictor and S. Tarantola,
© 2008 John Wiley & Sons, Ltd

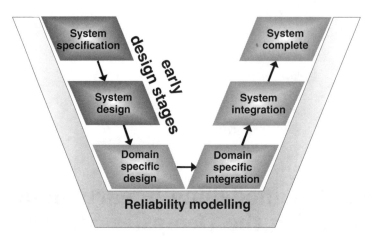

Figure 12.1 The V-Model (overview).

methodology is not intended to validate the system against this threshold. In fact, the goal is simply to get a first, 'quick and dirty' prediction of the reliability of a technical system, to see if the targets *could* be met.

Dempster-Shafer is an alternative non-probabilistic way to represent uncertainty which was experimented in this 'frontier' case study, although other elaborated uncertainty treatments (such as a level-2 probabilistic setting, with probability distributions on the reliability parameters) could also have been considered. The very peculiar context is that information available to build the uncertainty model is very scarce and the main goal is to highlight to the decision-maker the extent of ignorance remaining in the performance of the design. Although the theoretical foundations supporting the use of DST techniques in uncertainty modelling are less pervasive and more controversial than their probabilistic correspondents, and while industrial applications of those still appear as 'frontier cases' when this book comes into press, DST offers a framework where one may code a low level of specification of the uncertainty model (such as intervals instead of a precise pdf) while preserving a conservative, i.e. larger, resulting output uncertainty. A brief explanation of some features of DST is introduced in Appendix 3: detailed discussion of the theoretical foundations, implementation challenges and comparison to other uncertainty settings may be found in Dempster, 1967; Shafer, 1976; Helton and Oberkampf, 2004.

12.2   The study model and methodology

12.2.1   The system

The system under investigation is currently being developed as a member of the ZF-AS Tronic series (Figure 12.2). ZF-AS Tronic was developed especially for trucks with EDC (Electronic Diesel Control) engines and CAN (Controller Area Network) communications. The transmission system combines ZF mechanic technology with

Figure 12.2   Automatic transmission from the ZF-AS Tronic product line: (1) Transmission actuator, (2) Gearbox, (3) Clutch actuator. Reproduced by permission of ZF Friedrichshafen AG.

modern electronics. The integrated modular design simplifies both installation and maintenance and provides protection from outside influences. If the truck needs to perform other work during or after the run, it needs a power take-off. This is also possible with AS Tronic. The transmission can be equipped with one or more clutch-dependent power take-offs, even after it has been installed in the truck. The outputs can be shifted independently of each other. A speed-dependent power take-off system to drive auxiliary steering pumps is also available.

AS Tronic handles gear selection and clutch and shifting manoeuvres. The 12-speed gear shifts electropneumatically. Engine power is always transmitted optimally. The ZF-MissionSoft driving programme keeps the motor at an efficient engine speed. The driver can correct the automatic gear selection or switch to manual operation at any time and set the gear using the touch lever. These fully integrated transmissions are mainly used in buses, trucks and other special-purpose vehicles.

12.2.2   The system fault tree model

Being only a sub-system in the power train, which is again a sub-system in a modern car, it is difficult to specify exactly which failure modes may lead to a safety-critical situation for the driver. However, certain conditions can be identified as a critical failure, such as a failure of the clutch system or an uncovered failure in the gearbox electronics. In our study, the level of detail in the fault tree (Figure 12.4) is quite low. Sub-systems are not separated into different components, as the structure of the system has not yet been fully determined.

The system is composed of three different sub-systems. The clutch system mainly contains mechanical components that are highly reliable. Therefore it is not planned to add any additional sensors for the discovery of safety-critical situations. The gearbox with its set of mechanic and hydraulic components will include several sensors that may detect failures or near-failure conditions. The gearbox electronics, as the interface with the driver, are modelled as a separate sub-system. The expert estimates are idealized values and do not reflect the real failure behaviour of the AS Tronic.

Table 12.1 Tolerated target failure measure for safety-critical functions (high demand or continuous mode).

SIL	4	3	2	1
PTF	$[10^{-9}, 10^{-8}]$	$[10^{-8}, 10^{-7}]$	$[10^{-7}, 10^{-6}]$	$[10^{-6}, 10^{-5}]$

12.2.3   The IEC 61508 guideline: a framework for safety requirements

According to IEC 61508, it has to be shown that the probability of a critical failure for a safety-critical function is below a certain threshold (target failure measure) in order to comply with a certain SIL. Systems are classified according to frequency of their use into high and low demand. For systems with high demand such as the transmission under investigation, Table 12.1 shows the target failure measure (failure probability per hour of service). IEC 61508 provides intervals for these target failure measures. This motivates the use of uncertainty preserving methods. If it can be shown that the predicted failure measure p_s might be lower than the lower p_{TF} bound and is almost definitely below the higher p_{TF} bound, the safety arguments are much stronger.

IEC 61508 moreover defines limits on the SIL that can be reached by a system executing the safety function, dependent on data quality and fault-tolerance. (Sub-)systems are classified according to the quality of available data. A (sub-)system is of the (preferable) type A if:

(1) the behaviour of all components in case of failure is well-defined;

(2) the behaviour of the sub-system in case of failure is well-defined; and

(3) reliable failure data from field experience exists.

A system is of type B if one of the given conditions is not true. In an early design stage, type B can be considered to occur much more often. Even if criteria 1 and 2 above are well-defined, it is unlikely that enough data exists to fulfil the last criterion. Due to the recursivity of this definition, systems containing one type-B sub-system also have to be considered as type B. However, this has a strong impact on the safety requirements. Systems of type A can reach higher SILs than type B.

The IEC plays a large role in mechatronic reliability assessment. Even if originally intended for use with electronic/programmable systems, it has rapidly gained importance in areas involving mechanical components as well.

It may be desirable to be able to classify a system as type A. If exact failure data are missing, yet the (sub-)system needs to be classified as type A, it may be reasonable to work with a conservative uncertainty model. If the uncertainty model can be considered reliable (regarding criterion 3), the system may be treated as type A. Therefore, very conservative uncertainty modelling, such as is implicitly done by DST, can help to provide an argument for reclassifying the (sub-)system from type B to type A.

In contrast to purely probabilistic uncertainty treatment, it is possible, with DST, to include uncertainty in form of intervals. This may be seen as an alternative answer to uncertainty modelling of the reliability parameters when available failure information is very scarce, while in a probabilistic approach the possibly tricky choice of a pdf would require additional assertions, such as the maximum entropy principle (see comments in Section 16.2.2).

12.3 Underlying framework of the uncertainty study

12.3.1 Specification of the uncertainty study

Figure 12.3 illustrates the common framework and its application to the case study.

- **Pre-existing model and variables of interest**

The pre-existing model is a system fault tree with several components. Fault trees are one of the most popular reliability models due to their simplicity and good communicability. Formally, a fault tree is a Boolean System model, mapping component states (failed/working) into system states. However, fault trees, as they are typically used in reliability, are in themselves a level-1 probabilistic model, mapping the component failure probabilities into system failure probability. Therefore, most

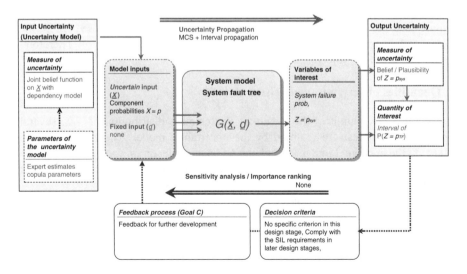

Figure 12.3 Overall uncertainty problem in the common framework.

reliability engineers consider the probabilistic level-1 uncertainty representation as the 'true' model, where the probability of failure per working hour of a component $i \in \{1, \ldots, n\}$ is described by a given failure probability value $p_i \in [0, 1]$. The transpose component probability vector p' represents the combined representation of all component failure probabilities:

$$p' = (p_1, \ldots, p_n) \tag{12.1}$$

The system failure probability p_{sys} is the variable of interest for the case study and can be obtained via the fault-tree model as:

$$\begin{aligned} G &: [0, 1]^n \rightarrow [0, 1] \\ G(p) &= p_{sys} \end{aligned} \tag{12.2}$$

The vector p represents the input into the pre-defined system model. Actually p is never perfectly known and in fact there are uncertainties on a large scale that can make an exact knowledge of p impossible. This uncertainty study will illustrate how to form an uncertainty model around p and to project this uncertainty onto the output, the system failure probability p_{sys}. Figure 12.4 illustrates the fault tree generated in this early design stage. In a level-1 probabilistic model $p_1 = 2 \cdot 10^{-8}$ and $p_{sys} = 5 \cdot 10^{-7}$ represent the probability of failure, in one hour, of component 1 (or the system).

- **Final goal(s) of the uncertainty study**

The goal of this study is to gain a first impression of the system failure probability with respect to the targets imposed by the SIL (Goal C-COMPLY). Indeed, at this early stage, the reliability model is far too crude to actually verify that the target failure measure will not be surpassed. However, it is important to see where the actual failure estimate is located with respect to this failure measure. Therefore, Goal C is a distant objective of the study.

- **Quantities of interest and decision criteria**

Besides the raw output of the Dempster-Shafer mass distribution of p_{sys}, several quantities of interest are returned by the uncertainty model. The expected value $E(p_{sys})$, the median $\text{Med}(p_{sys})$ and the 95% quantile on p_{sys}, $Q_{95}(p_{sys})$. For a preliminary testing of the reliability prediction against the SIL, the probability that p_{sys} exceeds the SIL lower and upper thresholds is also given.

In a purely probabilistic approach all these quantities are used to 'average out uncertainty' somewhat, reducing distributions to a point. Owing to the Dempster-Shafer uncertainty model chosen, uncertainty represented by intervals is not reduced to a single value. Any arbitrary distribution that is enclosed in the Belief/Plausibility functions is equally viable. All listed quantities therefore are only obtainable as intervals.

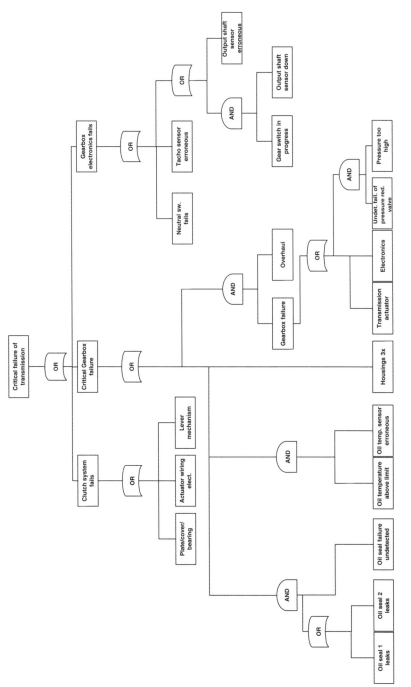

Figure 12.4 Early design stage fault tree of an automatic transmission.

12.3.2   Description and modelling of the sources of uncertainty

• Model inputs

The model inputs of the fault tree model are the component failure probabilities $p_1, \ldots, p_n$, which represent the failure probability vector p. A characteristic of fault trees is that the system function itself is very simple, but the number of inputs is generally quite large. The system fault tree in Figure 12.4 consists of 19 leaf components. Inputs other than p are not given.

• Sources of uncertainty involved: nature/types of uncertainty

The uncertainty model is composed of a level-1 probabilistic layer covering irreducible uncertainties (i.e. the component failure probabilities), on top of which a level-2 DST layer covers reducible uncertainties (i.e. the uncertainties from estimates of component failure probabilities drawn by reliability experts). Model uncertainty is not covered in this study. Nevertheless, a fault tree, though in common use, is only a rough approximation of the reality, used if more complex models are not available or are too expensive.

The probabilistic *level-1* model is a fault tree. The uncertainty represented by the probability values $p_1, \ldots, p_n$ is considered to be totally irreducible. It is assumed that a p_i value of 10^{-5}/h is equivalent to the chance of observing a failure with a probability of 10^{-5} every operation hour. By 'sitting and waiting' during some operating hours or by observing a larger population of systems, the number of observed failures rises according to this probability. Level 1 of the uncertainty model therefore has the probabilities $p_1, \ldots, p_n$, which are, even with perfect knowledge, probabilistic quantities. Its output is the failure probability p_{sys} of the system, a sharp probabilistic value representing the failure probability of the system.

Regarding probabilistic *level-2*, the probabilities $p_1, \ldots, p_n$ are never exactly known. In an early design stage they are estimated by reliability experts, provided by manufacturers or obtained by handbook calculations/similarity analyses. Several sources of uncertainty can be identified, represented by this second level. Expert uncertainty in providing reliability values is one crucial source of uncertainty. Another is expert conflict, i.e. several incongruent estimates for one probability value p_i. Other sources are the uncertainty in applying warranty data or test data from similar transmissions, or uncertainty because of imprecise information from the component manufacturer. Some of these uncertainties are eventually reducible, such as expert conflict (i.e. discussing the reasons for different opinions). If a link to the component manufacturer can be established, this uncertainty can also be reduced. Some uncertainties can be reduced in later project stages, i.e. with test data. But the real failure probabilities can only be obtained through in-service experience and field data. Another source of uncertainty is the unknown change of tolerances over time. These level-2 uncertainties in $p_1, \ldots, p_n$ cannot be neglected, because they are quite large and may have a strong influence on the final result. Furthermore, there is no repeatability in these uncertainties. Being induced by lack of knowledge, uncertainty in p_i persists until more detailed information is gained. If p_i is modelled by a uniform distribution $U(10^{-6}, 3 \cdot 10^{-6})$, it cannot be expected that for 10000 systems or 10000 working hours p_i will stabilize at $2 \cdot 10^{-6}$ (the average value of $U(10^{-6}, 3 \cdot 10^{-6})$). Therefore it is necessary to build a level-2 uncertainty model on top of level 1 to separate reducible uncertainty from irreducible.

- **Building of the uncertainty model**

The uncertainty representation chosen is the framework of Dempster-Shafer, which allows for the simultaneous representation of uncertainty with intervals and distributions (see Appendix 3 for a brief introduction of some features of DST). The uncertainty in p is represented by a joint belief/plausibility assignment (BPA) $m_{1,\ldots,n}$ which is the analogous concept to a joint distribution of the inputs. The joint BPA is obtained from the marginal BPA's $m_1, \ldots, m_n$ representing the component BPAs and a copula that models the dependency between the component estimates. It is entirely possible that a dependency model might be useful in this second layer of uncertainty, because components may come from the same manufacturer or because the same component may be found in different sub-systems. The manufacturer may have put an unknown margin between the specified failure probability and the real one, which would be similar for similar components. Therefore if the failure probability of one component is low, the other may also have a good chance of being low. Other causes of dependency in level 2 might be that the information has been obtained by the same elicitation method or that both components deal with similar technology.

In the system it is assumed that a correlation of +0.5 exists between events 4 and 5 (oil leak failure), events 18 and 19 (output shaft sensor failure modes), events 16 and 19 (tacho sensor and shaft sensor erroneous) and events 7 and 8 (failure

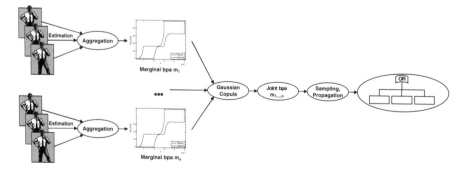

Figure 12.5 Illustration of the uncertainty model around the system fault tree.

modes concerning oil temperature). This dependence is modelled at the second uncertainty level. Although events 4 and 5 are not likely to occur simultaneously, it is more likely that if one failure probability is high, the other is also high.

Quantification of the uncertainty sources

As mentioned, DST can easily accommodate expert estimates. Therefore different ways of predicting failure probabilities are provided. The experts were given the possibility of predicting p_i using several methods:

- estimating a sharp value for p_i;
- estimating an interval that might contain p_i;
- estimating mean μ and standard deviation σ of a normal distribution (censored to $[0,1]$) describing the uncertainty in p_i.

Experts were also allowed to provide several estimates for the same component, which were aggregated using weighted mixing, a method for aggregating several mass functions to a single mass function. Intervals and sharp values were directly converted to a focal element with mass 1. Estimates of μ and σ could be given as values or intervals. The estimates were converted by sampling the inverse cdf of the distribution function.

The DST approach is tested here on a research study in which no data are available. In the study it is assumed that three experts (expert 1, expert 2 and expert 3) provide estimates of the failure probability. The values are idealized and do not reflect real failure probabilities. Table 12.2 shows these estimates on the basic fault tree events. While some components, such as component 6, contain only one estimate, component 10 has imprecise distribution and interval estimates which are aggregated to one marginal distribution. Figure 12.6 illustrates three different marginal BPAs of input components. It can be seen that interval estimates lead to an instantaneous increase in the BPA functions. Figure 12.6c is a pure distribution estimate without any uncertainty, whereby BPA's collapse to a cdf.

12.4   Practical implementation and results

Fault tree models involve very little computation. Thus, 10^6 samples were generated from the joint BPA and were propagated through the system function using a strategy similar to Monte Carlo sampling (see Appendix 3). All calculations performed and all plots shown were generated using the imprecise probability toolbox for MATLAB. This free and open-source tool is available at http://www.uni-due.de/il/software.

Figure 12.7 shows the prediction of Belief and Plausibility of the system failure probability. The vertical lines represent the upper and lower failure probabilities required by SIL 2. It can be seen that the system failure probability exceeds the stricter SIL level, but not the SIL upper bound, 10^{-6}. The uncertainty contributed by the interval width is definitely quite high.

Table 12.2 Expert estimates of the failure probability of different components.

Basic Event	ID	Source	Failure prob./h of service
Plate/Cover/	1	Expert 1	$[1 \cdot 10^{-8}, 3 \cdot 10^{-8}]$
Bearing		Expert 2	$4 \cdot 10^{-8}$
		Expert 3	$\mu = 1 \cdot 10^{-7}$, $\sigma = [1 \cdot 10^{-8}, 2 \cdot 10^{-8}]$
Actuator wiring	2	Expert 1	$\mu = [1 \cdot 10^{-8}, 2 \cdot 10^{-8}]$, $\sigma = 1 \cdot 10^{-9}$
elect.		Expert 2	$4 \cdot 10^{-8}$
		Expert 3	$\mu = 2 \cdot 10^{-8}$, $\sigma = 1 \cdot 10^{-9}$
Lever mechanism	3	Expert 3	$[0.1 \cdot 10^{-8}]$
Oil seal 1 leaks	4	Expert 1	$[1 \cdot 10^{-7}, 3 \cdot 10^{-7}]$
		Expert 2	$\mu = 3 \cdot 10^{-7}$, $\sigma = [1 \cdot 10^{-7}, 2 \cdot 10^{-7}]$
		Expert 3	$\mu = [3 \cdot 10^{-7}, 5 \cdot 10^{-7}]$, $\sigma = 1 \cdot 10^{-8}$
Oil seal 2 leaks	5	Expert 1	$[1 \cdot 10^{-7}, 3 \cdot 10^{-7}]$
		Expert 3	$\mu = 2 \cdot 10^{-7}$, $\sigma = 1 \cdot 10^{-8}$
Oil seal failure undetected	6	Expert 2	$1 \cdot 10^{-9}$
Oil temp.above limit	7	Expert 2	$[0.01, 0.03]$
Oil temperature sensor	8	Expert 1	$3 \cdot 10^{-7}$
		Expert 3	$\mu = 4 \cdot 10^{-7}$, $\sigma = 1 \cdot 10^{-8}$
erroneous			
Housings 3x	9	Expert 2	0
		Expert 3	$[0, 1 \cdot 10^{-9}]$
Transmission actuator	10	Expert 1	$[1 \cdot 10^{-8}, 3 \cdot 10^{-8}]$
		Expert 2	$\mu = 7 \cdot 10^{-8}$, $\sigma = [3 \cdot 10^{-9}]$
		Expert 3	$[2 \cdot 10^{-8}, 4 \cdot 10^{-8}]$
Electronics	11	Expert 2	$\mu = 3 \cdot 10^{-6}$, $\sigma = 5 \cdot 10^{-7}$
Undet. fail. of	12	Expert 1	$[7 \cdot 10^{-8}, 1.5 \cdot 10^{-7}]$
pressure red.		Expert 2	$[1 \cdot 10^{-7}, 2 \cdot 10^{-7}]$
valve		Expert 3	$4 \cdot 10^{-7}$
Pressure too	13	Expert 3	0.02

(*continued overleaf*)

Table 12.2 *(continued)*

Basic Event	ID	Source	Failure prob./h of service
Overhaul fails	14	Expert 1	[0.003,0.005]
		Expert 2	$\mu = [1 \cdot 10^{-3}, 5 \cdot 10^{-3}]$, $\sigma = 2 \cdot 10^{-4}$
		Expert 3	$\mu = 3 \cdot 10^{-3}$, $\sigma = [1 \cdot 10^{-4}, 2 \cdot 10^{-4}]$
Neutral sw. fails	15	Expert 1	$6 \cdot 10^{-8}$
		Expert 2	$[0, 1 \cdot 10^{-7}]$
Tacho sensor erroneous	16	Expert 2	$5 \cdot 10^{-7}$
		Expert 3	$[2 \cdot 10^{-7}, 4.5 \cdot 10^{-7}]$
Gear switch in progress	17	Expert 3	0.04
Output shaft sensor down	18	Expert 2	$\mu = 2 \cdot 10^{-7}, \sigma = 4 \cdot 10^{-8}$
Output shaft sensor erroneous	19	Expert 1	$\mu = 3 \cdot 10^{-8}, \sigma = 1 \cdot 10^{-8}$
		Expert 2	$\mu = 4 \cdot 10^{-8}, \sigma = 1 \cdot 10^{-8}$
		Expert 3	$\mu = 7 \cdot 10^{-8}, \sigma = 2 \cdot 10^{-8}$

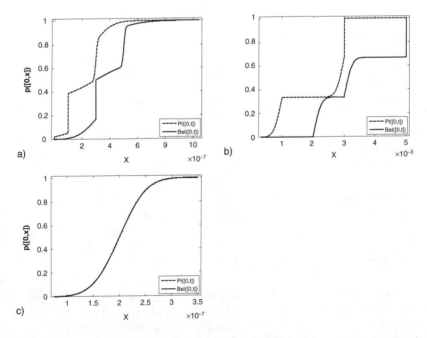

Figure 12.6 Visualization of the belief and plausibility functions of the aggregation of basic events: (a) oil seal 1 leaks; (b) overhaul fails; and (c) output shaft sensor down.

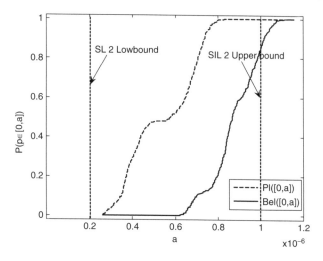

Figure 12.7 Results of the fault tree analysis and the illustrated bound, to comply with SIL2.

Table 12.4 gives some characteristics of the resulting BPA. Expectation value and median are approximately at the same level. These values are defined as the classical probabilistic quantities in the level 1 setting. However, the uncertainty over the parameters of that level 1 probabilistic setting is modelled by DST intervals,

Table 12.3 Table of study characteristics.

Final goal of the uncertainty study	Gain a first impression of the system failure probability. Test the system failure probability against SIL thresholds.
Variables of interest	System failure probability.
Quantity of interest	Interval of the exceedance probability.
Decision criterion	None
Pre-existing model	System fault tree: Level 1 probabilistic Boolean reliability model.
Model inputs and uncertainty model developed	Failure probabilities of system components are estimated by experts and modelled by a DST belief function.
Propagation method	Monte Carlo
Sensitivity analysis method	None
Feedback process	None

Table 12.4   Statistics on the resulting belief function.

Data source	Failure probability
$E(p_{sys})$	$[5.2 \cdot 10^{-8} \; 8.1 \cdot 10^{-7}]$
$Med(p_{sys})$	$[5.5 \cdot 10^{-7} \; 8.1 \cdot 10^{-7}]$
$Q_{95}(p_{sys})$	$[7.5 \cdot 10^{-7} \; 1.0 \cdot 10^{-6}]$
$Bel/Pl(p_{sys} < 10^{-6})$	0.92/1
$Bel/Pl(p_{sys} < 10^{-7})$	0/0

which results in probabilistic quantities given in bounds. These bounds are narrower for the median than for the expectation value.

12.5   Conclusions

The DST approach was experimented in a pilot case study approach in order to explore the potential of modelling both probabilistic distributions for the reliability randomness and interval descriptions for the expert discrepancies or lack of knowledge.

Because of the conservative uncertainty treatment inherent in DST the results could be exploited further, even in a sceptical environment. Experts have the option of describing critical uncertainties through intervals without the need to justify a distribution assumption.

When dealing with a second-order DST uncertainty treatment, it is necessary to communicate the meaning of the uncertainty expressed in the system output. The uncertainty is not generated by a random process (which is inherently modelled by the probabilistic fault-tree model). It stems from expert uncertainty and conflict between estimates. It cannot simply be 'averaged out' if more than one system is produced. This important difference has to be noted if results are to be used in a decision process.

In conclusion, the method looks especially interesting in reliability and safety studies during the first design stages; benchmarking with alternative descriptions of uncertainty, including level-2 probabilistic settings (as experienced in Helton and Oberkampf, 2004) could of course also enrich the findings.

References

Dempster, A.P. (1967) Upper and lower probabilities induced by a multivalued mapping, *Annals of Mathematical Statistics*, **38**, 325–339.

IEC (2001) *IEC 61508 Functional Safety of electrical/electronic/programmable electronic safety-related systems*, Parts 1–7. IEC, Geneva, Switzerland.

Helton, J. C. and Oberkampf, W. (2004) Alternative representations of epistemic uncertainty, *Reliability Engineering & System Safety,* **85** (1–3).

Jäger, P. and Bertsche, B. (2004) 'A new approach to gathering failure behavior information about mechanical components based on expert knowledge'. *Proceedings of the 2004 Reliability and Maintainability Symposium (RAMS).* Los Angeles.

Shafer, G.(1976) *A Mathematical Theory of Evidence,* Princeton University Press.

VDI (2004) *VDI 2206: Design methodology for mechatronic systems.* Beuth, Berlin.

Bundesrepublik Deutschland (2005) *V-Modell XT Dokumentation.* Available online at: http://www.kbst.bund.de

Part III

Methodological Review and Recommendations

13

What does uncertainty management mean in an industrial context?

'Complexity lies within the entanglement that does not allow us to tackle things separately, it severs what binds groups, and produces a crippled knowledge. The problem of complexity further appears since we are part of a world, which is ruled not only by determination, stability, repetitions, or cycles, but also by outbursts and renewal. Throughout complexity, there are uncertainties, either empirical or theoretical, but, most of the time, both.'

Edgar Morin, Philosopher
Translated from a French interview by François Ewald
in Magazine Littéraire, no. 312, July–August 1993.

13.1 Introduction

Many industries are situated in a so-called 'complex' environment in which organizations and knowledge are constantly being challenged. Complexity lies in the interplay of disciplines and their conflicting constraints. For instance, the management of a new aircraft project, the long-term operation of a power plant or the selection of oil drilling projects all present huge technical difficulties, in terms of multi-physics coupling between different disciplines: aerodynamics to optimize in-flight performance; mechanics to ensure structural robustness, weight performance of an aircraft or control of vibration in the steam turbines; geophysics to control drilling performance; thermics around the engine or thermal hydraulics inside a nuclear reactor; electromagnetics for direct and indirect lightning effects; acoustics to minimize environmental impact around airports; ageing processes under irradiation to optimize nuclear lifetime management; complex chaining between potential component failures to control

Uncertainty in Industrial Practice Edited by E. de Rocquigny, N. Devictor and S. Tarantola,
© 2008 John Wiley & Sons, Ltd

overall reliability, etc. This type of complexity is also accompanied by uncertainty: it involves, for example, dealing with supply chain management, commercial strategies, currency rates, political contexts, environmental regulation, corporate image, public opinion, etc. By 'uncertainty' is meant firstly macroscopic uncertainty, such as the international economic context: the '9/11' catastrophe was unpredictable and had a huge impact on the aeronautical industry; the Chernobyl accident, although less unpredictable, to a certain extent, similarly had a great impact on the energy sector. 'Uncertainty' also refers to a more specific (or microscopic) context, such as the electronic properties that improve or diminish the performance of an antenna as in *Electromagnetic Interferences in Aircraft* (Chapter 5).

Parts I and II exemplified the type of quantitative uncertainty treatment that is now being developed across a wide range of industrial activities, from design concept to in-service operations. This is a testament to the generic features of quantitative uncertainty treatment and to the possibility of developing a transverse approach in spite of the diversity of technical domains. The case studies also illustrate the influence of different constraints – time, cost, access to information – on the possibility of obtaining reliable results. Part III of the book will review a number of generic methods available to practitioners – the choice of algorithms or methodologies for each step will prove to be dependent on the key features and practical constraints faced in each situation. Therefore, the goal of this introductory chapter is to detail somewhat more systematically the industrial contexts, covering the ones characterizing the case studies of Part II and beyond, so that the difficulties encountered in undertaking uncertainty studies may be better anticipated.

13.2   A basic distinction between 'design' and 'in-service operations' in an industrial estate

For the sake of simplicity, two major phases of the life-cycle of a product or a system are distinguished in this section. Even if it is obvious that this short-cut gives only a poor representation of the complexity of an industrial environment, it is intended to help illuminate the different questions and contexts from a practical point of view. A distinction is proposed between: the early engineering stages when the product, system or utility does not yet exist, called the *design phases*; and the later phases which re-group all the engineering activities of manufacturing and operations of the product/system/utility, called the *in-service operations*.

13.2.1   Design phases

During the design phases, assumptions have to be made in order to assess future technical performance and adequacy to anticipated performance and market needs. The criteria or requirements applicable to the output variables of interest are partly negotiable (between the various actors) and partly imposed by a regulatory process.

The following case studies can be gathered within this category: *Hydrocarbon exploration* (Chapter 4), *Dyke reliability* (Chapter 10), *Aircraft fatigue modelling* (Chapter 11), *(Automotive) Reliability in early stages* (Chapter 12).

Within this context, engineers somehow tend to design according to fixed specifications. Depending on the industry, engineers have to conform to norms, standards (as for instance Eurocodes) or 'mil specs'. In many circumstances, traditional concepts of best practice lead designers to particular solutions. In short, engineering does not typically design for a *range* of possibilities. It designs according to fixed scenarios, trying to fulfil pre-determined criteria. Part of the associated uncertainty is due to the pre-existing model itself. Indeed, even if the most up-to-date technological or scientific knowledge in the discipline is taken into account in the pre-existing models, this may not be sufficient for the purpose, as innovation is underway. At the same time, lack of data and/or an unclear definition of the criteria also bring along a great deal of uncertainty.

However, the fact that engineers design according to fixed criteria does not imply that they may ignore uncertainty. The management of risk is a foundational issue in the design, development and extension of technology. Much engineering practice is thus directed towards failure-driven risk management, that is, towards the elimination of potential unacceptable consequences. This tendency can lead to over-conservative strategies through the definition of worst-case scenarios and additional unrealistic margins or safety coefficients.

Clarifying the uncertainty management steps proposed within this book could help to make these margins more explicit by differentiating the different elements of an analysis: data, system model, criteria. This could also facilitate the emergence of option-exploring approaches: the use of the same tools could be used to investigate new technological solutions that are typically neglected because of overly stringent demonstration processes. The engineering practice would thus evolve towards a compromise between risk control and the fulfilment of functional requirements and market needs.

Two case studies illustrate these points. Within *Dyke reliability*, the variable of interest is slope instability and the technical objective is to provide partial safety factors that would be useful to the design engineers, taking into account various sources of uncertainty. These factors should ensure the required safety level. *Aircraft fatigue modelling* is also a good example: current industrial practice involves SN curves materialising the fatigue limits used in conjunction with safety factors, in order to take into account (among other effects) the scatter inherent in fatigue experiments.

13.2.2   In-service operations

Within in-service operations or more generally 'downstream' stages, operators hope to optimize the performance of systems (e.g. a product, system or utility) in environments which are characterized by changing regulations and economic constraints. The following case studies are situated within this context: CO_2 *emissions* (Chapter 3), *Electromagnetic interferences in aircraft* (Chapter 5),

Airframe maintenance contracts (Chapter 7), *Radiological protection and maintenance* (Chapter 9).

On a day-to-day basis, operational engineers usually follow procedures that were defined upstream of operations, during the design process or the maintenance programme updates. These tools are defined to balance maintenance costsand strategies, in compliance with regulations and economic performance. When uncertainty studies are undertaken for this type of operation, a pre-existing model has already been defined, data or expertise is already available for the inputs, and a criterion is known. For example, *Radiological protection and maintenance* aims to define simple procedures under the constraint that the operator receives controlled doses; in CO_2 *emissions*, the goal is to compare different technologies to measure the CO_2 emissions from a thermal power plant in order to provide the easiest and most reliable configuration of a measurement chain.

Uncertainty management within this context typically makes it possible to adapt maintenance policy, better to balance maintenance costs, regulation and efficiency and to anticipate future demands and evolution at a reasonable cost.

13.3   Failure-driven risk management and option-exploring approaches at company level

Various sources (in particular de Neufville, 2004; and de Weck and Eckert, 2007) point out that the traditional engineering design approach to uncertainty often neglects two other important aspects of uncertainty management, in spite of their recognition in the earliest stages of economic research (Knight, 1921). First, the focus on *technological failure* leads to the neglect of uncertainties which may create opportunities. Rather than fully managing uncertainties, the traditional approach thus tends to deal mainly with worst-case failure-driven situations. Second, '*technological failure*' usually means that failures due to economics or other causes, including human factors, are disregarded. A fully responsible approach to the design of engineering systems should take care not just that the product performs as specified, but also that it anticipates market or public needs, which means that several other outputs and scenarios should be taken into consideration. Notwithstanding the managerial or strategic complexity that may result at company level, quantitative uncertainty management could technically involve a wider '*option-exploring approach*' rather than mere technological failures. This means expanding a failure-driven risk approach to a 'risk and opportunities' approach on a multi-disciplinary level. Two case studies anticipate these future moves and illustrate this view of uncertainty in a complex environment. Regarding opportunities, *Airframe maintenance contracts* illustrates both risks and opportunities, as it is intended to support market activities when negotiations are involved. In *Hydrocarbon exploration* different scenarios are investigated with economic constraints, leading to the ranking of different exploration scenarios.

13.4 Survey of the main trends and popular concepts in industry

Recent developments have created major new opportunities to improve procedures and practice when dealing with the uncertainties involved in design and in-service operations. Even though these concepts may not yet be widespread across business, they are becoming increasingly popular with industrial decision-makers.

As mentioned above, significant *conceptual reformulation* refers to the shift in emphasis from the 'failure-driven risk approach' to the 'option-exploring approach' in uncertainty management. What is new is the recognition that the performance of an engineered system has to be considered in its larger commercial and political environment. Designers today need not only to bullet-proof their designs against technical and human failures, but also to enable them to evolve and adapt to new circumstances. These approaches should be evaluated on the basis of flexibility and business agility as well as against safety constraints.

New advances in information technology (development of models, acquisition of computer power, etc.) make it possible to conceive of a much more coherent uncertainty management approach. For instance, the development and validation of best-estimate system models or the use of data-mining techniques are much more affordable nowadays. In short, as these advances continue, there will a process of change in the way uncertainty is analysed and managed.

Concerning recent methodological advances, two popular concepts should be highlighted. First, the notion of *robust design* (Phadke, 1989) encompasses a set of design methods for improving the consistency of system function across a wide range of conditions. The basic practice is to introduce noise factors in experiments so that systems can be made less sensitive to variations in customer-use conditions and internal degradation. By this means it is possible to improve quality without raising manufacturing costs.

A second popular concept is *real options analysis* (Dixit and Pindyck, 1994). Financial specialists have developed a theory and means of evaluating financial flexibility known as *option analysis*. The development of options analysis for real systems promises to enable the engineering profession to calculate the value of flexibility, similarly to the 'volatility' concept popular in finance. It is important to note that the options analysis developed in finance generally is not – and actually cannot – be applied directly to engineering, because financial analysis assumes that the elements of concern (the assets) are traded in efficient markets characterized by widespread trading, complete information, and an extensive history that provides reasonable statistics. The uncertainty associated with an engineering system, however, often relates to new technologies, markets or regulations for which historical data are unavailable or scarce. Beyond the research already done to transfer these methodologies to engineering, industrial practice has to test innovations in design and management of engineered systems over time. Undertaking a 'Real Options Analysis' approach at company level when assessing rival solutions involves methodologies that will not be detailed in this book. However, the way the case studies are examined and the techniques involved are fully compatible with this larger view: the corresponding decision-aid methods (for instance, Barbera *et al.*, 1998; or Bedford and Cooke, 2001) do rely on uncertainty quantification in the sense of assessing the distribution of plausible outcomes (i.e. variable of interest), be it in a standard probabilistic or extended setting.

13.5   Links between uncertainty management studies and a global industrial context

All the case studies presented within this book are intended to provide a quantified analysis in order to aid the decision-maker. The studies were undertaken in specific contexts: regulation (*CO₂ emissions, Dyke reliability, Electromagnetic interferences in aircraft*); investment (*Hydrocarbon exploration, Airframe maintenance contracts*); maintenance (*Radiological protection and maintenance*); and transport strategy (*Spent nuclear fuel behaviour* [Chapter 8]).

The sources of uncertainty range from more or less quantifiable uncertainties, according to how well the issues are understood. Each product or system bears its own uncertainties with it, which arise primarily from the '*inside*': this refers to the typical system boundary shown in the following Figure 13.1. Uncertainties *inside* the dashed box can be influenced by the system designer or company to a

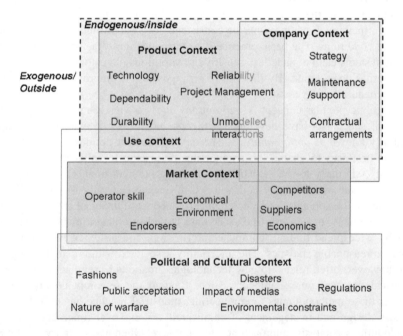

Figure 13.1   Links between uncertainties and industrial contexts. Based on de Weck O., Eckert C. 2007.

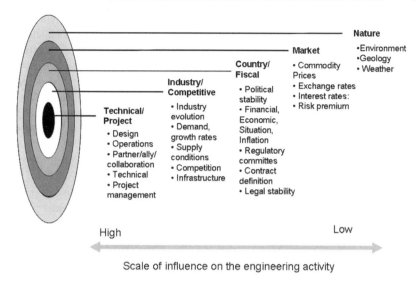

Figure 13.2 Different layers of uncertainty (inspired from [Miller R., Lessard D.R. 2001]).

greater extent. Uncertainties *outside* the system boundary can be influenced by the designers or company to a lesser extent. Many other uncertainties are outside the dashed box of a company's direct control, as they arise in the market-place where the product operates, in the way it operates and in the cultural and political context at the time of its design and use. *Outside* uncertainties are often much harder to predict or control.

13.5.1 Internal/endogenous context

In each development process there is an element of technical risk within the *product context*, as most products have an element of novelty in themselves or at least in their design. These *technical uncertainties* are monitored during the design process and should be resolved step by step, from conceptual phase analysis to detailed specifications. However, even the re-use of existing ideas involves considerable uncertainty. A component that works well in one product might not do so in another simply because a slightly different demand is made of it: its tolerance margins may be exceeded or the component may be placed in a new context and have to interact with unfamiliar components.

Un-modelled interactions between parts of a system frequently catch companies by surprise, when changes propagate through a system or unexpected failures occur. How well a system behaves during a change process depends on the exact state of all components (hardware, software, human, etc.), which are rarely well understood. This also affects the reliability of a component over its life cycle. Issues of reliability and robustness are now increasingly addressed by means of quality management strategies. Companies are now investing considerable effort in understanding the failure mechanism that could cause problems later in the product's lifecycle. This is part of the attempt to control the failure of components. While catastrophic failures are, of course, highly undesirable and often very costly, the life cycle is already significantly affected by the process of ageing and wear of a component. In *Electromagnetic interferences in aircraft*, the systems under consideration had been designed separately, since they originated in different industries with different needs, and therefore observed different regulations: the goal is thus to manage their interactions.

Some uncertainties arise from the *business context* within which the product is designed. Indeed, each company develops its own product strategies, which can affect particular products, by redirecting resources to and from the design process. The product is also strongly affected by the contractual arrangement under which it is designed, which can require properties that are difficult to achieve, or entail late changes to the product. *Airframe maintenance contracts* considers, for instance, different scenarios of investment/demand in order to evaluate potential returns on investments in a variable business context.

13.5.2   External/exogenous uncertainty

There is often a great uncertainty in the way a product is used and in the conditions under which it has to operate. The operational environment of products can change, requiring reliable operation in unfamiliar circumstances, climates or weather conditions. Similarly, the behaviour of the operators of the product is uncertain. This can be reflected in changes in maintenance contracts or in requiring companies to maintain products over unanticipated time-spans. Furthermore, industry may make incorrect predictions about the habits of their potential clients/customers. While a company can counteract this with inclusive design strategies, much is unknown about the capabilities and interests of different user groups. In *Radiological protection and maintenance*, the real behaviour of the operator, as compared to formal procedures, is one of the sources of uncertainty.

Markets contain a huge amount of uncertainty. The degree and diffusion of change in the market depends on the nature and the life span of the product. Demand profiles for a product may change very quickly, as environmental conditions change, or as the product is affected by forces outside the control of the company. The nature and timing of rival products also introduces great uncertainty into the market. If another player releases a comparable product earlier, they may conquer the market. Alternatively, innovations brought to the market by competitors can change the demand profile very rapidly.

Market forces are also at work at a deeper level through changes in the economy on a deeper level. Fluctuating exchange rates may have a huge impact on the cost of manufacturing as well as on the ability to export. Demand and supply forces push many products to the limits of financial viability. For instance, *Airframe maintenance contracts* illustrates how a supplier may try to anticipate the return on investments at different horizons, involving exogenous factors.

The market is strongly influenced in turn by wider *political and cultural contexts*, which may generate very tangible uncertainties for specific products. Changing regulations may require major changes both in the design of new products and the operability of existing products, as illustrated by CO_2 *emissions* in the context of the Kyoto protocol and growing public awareness, and by the definition of the different scenario in *Nuclear waste repository* (Chapter 6).

13.5.3   Layers of uncertainty

Another useful simplified mental model is to think of these uncertainties as occurring in layers around each other (see Figure 13.2). This representation shows an inner layer comprising the direct technical/project risks, roughly corresponding to the product context shown in Figure 13.1. The next outer layer is the industry/competitive layer corresponding to the corporate context in Figure 13.1. The next is the country/fiscal layer, followed by the market layer. This again is similar to Figure 13.1, whereby the ordering of the layers differs slightly. The outermost layer corresponds to natural events such as large-scale weather phenomena or geological formations, which can be particularly important for utilities, for instance in the oil or gas industries or in nuclear waste management.

A key to understanding the layers is the fact that the degree of influence decreases sharply from the inner to the outer layer. The possibility of mitigating risks or exploiting opportunities arising from these uncertainties thus also decreases. A company may have almost complete control over the choice of its technical architecture, supplier and operational strategy. Conversely, it may only have limited influence on future regulations (e.g. through lobbying efforts) and it has no influence at all on the occurrence of natural disasters. This does not mean that nothing can be done about the consequences of the outer layer events (consider procurement of insurance, redundancies in the architecture, protection mechanisms, etc). It means rather that their occurrence in the first place cannot generally be influenced: this is an aspect of the more general issue of reducibility which will be detailed in the following chapter. Note finally that individual companies may not be able to influence aspects such as regulation or marketing fashions, but collectively they may wield a great deal of power.

13.6   Developing a strategy to deal with uncertainties

From a practical point of view, it should be asked whether sources of uncertainty are reducible or not and whether the cost of different reduction strategies can be estimated. The decision-maker is concerned to decide whether there is a point to acting. Within the overall framework, this means that the criterion has already been more or less fulfilled by the end of the uncertainty study (see CO_2 *emissions*). In the case of non-compliance with a criterion, additional work is needed to define more accurate models or to collect additional data to evaluate the risk. This is an iterative process following the common steps of an uncertainty study (see *Electromagnetic interferences in aircraft*), taking into account the time-scale and the possible uncertainty management strategies.

The *time-scale* on which engineers and managers might choose to manage uncertainty could range from the very short to the very long. Thus, it might be reasonable to think about decisions which are:

- *Short-term.* These decisions concern immediate issues. Being limited to the available data and pre-existing models, the objective is to treat the information in the best possible way in order to aid the decision-maker. Obviously, the results are conditional on the quality of the available information, and could imply a mid-term effort to develop a new model and/or collect new information. Related goals are usually operational, i.e. Goals C (Comply) or S (Select).

- *Mid-term.* Unlike short-term decisions, these involve situations in which a support is conceivable, such as the development of refined models or the collection of additional data to improve the understanding of the global phenomenon. Goals U (Understand) and A (Accredit) support Goals C and S.

- *Long-term.* These concern long-term issues, including the influence of societal, political and market evolutions. These may also be the source of new opportunities. But little has been done at this level in the engineering field, since analysts need to consider a wide range of factors that have traditionally remained beyond the concern of engineers. The long-term aspect is not illustrated in this book.

Different *uncertainty management strategies* can be considered: either to reduce the uncertainty itself, or to enable the system to respond better to it. This latter strategy, the classical purpose of robust design, can consist in enhancing the capabilities of the system, either by modifying/strengthening it against an external factor or by making it more flexible so that it can adjust to the external factor. The choice of strategy depends on the most influential factors as well as on the time and cost constraints of the project. Figure 13.3 illustrates the level of reducibility that may be found in the various case studies, the other axes of the figure being identical to those introduced in Part I (position of the case study in the industrial lifecycle and maturity of the study). For example, within the context of *Electromagnetic interferences in aircraft*, different strategies are examined to manage uncertainty:

- forbidding the use of PEDs during take-off and landing (reducing one source of uncertainty);

- testing shielded windows and electromagnetic absorbers (strengthening passively);

- use of other radio navigation systems, such as GPS systems (modifying the system).

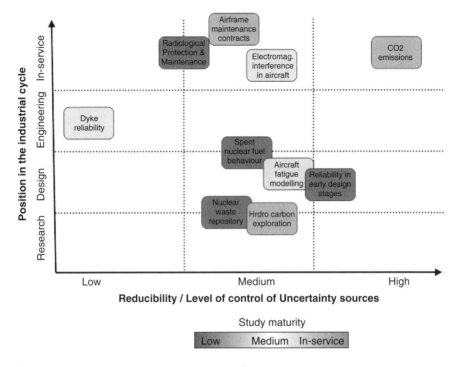

Figure 13.3 Level of control over uncertainty and position in the industrial cycle for case studies of the book.

These proposals are, of course, not new. Engineers should already be employing these concepts. The main interest of uncertainty management lies in the possibility of better definition and control of the margins, as well as of the underlying hypotheses, risks and opportunities in a design or in-service operations strategy.

References

Apostolakis, G. and Wu, J. S. (1993) *Reliability and Decision-making*, London: Chapman & Hall.

Aven, T. (2003) *Foundations of Risk Analysis*, Chichester: John Wiley & Sons, Ltd.*

Barberà S., Hammond P. J. and Seidl, S. (1998) (Eds) *Handbook of Utility Theory, Volume 1 Principles*, Dordrecht: Kluwer Academic Publisher.

Barberà S., Hammond, P. J. and Seidl, S. (1998) (Eds) *Handbook of Utility Theory, Volume 2 Extensions*, Dordrecht: Kluwer Academic Publisher.

Bedford, T. and Cooke, R. (2001) *Probabilistic Risk Analysis – Foundations and Methods*, Cambridge: Cambridge University Press.*

De Neufville, R. (2003) *Architecting/Designing engineering Systems using Real Options*, MIT report ESD-WP-2003-01.09 (Available online at: http://esd.mit.edu).

---

*Also suitable for beginners.

De Neufville, R. (2004) Uncertainty management for engineering systems planning and design, *MIT Engineering Systems Monograph* (Available online at: http://esd.mit.edu/symposium/pdfs/monograph/uncertainty.pdf).

Dixit, A. K. and Pindyck, R. S. (1994) *Investment Under Uncertainty*, Princeton University Press.

Granger Morgan, M. and Henrion, M. (1990) *Uncertainty – A Guide to Dealing with Uncertainty in Quantitative Risk and Policy Analysis*, Cambridge: Cambridge University Press.*

Knight, F.H. (1921) *Risk, Uncertainty and Profit*, Hart, Schaffner & Marx.

Kumar, A., Keane, A. J., Nair, P. B. and Shaphar, S. (2006) Efficient robust design for manufacturing process capability, in *6th ASMO-UK/ISSMO International Conference on Engineering Design Optimization*, Oxford, UK, 3–4 July, 2006, 242–250.

Miller R., Lessard D.R., Strategic Management of Large Engineering Projects: Shaping Institutions, Risks, and Governance, MIT Press, 2001.

Phadke, M. S. (1989) *Quality Engineering Using Robust Design*, New Jersey: Prentice Hall.

Quiggin, J. (1982) A theory of anticipated utility, *Journal of Economic Behavior and Organization*, **3**, 323–343.

Savage, L. H. (1954, 1972) *The Foundations of Statistics*, New York: Dover Publication Inc.

Trigeorgis, L. (1996) *Real Options – Managerial Flexibility and Strategy in Resource Allocation*, Cambridge, MA: MIT Press.

de Weck, O. and Eckert, C. (2007) *A Classification of Uncertainty for Early Product and System Design*, MIT report ESD-WP-2007-10 (Available online at: http://esd.mit.edu/WPS).

Zarembka, P. (1974) *Frontiers in Econometrics, Economic Theory and Mathematical Economics*, New York: Academic Press.

---

*Also suitable for beginners.

14

Uncertainty settings and natures of uncertainty

This Chapter discusses the quite classical theoretical distinctions made in the literature on the natures of uncertainty (reducible or irreducible, variable or uncertain, error, etc.). The central point to be made is the following: although these classical distinctions appear as both natural and useful in industrial studies, as already exemplified in some of the case studies of Part II, a number of difficulties and controversies quickly emerge when considering the formal differentiation of treatment according to those distinctions. The practitioners' point of view of this book should also be understood in the wake of what was discussed in Section 1.4: different mathematical settings (such as level-1 or level-2 settings, probabilistic or mixed deterministic, etc.) may be considered to represent uncertainty and are actually encountered in various industrial regulations. There are even multiple interpretations of the same settings. Therefore a panel of differing settings introduced below has been deemed acceptable, provided that a clear explanation is given on how the setting relates to the natures of uncertainty involved.

14.1 A classical distinction

It appears quite typical in the literature to discuss the nature of uncertainty and to draw out a number of theoretical distinctions (*cf.*, for instance, Granger Morgan and Henrion, 1990, or Paté-Cornell, 1996). These will briefly be reviewed below, with some additional comments on the practical difficulties or ambiguities that immediately arise in industrial uncertainty studies when considering those distinctions.

Uncertainty in Industrial Practice   Edited by E. de Rocquigny, N. Devictor and S. Tarantola,
© 2008 John Wiley & Sons, Ltd

- Irreducible' or 'aleatory' (or random, stochastic, objective, inherent) *vs.* 'reducible' or 'epistemic' (or lack of knowledge, ignorance, subjective, imprecision) is the most classic:

- Irreducible uncertainty refers to events which remain unpredictable whatever the amount of data available; in many cases, however, the regularity of their frequential behaviour in long series is classically considered to be amenable to probability calculations (for example, weather patterns or natural risks), which explains the alternative classical denomination of 'aleatory' for many examples of irreducible uncertainty. Reducible or epistemic uncertainty refers to types of uncertainty which can be directly reduced by an increase in the data available. This may include situations in which there is somehow a deeper lack of knowledge of the uncertainty, less (or even not at all) amenable to probabilistic treatment or estimation.

- Some authors reserve the word 'uncertainty' for the reducible or epistemic kinds, i.e. those based on lack of knowledge, while 'risk' encompasses the irreducible and aleatory kind in its purest sense, i.e. unpredictability associated with complete knowledge of the probability distributions[1]. In practical applications, this kind of distinction implies that an aleatory component never arises without some associated epistemic component: indeed, probabilities in non-academic examples are never known perfectly due to data scarcity.

- For many, there is a clear and definitive distinction to be made between uncertainty which is irreducible or reducible in the light of an increase in data/knowledge. However, it is generally necessary in practice further to distinguish uncertainties that are *theoretically reducible* from those that are *industrially irreducible* because of operational or economic constraints, data scarcity or modelling limitations. *For practitioners, the reducibility issue may therefore be more of a context-dependant feature or even of a modelling choice* (as will be explained in Section §14.2).

- 'Variability' *vs.* 'uncertainty'

- This distinction, close but not completely equivalent to the preceding one, is used more specifically when the system inputs mix a *population* of objects (or scenarios), a *spatial* distribution of properties within a system, or even a *temporal* distribution of properties affecting a system: in which case, the variation in properties from one object (or part of the system, or instant in time, respectively) to another within the population considered is distinguished from any uncertainty attached to the properties of a given object (or part, or instant), such as lack of knowledge, measurement error, etc.

- In practice, *temporal* variability is often identified with aleatory uncertainty, since the goal is generally the control or understanding of the uncertain

---

[1]This view is closely related to the analysis of human behaviour towards uncertainty in decision theory, risk aversion or uncertainty aversion, etc., which will not be reviewed in this book. See: Knight, 1921; Savage, 1954; Quiggin, 1982.

behaviour of an industrial system at any time in the future. However, it may be necessary to further distinguish a temporal variability which is known or fairly predictable (such as a seasonal pattern) from the uncertain realizations in time most appropriately supposed to be stationary: this is generally implicit when considering the uncertain point events of a system.

- *Spatial* variability may not always play the same role and may not be as clearly identified with an irreducible or aleatory component. If the behaviour considered involves all parts of a system at once, such as an oil basin or a dyke system, there is just one global realization of the spatial distribution for the lifetime of the system, but it will be poorly known due to the scarcity of realistic measurements. In that case, variability could be seen rather as epistemic in nature, or theoretically reducible, although irreducible in practice. It is, however, closely mixed up with the limitations of measurement processes, or even with the subsequent modelling simplifications, etc.

- 'Epistemic uncertainty' *vs.* 'error', is a finer distinction.

- This depends on whether ignorance or subjective uncertainty is 'inevitable' or 'deliberate', in spite of the availability of knowledge. This is the case, for instance, if practical constraints or choice result in the use of a simpler model than might have been used, or the use of a numerical algorithm which is faster but less accurate.

- This depends on the acceptability of a certain limited degree of error, defined by the practitioner, to allow the use of a simpler model. In this case, the degree of deliberate error will somehow be controlled by an additional margin; this may also include, more insidiously, the occurrence of involuntary study errors, in spite of quality assurance.

- Uncertainty which is referred to as 'parametric' (associated with model inputs according to the level of information available on those inputs) vs. 'modelling uncertainty', concerning the adequacy of the model to reality (structure, equations, discretization, numerical resolution, etc.).

- Although the difference is not so clear-cut in theory, because it depends on the conventional definition of a 'model', this distinction is quite instrumental in practice, since uncertainty analysis will result in the consideration of parameters as inputs of a model, on which a series of runs will be ordered for uncertainty propagation or sensitivity analysis.

- According to the physical-industrial domain considered, traditions differ as to the inclusion or not of model uncertainty alongside the uncertainty in 'parametric' inputs. It appears quite common in physical fields where calibration of meso-scale models is common, such as in fluid mechanics: considering modelling parameters as uncertain inputs is commonly taken to cover both epistemic uncertainty regarding the un-modelled details of the phenomena involved (such as wavelets, micro-variability of riverbed conditions, etc.) and aleatory uncertainty, such as intrinsic turbulence structures, not predicted by average models. In other fields, such as risk analysis, there may be a large theoretical controversy over the acceptability of including quantified model uncertainty.

– Note that in general, modelling uncertainty refers to a level of uncertainty over which there is less control than what might be had over a deliberate model simplification with an error margin.

These distinctions are often closely related to the theoretical foundations and interpretation that is being made on uncertainty and the use of probabilistic settings, as already mentioned in Section 1.4. For instance, in what was called the *functional analysis* view on uncertainty exploration in numerical models, distinctions according to the nature of uncertainty are rather unimportant as uncertainty modeling and sensitivity analysis may be viewed as a formalized and efficient space-exploring mathematical tool to understand upstream models. Double probabilistic settings interpreted in the *probability of frequency* approach (Apostolakis and Wu, 1993) would clearly distinguish two natures, typically aleatory vs. epistemic components, or variability vs. epistemic components. A predictive Bayesian approach (Aven, 2003) would limit the description of uncertainty to observable quantities, discarding in particular modeling uncertainty (or uncertainty of pure model parameters), and eventually considering that all uncertainty is 'epistemic'.

14.2   Theoretical distinctions, difficulties and controversies in practical applications

As was explained in some of the examples in Part II, commenting on the different types or natures of uncertainty involved in a case study is both natural and useful as an introduction to uncertainty treatment. It may also be viewed as a sort of 'check-list' to make sure that a large enough inventory has been undertaken before conducting the uncertainty analysis.

Many classifications have already been produced in the literature: however, no classification appears perfect, and it is often adherent to a given theoretical inspiration and interpretation of probability. In practice, types of uncertainty often overlap and sources are mixed in a complex way. Table 14.1 is therefore intended to analyse the natures most commonly encountered in the examples of Part II rather than to establish a new authoritative classification: the structure of this table follows a general distinction according to reducibility (which is viewed as more practical than the classical 'aleatory vs. epistemic' distinction often encountered in the literature), although this does not preclude the limitations already mentioned on this theoretical distinction. Indeed, for some it was deemed preferable to maintain a 'mixed' category.

Note that two types of reducible uncertainty have deliberately been made explicit in the table, since these are easily identified in practical applications:

Table 14.1   Review of the natures of uncertainty encountered in Part II case studies.

General types	Details	Examples
Mostly irreducible	Inherent/ variability in time	High Water Levels (Dyke reliability) Failure/repair events, maintenance costs, flight characteristics (Airframe maintenance) Scenario uncertainty (Radioactive waste repository) Failure time of components (Early design)
Mostly reducible	Statistical uncertainty (or *estimation* uncertainty) -shape -parameters	High water pdf parameters or shapes (Dyke reliability) Component reliability parameters (Early design, Airframe maintenance)
	Propagation uncertainty*	FORM accuracy in predicting failure probability (Spent nuclear fuel behaviour, Electromagnetic Interference in Aircraft, Dyke reliability) Variability of 95th dose quantile predictions through limited Monte Carlo trials (Radioactive waste repository)
Particularly mixed natures	Measurement uncertainty	Coal beltweighers (CO_2 emissions) Basin geophysical properties (Oil exploration)
	Model uncertainty	Analytical simplifications of the reliability model (Airframe maintenance) Fluid behaviour in the oil basin (Oil exploration) Electromagnetic coupling (Electromagnetic Interference in Aircraft) Underground source term and flow model (Radioactive waste repository) Heat rate model approximations (CO_2 emissions) Soil stability model simplifications (Dyke reliability) Basin geophysical properties (Oil exploration)

(continued overleaf)

Table 14.1 (*continued*)

General types	Details	Examples
	Variability in space	Geometrical simplification (Radiological protection and maintenance)
		Dyke soil properties (Dyke reliability)
		Basin geophysical properties (Oil exploration)
		Coupling characteristics (Electromagnetic Interference in Aircraft)
		Deep geological properties (Radioactive waste repository)
	Variability in population	Failure/repair events, flight characteristics, maintenance costs (Airframe maintenance)
		Emission characteristics (Electromagnetic Interference in Aircraft)
		Fuel carbon rates (CO_2 emissions)
		Waste inventory (Radioactive waste repository)
		Radiological source composition (Radiological protection and maintainence)

*also referred to as sampling uncertainty or variability, confidence in the results.

- Classical 'statistical' uncertainty comes from the inadequacy of data to properly estimate the pdf of the sources of uncertainty. This is even more the case when expertise complements or completely replaces the use of data for the statistical modeling of the sources of uncertainty.

- 'Propagation uncertainty', less often mentioned in the literature, arises from the limitations of the propagation methods, such as relatively small sizes of Monte Carlo sampling or approximations underlying the use of FORM/SORM methods. This becomes all the more important when large scientific computing constraints are encountered in complex industrial models.

Note again that the 'reducibility' of some of the uncertainties does not equate to an 'epistemic' (or theoretical reducibility) nature: reducibility also involves some industrial/practical constraints, and even a cost-benefit perspective. There may be a continuum between 'strictly irreducible' or 'reducible': in some cases, 'epistemic' uncertainty cannot be reduced: e.g. in very expensive measurements, or even in experiments which necessarily damage equipment. This is one of the reasons for which, in the table above, it was deemed preferable to address the nature of reducibility rather than the theoretical 'epistemic/aleatory' natures, and to leave a category for 'mixed natures'.

Measurement uncertainty is another example of a practical source of uncertainty within which it is hard to distinguish epistemic and aleatory components: the *Guide to Expression of Uncertainty in Measurement* (GUM, CEN, 1999) has in fact eliminated any mention of epistemic/aleatory structures as a mandatory distinction in analyses and treatments. In the case of oil exploration, for instance, it seems that measurement error in seismic velocities is very much bound up with spatial variability and even with modelling uncertainty, since the basin geometry model is somewhat simplified.

Note also that a distinction may be found between the uncertainty attached to the definition and occurrence of scenarios (which could be referred to as *scenario uncertainty*), and uncertainty issuing from the phenomena and consequences of a given scenario. This may be the case in certain contexts with exceedingly poor information or only rather controversial expert opinion for the assessment of the probability distributions of the occurrence of scenarios (consider, for example, extreme seismic events or very long-term intrusions into a deep repository). Although it is not a theoretical necessity to define intrinsic difference from phenomenological uncertainties, in this kind of context the distinction may facilitate a quantitative study through the randomization of the phenomena/consequences alone (*cf. Radioactive Waste Repository*). An essential point is then to make clear within the definition of the study that quantities of interest (such as exceedance probabilities or expected doses) should be understood *conditional* to the occurrence of these scenarios.

14.3   Various settings deemed acceptable in practice

The previous paragraphs have demonstrated that while some distinctions between natures of uncertainty appear quite natural and useful in the course of a study, finding a definitive classification suitable to every example is a difficult task. Indeed, natures of uncertainty tend to be quite confused in real examples. The key question, beyond a classification, is whether or not specific treatments should be applied to only certain types of uncertainty: many have argued in the literature for a duly separated treatment of reducible (or epistemic) and irreducible (or aleatory) components, although the best mathematical setting for this separation (between probabilistic or more complex settings) is still a matter of discussion (see Helton and Oberkampf, 2004)

In some cases it is difficult to make proper distinctions. Lack of data may make it difficult to separate, for instance, the reducible/epistemic and irreducible/aleatory components of a source, therefore most sources of uncertainty are randomized together: this was the case in *Electromagnetic interferences in aircraft, CO_2 emissions, Hydrocarbon exploration* and *Spent nuclear fuel behaviour*. In others, such as *Reliability in early stages* or *Airframe maintenance contracts*, a distinction is implemented practically: aleatory lifetime variables on one hand, and, on the other, epistemic uncertainty in the value of their parameters (such as MTTF) which is modelled either in a probabilistic setting (*Airframe maintenance contracts*) or in a DST setting (*Reliability in early stages*).

A number of settings can therefore be distinguished, as already listed in Part I:

• **Standard probabilistic setting**

In this setting, a number of model inputs of factors (within the vector $\underline{x}$) are given an uncertainty model which is a random variable with a pdf for each, or more generally a random vector with a joint pdf if non-independence is modelled. A number of other model inputs or factors are generally fixed at conventional deterministic values (within the vector $\underline{d}$), so that this setting can in fact be considered a 'mixed deterministic – probabilistic' setting.

This means generally that no explicit separation has been made between natures of uncertainty; whatever their natures, all sources of uncertainty pertaining to vector x are randomized together. A number of interpretations may be made of that setting: frequentist, subjective or purely on a functional analysis basis (see Section 1.4).

• **Standard probabilistic setting, with additional level-2 deterministic treatment**

This setting is a slight variation of the preceding. In some practical cases the (epistemic) uncertainty in the values of the pdf parameters, i.e. the level-2 uncertainty in the level-1 uncertainty model, is deemed sufficiently important.

This is the case, for instance, in metrology, where the samples that were used to estimate the pdfs' parameters are small enough for the statistical variability to be considered non-negligible. This can be also the case when limited expertise is available to quantify the sources of uncertainty, or when doubt or disagreement about the expert values is considered significant.

In that case, a potential response, used in some industrial cases, is to take a margin on the values of the pdf parameters. This sort of practice makes sense essentially when the system model $G(.)$ is monotonic, so that, for instance, a larger variance in some uncertain model factors implies a larger variance in the quantity of interest.

This setting introduces two levels of treatment of uncertainty, the second being of a clearly epistemic nature: but generally it does not preclude the possibility that both aleatory and epistemic mixed components may have been considered in the first level, as was the case in CO_2 *emissions*.

• **Double probabilistic setting**

As in the first two settings, the model inputs are also given random variables (level-1 pdfs) to represent the uncertainty attached to them. But in addition, the parameters of their pdfs are considered sufficiently uncertain to be modelled within a probabilistic setting (level-2 pdfs): this second level replaces the 'deterministic penalization' of the pdf parameters considered in the previous setting.

This double probabilistic setting corresponds to the settings explicitly separating the two components, often named 'aleatory' and 'epistemic', which became popular through nuclear studies in the USA in the 1990s. As mentioned in Section 1.4, note that numerous authors have further developed Bayesian interpretations or modifications of this kind of double probabilistic setting (see Helton and Burmaster, 1996; Apostolakis, 1999; Aven, 2003), although these developments will not be reviewed here.

Propagation of uncertainty becomes more complex in this setting than in the previous two; in theory, a double level of propagation algorithms is needed. Often, as was the case in *Airframe maintenance contracts* or *Reliability in early stages*, the first-level propagation is somehow solved or approximated through an analytical calculation; hence, only one Monte Carlo or other propagation algorithm is necessary for the second level.

- **Mixed probabilistic – DST setting**

Similarly to the above, the model inputs are probabilized with level-1 pdfs. The uncertainty in the values of their pdfs is then modelled within a setting that is neither deterministic nor probabilistic, such as with Dempster-Shafer Theory. This is probably one of the most advanced settings: it involves a rather complex mathematical structure at the forefront of industrial practice. The theoretical foundations of that last setting are distinct to that of probabilistic settings (see Appendix C, [Dempster, 1967] or [Helton *et al.*, 2006]), and raise a number of controversies and implementation challenges.

It is essential to note the following, in relation to double probabilistic settings (mixed probabilistic and DST, respectively). They may be viewed as standard probabilistic settings (level-1 DST or purely DST setting, respectively) if the system model is conventionally considered to represent the probabilistic reliability model rather than the underlying deterministic causal reliability model. *Airframe maintenance contracts*, for instance, can be viewed as a 'standard probabilistic' uncertainty study if the system model inputs include reliability parameters such as time-to-repair or time-to-failure. But, in fact, the underlying model already includes some aleatory processes such as uncertain failure or repair dates. If one defines the system model as the purely deterministic causal process, i.e. a balance sheet of costs due to known repair dates, then the study becomes double probabilistic.

Table 14.2 recapitulates the types of settings encountered, adding a deterministic setting by which the model inputs would be 'penalized' (i.e. taken at lower or higher values covering the uncertainty, according to the variable of interest) in a single calculation of the pre-existing model. This deterministic setting is not used in any of the cases presented in Part II. However, in industrial practice it may be in the background of some of the case studies, and therefore constitutes an implicit common reference when practices or regulations in the field of uncertainty are in question.

Table 14.2 Different settings of uncertainty treatment encountered in the case studies.

Name of the setting	Level 1	Level 2	Type of treatment	Examples/ comments
Deterministic	x model inputs are taken at 'penalized' values	None	Deterministic: one calculation at the penalized values*	None in the case studies, however in the background of Electromagnetic Interferences in Aircraft or Dyke Reliability
Standard probabilistic	x model inputs are randomised together in the uncertainty model (along pdfs); some d model inputs remain fixed	Uncertainty model parameters (i.e. pdf parameters) are fixed	Probabilistic (single level)	Electromagnetic Interferences in Aircraft, Nuclear Waste Repository, Radioprotection and Maintenance, Hydrocarbon Exploration, Dyke Reliability, Spent Nuclear Fuel Behaviour
Probabilistic with level 2 deterministic	Idem	Epistemic uncertainty on uncertainty model is modelled through 'penalized' fixed parameters in the pdfs	Generally probabilistic (single level) But may become mixed probabilistic – deterministic if no monotonicity ...	CO_2 emissions

Table 14.2   (*continued*)

Name of the setting	Level 1	Level 2	Type of treatment	Examples/ comments
Double probabilistic (with level 2 probabilistic)	*Idem*, provided *x* refers to the underlying model inputs, before any aleatory setting (e.g. reliability lifetimes)	pdf parameters become also random variables	Probabilistic (double level)	Airframe Maintenance Contracts
Mixed probabilistic – DST	*Idem*	pdf parameters receive a DST distribution	Probabilistic + DST	Reliability in Early Stages

*When the monotonicity of $G(.)$ is dubious as regards some components of $\underline{x}$, calculating $G(.)$ at one 'penalised' value is not enough. A more rigorous deterministic treatment requires the bounding of $G(.)$ over the input ranges between lower and higher values for the components considered, e.g. through full-scale optimisation or other techniques from the field of *interval computations*.

Hence, the examples of Part II show that, rather than one unique approach, a variety of approaches to the separation or non-separation of mathematical treatments according to the natures of uncertainty can reasonably be derived. In fact, the essential recommendations are to:

- List the sources of uncertainties identified and their nature. Discuss in particular the *practical* reducibility issue.

- Detail clearly how they are modelled according to a separate or unified uncertainty model, be it standard probabilistic, mixed with level 2 deterministic or probabilistic of DST, etc.

Those two points are the minimal basis for a justifiable choice of method of treatment, and for a clear communication of the results. Deeper theoretical understanding of decision-theory justifications remains obviously desirable to consolidate the choices, as represented by some salient references mentioned hereafter.

References

Apostolakis, G. and Wu, J. S. (1993) *Reliability and Decision-making*, London: Chapman & Hall.

Apostolakis, G. (1999) 'The distinction between aleatory and epistemic uncertainties is important: an example from the inclusion of aging effects in the PSA', in *Proceedings of PSA '99*, Washington DC.

Aven, T. (2003) *Foundations of Risk Analysis*, Chichester: John Wiley & Sons, Ltd.*

Avizienis, A., Laprie, J-C., Randell, B. and Landwehr, C. (2004) Basic concepts and taxonomy of dependable and secure computing, *IEEE Transactions on Dependable and Secure Computing*, 1(1), 11–33.

Barberà, S., Hammond, P. J. and Seidl, S. (1998) (Eds) *Handbook of Utility Theory, Volume 1 Principles*, Dordrecht: Kluwer Academic Publisher.

Barberà, S., Hammond, P. J. and Seidl, S. (1998) (Eds) *Handbook of Utility Theory, Volume 2 Extensions*, Dordrecht: Kluwer Academic Publisher.

Bedford, T. and Cooke, R. (2001) *Probabilistic Risk Analysis – Foundations and Methods*, Cambridge: Cambridge University Press.*

Cooke, R.M. (1991) *Experts in Uncertainty: Opinion and subjective probability in science*, Oxford: Oxford University Press.

Daudin, J-J. and Tapiero, C. (1996) *Les Outils et le Contrôle de la Qualité*, Economica.

De Finetti, B. (1974) *Theory of Probability, volumes I and II*, New York: John Wiley & Sons, Ltd.

De Neufville, R. (2003) *Architecting/Designing Engineering Systems Using Real Options*, MIT report ESD-WP-2003-01.09 (Available online at: http://esd.mit.edu).

De Neufville, R. (2004) Uncertainty management for engineering systems planning and design, *MIT Engineering Systems Monograph* (Available online at: http://esd.mit.edu/symposium/pdfs/monograph/uncertainty.pdf).

Devictor, N. and Bolado-Lavín, R. (2006) *Uncertainty and Sensitivity Methods in Support of PSA level 2*, EUR21762EN.

Dixit, A.K. and Pindyck, R.S. (1994) *Investment Under Uncertainty*, Princeton University Press.

Forrest, W. B. III (2003) *Implementing Six Sigma: Smarter solutions using statistical methods* (2nd edition), Chichester: John Wiley & Sons, Ltd.

Gelder, P.H.A.J.M. van (1999) *Statistical Methods for the Risk-based Design of Civil Structures*, TU Delft report (Available online at: www.hydraulicengineering.tudelft.nl/public/gelder/homepg.htm).

Granger Morgan, M. and Henrion, M. (1990) *Uncertainty – A Guide to Dealing with Uncertainty in Quantitative Risk and Policy Analysis*, Cambridge: Cambridge University Press.*

---

*Also suitable for beginners.

Helton, J.C. (1994) Treatment of uncertainty in performance assessments for complex systems, *Risk Analysis*, **14**, 483–511.

Helton, J.C. and Burmaster, D.E. (1996) (Eds) Treatment of aleatory and epistemic uncertainty, Special Issue of *Reliability Engineering & System Safety*, **54**(2–3).

Helton, J.C. and Oberkampf, W. (2004) (Eds) Alternative representations of epistemic uncertainty, Special Issue of *Reliability Engineering & System Safety*, **85**(1–3).

Helton, J.C., Cooke, R.M., McKay, M.D. and Saltelli, A. (2006) (Eds) sensitivity analysis of model output: SAMO 2004, Special Issue of *Reliability Engineering & System Safety*, **91**(10–11).

Kane, V. E. (1986) Process capability indices, *Journal of Quality Technology*, **18**(1), 41–52.

Knight, F.H. (1921) *Risk, Uncertainty and Profit*, Hart, Schaffner & Marx.

Kumar, A., Keane, A.J., Nair, P.B. and Shaphar, S. (2006) 'Efficient robust design for manufacturing process capability', in *6th ASMO-UK/ISSMO International Conference on Engineering Design Optimization*, Oxford, 3–4 July, 2006, pp.242–250.

Kurowicka, D. and Cooke, R.M. (2006) *Uncertainty Analysis with High Dimensional Dependence Modelling*, Chichester: John Wiley & Sons, Ltd.

Law, A.M. and Kelton, W.D. (2000) *Simulation Modelling and Analysis* (3rd edition), London: McGraw Hill.

Oberkampf, W.L., De Land, S.M., Rutherford, B.M., Diegert, K.V and Alvin, K.F (2002) Error and uncertainty in modelling and simulation, Special Issue of *Reliability Engineering & System Safety*, **75**(3), 333–357.

Paté-Cornell, M.E. (1996) Uncertainties in risk analysis: six levels of treatment, *Reliability Engineering & System Safety*, **54**(2–3), 95–111.

Phadke, M. S. (1989) *Quality Engineering Using Robust Design*, New Jersey: Prentice Hall.

Quiggin, J. (1982) A theory of anticipated utility, *Journal of Economic Behavior and Organization*, **3**, 323–343.

Savage, L.H. (1954, 1972) *The Foundations of Statistics*, New York: Dover Publication Inc.

Trigeorgis, L. (1996) *Real Options – Managerial Flexibility and Strategy in Resource Allocation*, Cambridge, MA: MIT Press.

de Weck, O. and Eckert, C. (2007) *A Classification of Uncertainty for Early Product and System Design*, MIT report ESD-WP-2007-10 (Available online at: http://esd.mit.edu/WPS).

Schonberger, R.J. (2007) *Best Practices in Lean Six Sigma Process Improvement*, Chichester: John Wiley & Sons, Ltd.

Zarembka, P. (1974) *Frontiers in Econometrics, Economic Theory and Mathematical Economics*, New York: Academic Press.

15

Overall approach

15.1 Recalling the common methodological framework

Suppose now, as a starting point, that a given pre-existing model (or system model) $G(.,.)$ has been defined for a given purpose, with the corresponding (output) variables of interest $(\underline{z})$ and model inputs $(\underline{x},\underline{d})$. One or several of the four salient final goals has also been specified, thus setting the context for the quantitative uncertainty assessment (*cf.* Chapter 1).

The key requirements, as schematized in the common framework (cf. Figure 1.3 of Part I), are:

- to *specify* some measure of uncertainty, as well as corresponding quantities of interest (first step);

- to undertake *uncertainty modelling* (second step);

- to launch *uncertainty propagation* (third step) and *sensitivity analysis* (fourth step);

- to follow through with a *feedback process* after one step or another.

The Part II case studies have all illustrated variations of those requirements, and in Section 14.3, it has already been mentioned that those variations were partially due to the adoption of different settings for uncertainty treatment (for instance, more or less probabilistic, with one or two levels). Note that this Chapter will address only one- or two-level settings, which cover most industrial applications and known regulatory guidelines. Theoretically, three-level or even more complicated settings are conceivable, if one desired to separate further aleatory from epistemic uncertainty, and, within epistemic uncertainty, the estimation uncertainty (such as choice of pdf

Uncertainty in Industrial Practice   Edited by E. de Rocquigny, N. Devictor and S. Tarantola,
© 2008 John Wiley & Sons, Ltd

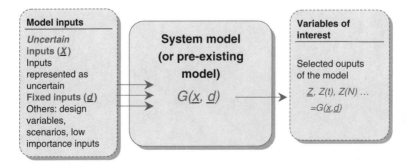

Figure 15.1   The pre-existing model and its inputs/outputs.

shapes or parameters) from the propagation uncertainty (for instance, as generated by the limitations of the CPU in Monte Carlo convergence or FORM approximations in the uncertainty propagation for a given uncertainty model). To the best knowledge of the authors, this tends rather to be limited to research options, being too complicated to be entertained in actual regulations or practice.

The following paragraphs will introduce a common positioning:

- developing a more explicit mathematical formulation of the measures and quantities of interest, according to the setting for uncertainty treatment;

- introducing the key steps for which recommendations will be made in Chapters 16–18;

- reviewing how the final goals connect with this mathematical formulation, within an applied study process.

The question of how the common methodological framework positions itself in relation to some classical settings encountered in systems analysis, identification or command and control will be discussed, as will its link to decision theory. Finally, a brief perspective will be given on model validation issues.

15.2   Introducing the mathematical formulation and key steps of a study

15.2.1   The specification step – measure of uncertainty, quantities of interest and setting

As was explained in Part I, any uncertainty setting requires a mathematical 'instrumentation' of the uncertain spaces containing the inputs/outputs, through the definition of measures of uncertainty (e.g. distributions in a probabilistic setting) as well as one or several quantities of interest (e.g. variance of the variable of interest, probability of exceeding a threshold, etc). These may form the basis for a formal decision criterion (or criteria). In fact, these concepts vary greatly according to the setting for uncertainty treatment.

To begin, suppose the uncertainty setting is one-level in one of the two simplest options: the deterministic setting (often implicit when no formal uncertainty treatment is considered) or the standard probabilistic setting (as in the majority of Part II case studies). Note that a common 'casual' interpretation of the deterministic setting is 'worst-case' analysis, taking the image of 'penalized' model inputs. But beyond monotonicity of the system model, full-scale optimization becomes necessary. Table 15.1 introduces the corresponding measures, the most frequent quantities of interest and the corresponding decision criteria.

When level 2 is considered, the measures of uncertainty or quantities of interest become formally a little more complex, as they may extend to both levels, such as the range of possible values for the coefficient of variation or the exceedance probability, or even to belief/plausibility functions over an exceedance probability. In the case of double-layer settings, a careful distinction must be made between:

- $\underline{X}$: the model inputs of the pre-existing model, which are considered uncertain. They are 'level-1 variables';

- $\underline{\theta}_X$: the parameters of the measure of uncertainty or distribution of $\underline{X}$. They are 'level-2 variables'. Also modelled as uncertain, they are given a distribution characterized by additional parameters (e.g. upper and lower bounds or parameters of a probabilistic or DST distribution).

Table 15.2 lists the main quantities of interest and potential decision criteria.

Remember, however, that level 2 can always be re-interpreted as level 1 if the model is re-written, whereby $\underline{Z} = G(\underline{X}, \underline{d})$, for instance, already incorporates the level-1 probabilistic model. This may improve the readability of the results, as in *Airframe Maintenance Contracts*.

Once the system model has been defined and the uncertainty setting, measures and quantities of interest specified (which prove to be the first step of the common framework), further elementary steps are: uncertainty modelling, i.e. the quantification of a measure of uncertainty in the model inputs; uncertainty propagation, i.e. the transformation into the measure of uncertainty in the model outputs, or at least the computation of given quantities of interest; and sensitivity analysis or importance ranking, as will be reviewed briefly in the following paragraphs.

15.2.2   The uncertainty modelling (or source quantification) step

Within the framework of this book, *uncertainty modelling* involves the estimation of a measure of uncertainty in $\underline{X}$, the uncertain model inputs (or more generally a combined $\underline{X}$ and $\underline{\theta}_X$ estimation in double-level settings), that is, *quantifying the sources of uncertainty* (Table 15.3). Chapter 16 will review the main methodological challenges and solutions around that key question: the following table simply introduces the typical features of the step, according to uncertainty setting.

Table 15.1 Measure of uncertainty, quantities of interest and decision criteria in the level-1 deterministic or standard probabilistic settings.

U. Settings	Measures* of uncertainty input/output	Quantities of interest	Potential decision criteria	Case study examples
Deterministic	Intervals (on $\underline{X}^i$, $\underline{Z}$) or more generally subsets (on $\underline{X}$, $\underline{Z}$)	Interval boundaries ($\underline{X}^i$ min, X^i max, $\underline{Z}$ max ...) or more generally limit functions ($\underline{Z}$ $max(N)$ ...)	$\underline{Z}$ $max < \underline{Z}_s$ or subset boundary constraints	None directly, but often in the historical background of some examples (e.g. electromagnetic interf. or dyke reliability)
Standard probabilistic	pdf (on $\underline{X}^i$, $\underline{Z}$) > or more simply limited characteristics of the pdf[†] > or more generally joint pdf (on $\underline{X}$, $\underline{Z}$)	Variance $var(\underline{Z})$, Expectation $E(\underline{Z})$ or other central dispersion quantities (standard deviation, coefficient of variation...) (or the whole distribution) Exceedance probability[‡] $P(\underline{Z} = G(\underline{X}, \underline{d}) > \underline{Z}_s)$ or quantile ($z^{95\%}$ ...) or probability of relative exceedance $P(G(\underline{X}, \underline{d}_1) > G(\underline{X}, \underline{d}_2))$	$CoV(\underline{Z}) < 3\%$ $P(\underline{Z} > \underline{Z}_s) < p_s$ $z^{95\%} < z_s$	Hydrocarbon exploration; nuclear waste repository; radioprotection and maintenance Electromagnetic interference in aircraft; spent nuclear fuel behaviour; dyke reliability; nuclear waste repository; radioprotection and maintenance.

*Measures should not always be understood in a mathematical sense; it may be the case in the standard probabilistic setting, but not always in other settings.

[†] Mean and variance, other moments, etc.

[‡] Recall that it is generally conditional on $\underline{d}$, so that variance or exceedance probabilities should be understood as conditional variance or probability.

Table 15.2 Measure of uncertainty, quantities of interest and decision criteria in the double-level mixed probabilistic–non-probabilistic settings.

Setting	Measure of uncertainty	Quantity of interest	(Potential) decision criterion	Case study examples
Probabilistic with level-2 deterministic	Joint pdf (on $\underline{X}$), the parameters of which, θ_X are given intervals of variation, or more generally, subsets.	Max. of probabilistic q.i. (variance $var(\underline{Z})$, standard deviation, coef. of variation, probability of exceeding a threshold $P(\underline{Z}>\underline{Z}_s)$ or quantiles $z^{a\%}$ ...) when θ_X varies within its interval or subset.	Max. $CoV(\underline{Z})<3\%$ Max. $P(\underline{Z}>z_s)<P_s$ Max. $z^{95\%}<z_s$	CO_2 emissions
Double probabilistic	Joint pdf (on $\underline{X}$), the parameters of which, θ_X, themselves follow a given joint pdf.	Level-2 quantile of a level-1 quantile, level-2 expectation of a level-1 quantile, etc.	$P[P(\underline{Z}>z_s)<p_s]>95\%$ $P[E(\underline{Z})<z_s]>95\%$	Airframe maintenance contracts
DST – probabilistic	Joint pdf (on $\underline{X}$), the parameters of which, θ_X, are given a joint plausibility and belief distribution.	Belief-Plausibility intervals for level-1 expectation, quantile or exceedance probability.	$Pl(P(\underline{Z}>z_s)<P_s)>P_{min}$ or $Be(P(\underline{Z}>z_s)<P_s)>Be_{min}$ ...	Reliability in early stages

Table 15.3   Main features for uncertainty modelling according to the uncertainty setting.

Uncertainty settings	Measure of uncertainty to be modelled	Typical estimation procedures
Deterministic	Intervals (on $\underline{X}^i$) or more generally subsets (on $\underline{X}$) or more simply point values.	Choose the boundaries according generally to physical constraints and/or expertise.
Standard probabilistic	pdf (on $\underline{X}^i$) or more generally joint pdf or more simply limited characteristics of the pdf (on $\underline{X}$)	Estimate pdf shapes and parameters, merging observations, expertise, and/or physical properties.
Probabilistic with level-2 deterministic	Joint pdf (on $\underline{X}$) the parameters of which θ_X are given intervals of variation or more generally subsets	This is generally a close extension to the estimation step of the standard probabilistic: it
"Double probabilistic	"Joint pdf (on $\underline{X}$) the parameters of which θ_X follow themselves a given joint pdf	involves describing in more detail the estimation uncertainty of the
"DST–probabilistic	"Joint pdf (on $\underline{X}$) the parameters of which θ_X are given joint plausibility and belief distributions / bpa	pdf parameters, again either through statistical estimation theory, expertise, and/or physical considerations.

15.2.3   The uncertainty propagation step

Once an uncertainty model has been developed, the computation of the quantity(ies) of interest involves the well-known *uncertainty propagation* step. This step is needed to transform the measure of uncertainty in the inputs $\underline{X}$ into a measure of uncertainty in the outputs $\underline{Z}$ of the pre-existing model. In a probabilistic setting, this means estimating the pdf of $\underline{z} = G(\underline{x}, \underline{d})$, knowing the pdf of $\underline{x}$ and given values of $\underline{d}$, $G(., .)$ being a numerical model. As will be reviewed in Chapter 17, this computational step is primarily dependent on the uncertainty setting:

- in a deterministic setting: scenario testing, design of experiments or interval computation techniques;

- in a probabilistic setting: Monte Carlo sampling, Taylor quadratic approximation, etc.

Within a given uncertainty setting, some propagation methods are theoretically virtually universal, since they generate the *complete measure* of uncertainty in the variable(s) of interest, hence enabling the computation of *any* quantity of interest: Monte Carlo sampling can, in theory, do this in a standard probabilistic setting. In practical cases, according to the main quantity(ies) of interest, and to the study characteristics, the virtually universal propagation method may not be practical (see Chapter 17); the propagation becomes a more or less difficult numerical step, involving a large variety of methods, such as Monte Carlo accelerated sampling techniques, simple quadratic sum of variances, FORM/SORM or derived reliability approximations, deterministic interval computations, etc. Prior to undertaking one of these propagation methods, it may also be desirable to develop a surrogate model (equivalently referred to as response surface or meta-model), i.e. to replace the pre-existing system model with another which produces comparable results with respect to the output variables and quantities of interest, but which is much quicker to compute.

15.2.4 The sensitivity analysis step, or importance ranking

The *sensitivity analysis* step (or *importance ranking*) refers to the computation and analysis of so-called sensitivity or importance indices of the components of the uncertain input variables $\underline{x}$ with respect to a given quantity of interest in the output $\underline{z}$. In fact, this involves a propagation step, e.g. with sampling techniques, but also a post-treatment specific to the sensitivity indices considered, typically involving some statistical treatment of the input/output relations that generate quantities of interest involving the measure of uncertainty both in the outputs and inputs. The wide variety of probabilistic sensitivity indices includes, for instance: graphical methods (scatterplots, cobwebs), screening (Morris, sequential bifurcations), regression-based techniques (Pearson, Spearman, SRC, PRCC, PCC, PRCC, etc.), non-parametric statistics (Mann-Whitney test, Smirnov test, Kruskal-Wallis test), variance-based decomposition (FAST, Sobol', Correlation ratios), and local sensitivity indices on exceedance probabilities (FORM).

Chapter 18 will discuss the practical challenges of this step, for which the choice of the most efficient methods has to be carefully made. It will be shown that it does not depend on the specificities of a physical or industrial context as such, but on the generic features identified above: the computing cost and regularity of the system model, the predominant final goal, the quantities of interest involved, the dimensions of vectors $\underline{x}$ and $\underline{z}$, etc.

15.3   Links between final goals, study steps and feedback process

Recall that uncertainty assessment in industrial practice is generally undertaken with one predominant final goal of the four identified in Part I (plus possibly a second important one). How do the mathematical formulation of the quantities of interest and the key steps of the study introduced in Section 15.2 relate to these goals? A first *formal* answer to the question is given below (see Table 15.4 for more detail):

- Understanding of influence or importance ranking (Goal U) essentially involves a sensitivity analysis step with respect to a given q.i. (at least implicitly, as variance is often the implicit q.i. for that goal).

- Model calibration, validation or simplification (Goal A) formally requires a more elaborate mix of propagation and sensitivity analysis or source modelling, with respect to a q.i. which is defined to control the validity of the model (e.g. variance or maximal value of a cost function defined on the variable of interest).

- Performance comparison or relative optimization (Goal S) involves the formal comparison, with respect to a given q.i., of two or more scenarios, designs, etc., indexed formally by corresponding values for $\underline{d}$, the fixed model inputs (i.e. two or more propagation steps for the corresponding $\underline{d}$).

- Compliance demonstration (Goal C) involves essentially the computation – through uncertainty propagation – of a quantity of interest (q.i.) and the testing of a decision criterion involving that q.i..

Table 15.4   Relating the final goals and the quantities of interest.

Goals	Handling the q.i.
U: Understand, rank importance	Which x^i contributes most to the q.i., e.g. $Var[G(X, d)]$, $P(Z < z_s\ d)$ and, more generally, how do $\underline{x}$ components interact to explain, or how do regions of $\underline{x}$ map, the q.i.
A: Accredit (calibrate/simplify/validate)	Calibrate $\underline{X}$ or θ_X with measurements; or downscale $\underline{X}$ or G, and finally validate the model, given control over a given q.i. (e.g. acceptable level of model-measurement variance).
S: Select by comparison in relative terms, optimize	Compare q.i. on selected values of d or choose d to minimize a q.i., e.g. $E(G(x, d))$, $G(x, d)^{95\%}$
C: Demonstrate compliance	Demonstrate a threshold on a q.i., e.g. $P(Z < z_s d) < 10^{-2}$, $z^{95\%} < z_s$, $CoV(Z) < 30\%$, etc.

As illustrated in Part II, a more iterative and complex process takes place in practice: none of those final goals stands alone in a one-way sequential process. Goals C or S are seldom fulfilled with just one set of modelling and propagation step(s). The real process may often involve in fact *a great deal of feedback*:

- simple uncertainty modelling at the beginning, before costly complete data collection or model computation;

- sensitivity analysis to identify which of the model inputs are the most influential, in order to steer data collection efforts, and potentially a model simplification to adapt the vector $\underline{X}$;

- return to propagation with a new uncertainty model;

- adaptation of the design or conventionally fixed variables $\underline{d}$ to lead to a final case (or portfolio of cases) that fulfil the criterion.

15.4   Comparison with applied system identification or command/control classics

Conceptual settings in systems analysis or automatic theory also consider the description of the system by a model around which variables play different roles, as illustrated in Figure 15.2. On the input side they often distinguish between:

(i) uncontrolled or environmental variables or simply inputs;

(ii) controlled or command inputs.

A rather clear distinction is made according to the intrinsic nature of the variables, and to their decision-making status. Note that a feedback loop may additionally link the output to the command variables.

The common framework described in this book focuses on a slightly different setting. It prefers to distinguish between:

(i) the variables or inputs that are deliberately modelled as uncertain $(\underline{x})$;

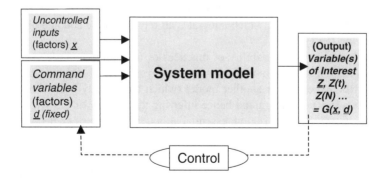

Figure 15.2 A simplified 'systems analysis' view of the system model.

(ii) other variables or inputs that are (at one point of a study) *conventionally fixed* ($\underline{d}$).

The case studies showed that during the process of a study, which involves a potential regulatory certification process, it may well happen that some naturally uncertain factors are conventionally fixed (e.g. uncertain scenarios for waste storage altering), either because the variability associated with them is too difficult to model or because it is not thought to contribute significantly to output uncertainty. Hence, in practice, fixed vector $\underline{d}$ may well combine two components: first, variables which are *intrinsically uncertain* but conventionally fixed; second, more traditional *design or 'command'* variables that are presumed to be well controlled by the industrial project. The intrinsically uncertain nature of the variables appears less important in practice than the conventional choices to fix them or not. The classic systems analysis convention may nevertheless be more convenient in many situations, e.g. when the system is well known and 'quite closed', or when a regulation of command/control loop is developed, etc.

15.5 Pre-existing or system model validation and model uncertainty

Validating the system model, in the broad sense of establishing an acceptable level of confidence in its use for a number of goals, is a very fundamental issue in industrial practice, particularly upstream of certification steps. In a sense, the whole issue of uncertainty treatment is linked to model validation: however, model validation may also be considered in a more restricted sense, as one of the four salient goals (Goal A-Accredit) of uncertainty. Certain other principles hold in general. First, a model is never completely validated, but rather only *corroborated* by *available* information until a better model is available or affordable. Model validation covers a wide range of practices, ranging from purely numerical benchmarks to rigorous experimental designs. In nuclear code development, model *qualification* may refer to a comparison to experimental results, step-by-step for each physical phenomenon involved in the system, whereas validation would require a comparison to an integrated experiment, which integrates all representative coupled phenomena characterizing the system.

Notwithstanding the variability of practices or of definitions involved in more or less rigorous processes, model validation always implies comparing the system model to data, expertise, or another model (which has already been validated) on a certain experimental design and hence inferring the validity of the system model *in a given context*. This restriction, essential although sometimes neglected in practice, should specify the input space domain ($\underline{X}$, $\underline{d}$), the type of application, variables of interest and the q.i. In fact, a given system model may be acceptably validated for predicting the central dispersion of a given output variable of interest while being unacceptably inaccurate for other outputs, or even for the tail distribution of the original variable.

Note also that model validation often involves a prior *calibration* (or *parameter identification*) phase of some of the system model parameters, to fit better to the data (or reference model results). A great number of inverse identification or data assimilation techniques (Beck & Arnold, 1977; Tarantola, 1987; Walter and Pronzato, 1994) may be mobilized for that purpose. This calibration phase may even be seen as a way of quantifying the uncertainty characteristics of certain model inputs which cannot be observed directly (see Chapter 16).

Therefore, besides supporting a model validation statement, the calibration phase may result in some quantification of the residual error remaining after the best fit to the observations (or reference model), exactly in the same way a statistical regression model comes with the estimate of its residuals. Variance in those residuals could sometimes be taken as a component of *model uncertainty* to be included in later uncertainty propagation: however, residuals offer a necessarily limited representation of model uncertainty, and it is sometimes hard to determine whether they are generated by observational measurement or intrinsic model uncertainty and how they behave outside the experimental design used in validation. Chapter 14 has already referred to the controversies affecting the concept of model uncertainty.

15.6   Links between decision theory and the criteria of the overall framework

A final remark must be made on the positioning of the overall framework of the present book with respect to decision theory. Choosing the appropriate decision criteria for decision-making under uncertainty is a large and specific domain of research: it involves in-depth understanding of the larger decision processes, and potentially a conceptual decision-theory approach to the decision-maker's preferences, attitudes towards risk and uncertainty (see, for instance, Barbera *et al.*, 1998), etc. As mentioned in Chapter 1 and Chapter 14, a long-standing literature has discussed the acceptability and rationale of probabilistic settings, the distinctions between risk and uncertainty, risk aversion and related decision theory issues (see Knight, 1921; Von Neumann & Morgenstern, 1944; Savage, 1972; Quiggin, 1982; or, nearer to industrial risk and uncertainty practice, Bedford and Cooke, 2001; Aven, 2003).

Notwithstanding the theoretical debates around choosing the right decision theory paradigms, the present book is focussed on quantitative modelling of industrial systems with a *given decision setting*. Its overarching approach is to treat the resulting uncertainty in the manner of supposedly known mathematical quantities, namely the quantities of interest and 'decision criteria' that were introduced above. Indeed, industrial regulations and practice show that there is always a need for a preliminary choice of a decision model, which may take various settings according to the paradigm (see Chapter 14 on deterministic, probabilistic Bayesian, nonprobabilistic settings, etc.). However, quantities to be computed later mostly accord with the list mentioned above whatever the decision model chosen. For example, risk-aversion aspects could imply the choice of different quantities of interest or thresholds for Goal C, while remaining within the framework. A utility function, for instance, could be involved to post-process the intermediate variable of interest and generate the final variable of interest, for which the q.i. would be the expected utility (Von Neumann & Morgenstern, 1944). Or alternatively, a non-linear transformation could post-process the distribution of utility in order to generate the new measure of incertainty and hence the rank-dependent expected utility (Quiggin, 1982), which may be viewed as another q.i. according to the book framework.

References

Aven, T. (2003) *Foundations of Risk Analysis*, Chichester: John Wiley & Sons, Ltd.*

Beck, J.V. and Arnold, K.J., (1977) *Parameter Estimation in Engineering and Science*, New York: Wiley.

Bedford, T. and Cooke, R. (2001) *Probabilistic Risk Analysis – Foundations and Methods*, Cambridge: Cambridge University Press.*

Barberà, S., Hammond, P.J. and Seidl, S. (1998) (Eds) *Handbook of Utility Theory, Volume 1 Principles*, Dordrecht: Kluwer Academic Publisher.

Barberà, S., Hammond, P.J. and Seidl, S. (1998) (Eds) *Handbook of Utility Theory, Volume 2 Extensions*, Dordrecht: Kluwer Academic Publisher.

Knight, F.H. (1921) *Risk, Uncertainty and Profit*, Hart, Schaffner & Marx.

Quiggin, J. (1982) A theory of anticipated utility, *Journal of Economic Behavior and Organization*, **3**, 323–343.

Savage, L.H. (1954, 1972) *The Foundations of Statistics*, New York: Dover Publication Inc.

Tarantola, A. (1987) *Inverse Problem Theory and Methods for Data Fitting and Model Parameter Estimation*, London: Elsevier.

Von Neumann J. and Morgenstern O. (1944) *Theory of games and economic behaviour*, Princeton, New Jersey: Princeton University Press.

Walter, E. and Pronzato, L. (1994) *Identification de Modèles Paramétriques à Partir de Données Expérimentales*, Collection MASC, Masson.

---

*Also suitable for beginners.

16

Uncertainty modelling methods

16.1 Objectives of uncertainty modelling and important issues

The overall methodological approach described in the previous section shows that the first step of an uncertainty study consists in defining the key goals of the study, the quantities of interest (minimum-maximum, central dispersion, exceedance probabilities, etc.), and the choice of the uncertainty settings (deterministic, probabilistic, DST, etc.). The outputs of this first step make it possible to identify the elements of the following equation:

$$\underline{Z} = G(\underline{X}, \underline{d})$$

where $\underline{Z}$ denotes the vector of variables of interest, G the pre-existing model, $\underline{X}$ the sources of uncertainty, and $\underline{d}$ the fixed (or control) variables. The current section addresses the next step of the uncertainty study, which is the building of the uncertainty model on the sources $\underline{X}$.

As presented below, the choice of a relevant uncertainty model depends on the available information on the uncertain model inputs (database, engineering or expert judgement). But there also exists an important interaction between this step of the uncertainty study and the other steps in the overall approach. The relevant method depends on the uncertainty setting and the quantity of interest considered; moreover, the uncertainty propagation method that will be used later in the study may require different levels of complexity in the uncertainty model.

- **Deterministic setting**

In a deterministic setting, a set of values must be chosen to define the range of variation for each of the uncertainty sources represented by uncertain model

Uncertainty in Industrial Practice   Edited by E. de Rocquigny, N. Devictor and S. Tarantola,
© 2008 John Wiley & Sons, Ltd

inputs X^i. In this chapter the emphasis will be placed rather on the probabilistic settings, which require more detailed uncertainty models.

- **Standard probabilistic setting**

$\underline{X}$ is modelled here as a random vector. It is then necessary to define its *probability distribution*, which is the mathematical tool allowing for the calculation of any interesting feature (mean value or expectation, median, standard deviation, probability of exceeding a threshold, etc.). One characterization of a probability distribution is the *cumulative distribution function* (cdf) F_X:

$$F_X(x^1, \ldots, x^n) = P[X^1 \leq x^1, \ldots, X^n \leq x^n]$$

In fact, defining the complete cdf is not always obligatory, depending on the couple (quantity of interest, propagation method) considered. Indeed, in particular cases, such as Taylor approximation for the computation of mean value and variance of the variables of interest (see Chapter 17), only partial characteristics are necessary, such as average values, variances and correlation coefficients between uncertain model inputs. This makes the problem simpler.

In the general case in which the entire cdf is required, the quantity of interest considered also influences the tools which should be used to build the uncertainty model. More precisely, the accuracy of the model is particularly crucial in some parts of the probability distribution, which may have a dramatic impact on the quantity of interest. For instance, assessing a very low probability of exceedance $P[Z^k > z_s]$ or a rare quantile z^α (with α close to 0 or 1) often requires a good level of confidence in the *distribution tails* of the uncertain model inputs' probability distribution. This issue may turn out to be less critical for the central dispersion of Z^k (mean and standard deviation). Some existing statistical tools place the focus either on distribution tails or on central dispersion: the final goal of the study will have to be remembered when choosing the most relevant tool.

Section 16.2 takes these considerations into account in order to adapt some recommendations to the standard probabilistic setting.

- **Level-2 probabilistic setting**

Apart from the *non-parametric methods* which will be described in a few paragraphs, building the cdf of $\underline{X}$ is often carried out with a *parametric approach*. The shape of the cdf F_X is then chosen from among a list of existing probability distributions (e.g. a Gaussian, Weibull or extreme value distribution, etc.), in which a few parameters θ_X – in other words, degrees of freedom – can be tuned to achieve a satisfactory description of the uncertainty sources. These parameters are chosen via statistical analysis or engineering judgement, and thus remain uncertain themselves to a certain extent. In a level-2 setting, an uncertainty model on θ_X also has to be determined, be it deterministic, probabilistic or non-probabilistic. Section 16.3 briefly presents some starting ideas on this issue.

16.2   Recommendations in a standard probabilistic setting

Some diagrams are presented below to clarify the major questions to be addressed when building the uncertainty model. Note that the following deals with certain statistical problems that may seem quite tedious to non-specialists. This is why it is necessary to emphasize beforehand that the complexity of the uncertainty model may be increased progressively: a first 'rough' (but sufficiently relevant) uncertainty model may be used to carry out the next steps of the overall methodology (uncertainty propagation and sensitivity analysis), which will reveal the model inputs on which the uncertainty model should be refined in a feedback step.

As indicated in Figure 16.1, the first crucial question to be addressed concerns possible dependencies among uncertain model inputs. If no physical or modelling reason linking the uncertain variables can be thought of, or if a fast and simple study is envisaged, then the situation becomes simpler, since each uncertainty source X^i can be studied separately. The sequel of the diagram can then be found in Figure 16.2.

Remember that even if such a simplification is an attractive option, *neglecting dependencies among uncertain model inputs may dramatically distort the study's results*. This is often particularly critical if the quantity of interest concerns a rare event. Thus, if the assumption of independence is judged tenuous, the reader is referred to Figure 16.3, which addresses the questions of detection, modelling and quantification of dependencies. Independence can clearly be questioned when dealing with physical parameters affected by spatial or time variability. In such cases, several model inputs, say $x^{i1}, \ldots, x^{in}$, may correspond to the same physical parameter, but at different instants or at different locations. A dependency could often exist between inputs in such cases.

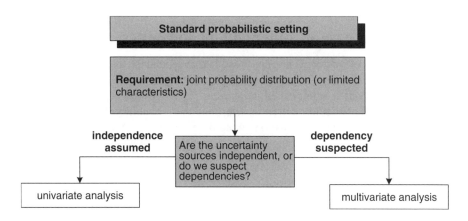

Figure 16.1 The crucial question of dependencies.

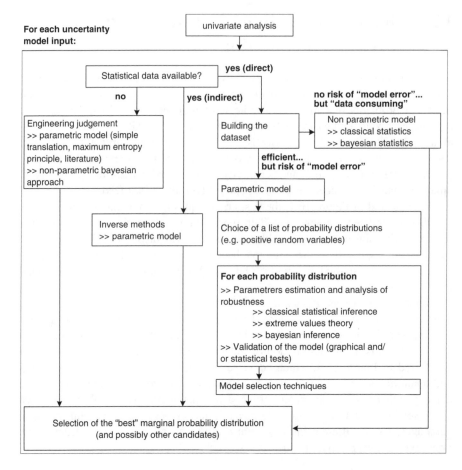

Figure 16.2 Diagram for independent uncertain model inputs.

16.2.1   The case of independent variables

The components of vector $\underline{X}$ are here assumed independent. It is then necessary and sufficient to choose a separate probability distribution for each X^i: this one-dimensional analysis is often called *univariate*.

The method that should be used to determine the probability distribution of X^i depends on the available information. According to the type and richness of this information, one may choose between two categories of approaches:

- **Non-parametric models**

In such models no prior hypothesis is made on the shape of the probability distribution. The relevant distribution is built on the basis of data alone (Scott, 1992). If this makes it possible to avoid 'model errors', note that the amount of data necessary to reach the desired accuracy may be important, particularly when interested in extreme values of X^i. Extrapolating beyond the fluctuation observed statistically is indeed somewhat limited in such an approach. For example, a non-parametric approach cannot be used to assess the 99.9% quantile of X^i with acceptable confidence if only one hundred data are available. On the other hand, the robustness of the non-parametric approach is quite attractive in similar situations when dealing with central dispersion or lower quantiles.

• **Parametric models**

The idea here is to make some hypothesis on the probability distribution of X^i: the cdf is chosen from among a list of pre-defined models that can be adjusted through a limited set of parameters denoted by $\underline{\theta}$, playing the role of degrees of freedom (Johnson, Kotz and Balakrishnan, 1994). Such an approach is far less costly than a non-parametric, in terms of data required – tuning a small number of parameters is easier than generating ad-hoc models. Moreover, choosing a distribution shape facilitates extrapolation beyond the statistically observed range of variation. To summarize, a parametric approach is more efficient, but only if its shape can be justified. There is a risk of model error:

• The low number of degrees of freedom does not always provide enough flexibility to describe satisfactorily the data on all the range variation.

• The extrapolation to extreme values may be quite misleading if the model shape is not relevant to the real unknown input distribution (Castillo, 1988).

Therefore, the two approaches should not be systematically opposed, but may be seen, on the contrary, as complementary: for instance, a non-parametric treatment of data can be helpful in identifying potentially judicious parametric models, or in rejecting obviously inappropriate ones.

More generally, let us now look in detail at the type of information which may lead to a parametric or a non-parametric approach.

16.2.2   Building an univariate probability distribution via expert/engineering judgement

Expert judgement (EJ) has been used since the late 1940s to estimate probability density functions and event probabilities, among other things. The lack of probabilistic background and the threat of bias in the estimates provided by experts triggered the design of protocols and techniques to help them to provide unbiased estimates. A typical set of steps in a generic EJ protocol is (see, for example, NUREG-1150): (i) selection of team project; (ii) definition of the questions to be studied; (iii) selection of experts; (iv) training; (v) definition of tasks; (vi) individual experts' work; (vii) elicitation of experts' opinions; (viii) analysis and aggregation of results; and (ix) documentation. Comprehensive information may be obtained in Mengolini *et al.* (2005).

Expert judgement is mostly used when information about uncertain model inputs is scarce. Nevertheless, simpler approaches are also commonly used in practice. Suppose, for instance, that the only available information on the uncertain model input X^i comes from experts or engineers, such as absolute bounds of X^i, estimates of the average value, etc. Parametric models can then be used to translate these judgements as accurately as possible into a probability distribution.

'Accurately', in this case, means 'matching the characteristics provided by the experts, without adding too much extra information in the choice of the distribution type'. For instance, if X^i is known *always* to be bounded by a and b, without any prior knowledge of the likelihood of values between a and b, then a uniform distribution seems a reasonable choice – all values between a and b are equally likely, and values outside $[a,b]$ are impossible. But if experts think that a value c belonging to $[a,b]$ is the best estimate possible, a triangular distribution could be used: the difference from the uniform distribution is that the likelihood of a value x belonging to $[a,b]$ increases linearly as x approaches c. Other 'translation rules' exist, such as those deriving from the *maximum entropy principle* (Cover and Thomas, 1991). However, they may be less intuitive: this principle tells us, for example, that a variable X^i, known always to be non-negative and with mean value m, should be modelled by an exponential distribution.

Note that the maximum of entropy is not the only solution to translate (or 'encode') engineering judgement into a model. An analysis of the literature dedicated to the physical field of interest may provide some ideas. For instance, in the field of material mechanics, the variable 'fracture toughness of a steel' is often considered random because of material heterogeneity. Physical considerations (such as the weakest link theory) then point to the Weibull distribution as one of the relevant choices (Wallin, 1984).

Finally, note that the principle of *Bayesian statistics* is precisely to take expert judgement into account (Robert, 2001). The possibilities are quite interesting, since the modelling of expert judgement is less 'absolute': for steel toughness, for instance, expert judgement may be translated as a prior 'set of possible Weibull distributions', instead of a given Weibull distribution with completely determined parameters. It also allows for an updating of the uncertainty model if data are collected afterwards. Other frameworks exploit this idea of 'imprecise probability distributions', such as the Dempster-Shafer theory, which will be discussed later. All these approaches are well suited to the description of *epistemic uncertainties* (lack of knowledge), but they remain quite sophisticated and generally more advanced than present practice in industrial uncertainty studies, as evidenced by the choices made in the case studies of Part II. In practice, their use depends on the stakes involved and the maturity of the uncertainty analysis.

- ● **Building an univariate probability distribution via 'direct' data**

Suppose now that a dataset $x^i_1, \ldots, x^i_n$, called 'realizations' of X^i, is available. In this situation numerous statistical tools can be used to build the most relevant probability distribution (Saporta, 1990; Dixon and Massey, 1983). However, remember the obvious danger of using these tools as a 'black-box' – every statistical method is based on hypotheses and can produce misleading results if used outside of its domain of validity.

To illustrate, note that the most common statistical methods presume that data are *independent and identically distributed (i.i.d.)*. This means that the experiments that have generated the dataset are independent from one another, and that the variability of the uncertainty source X^i has not evolved during these experiments. In the example of steel toughness, these assumptions could be invalid if data have been collected via mechanical tests carried out on very different types of materials. More generally, a tool adapted for i.i.d. data should be used cautiously if the dataset was built by aggregating different sources: data may be *non-identically distributed*. Another example is the measurement of a physical parameter (e.g. a temperature) at regular time intervals. If the length of the interval is too small, a dependency may appear among data: temperatures at time t and $t + 1$ cannot be assumed completely *independent*.

Although the i.i.d. hypothesis may not be valid, this does not mean that a rigorous statistical analysis is impossible. However, more advanced statistical tools have to be used, such as time series analysis (Box and Jenkins, 1970; Gouriéroux and Monfort, 1995). For the sake of simplicity, the framework of an i.i.d. dataset will be retained in the following.

Once the database has been built (i.e. definition of the i.i.d. sample $x_1^i, \ldots, x_n^i$), both parametric and non-parametric approaches are possible. As mentioned above, none of these approaches is consistently the best: a balance has to be found between a robust but 'data-consuming' non-parametric approach, and an efficient but potentially risky parametric approach. Choosing the parametric solution will be natural if physics or the literature point to a particular parametric model (e.g. the Weibull distribution for steel toughness). In general, it is not advisable to make an arbitrary choice of the parametric model; several models should be tried in order to sort the relevant from the non-relevant and to identify the 'best candidate'. More precisely, the following steps are to be carried out.

For each type of parametric model considered (e.g. a lognormal, normal and a Weibull distribution), characterized by the parameters $\underline{\theta}$, the first step is to find the values of $\underline{\theta}$ that make it possible to fit the dataset as well as possible. The most common methods for this are the maximum likelihood and the method of moments. This is an instance of the so-called *classical frequential inference*, opposed generally to the *Bayesian inference* already described in this section as an advanced tool. Another possible statistical framework is the *extreme values theory*, which can be referred to if the final quantity of interest implies extremely rare values (Coles, 2001).

In the framework of classical statistical inference, one generally denotes by $\hat{\underline{\theta}}$ the estimated parameters. These estimates inevitably undergo statistical fluctuation (i.e. estimation uncertainty), the amplitude of which depends on the sample size n, on the number of parameters composing $\underline{\theta}$ and on the estimation method used. Generally, this statistical uncertainty decreases proportionally to $n^{1/2}$. Apart from simple cases (e.g. normal or exponential distribution), asymptotic theoretical results have to be used to control this uncertainty via confidence intervals. Such results are only valid if n is large enough: no absolute threshold exists for the minimum size of the sample, but asymptotic results are commonly considered acceptable when n reaches a few dozens. For smaller samples, numerical *bootstrap* or re-sampling methods can be used (Efron and Tibshirani, 1993).[1]

Explicitly taking into account the statistical uncertainty in parametric models in the sequel of the uncertainty study is the object of level-2 probabilistic settings. But even if it is decided to remain in the standard probabilistic setting, assessing confidence intervals is always a prudent move to determine how robust (or not) the final uncertainty model is.

At this stage, *several* parametric models may have been adjusted. The question is then to select the best. Many graphical analyses (QQ-plots, histograms, empirical vs. theoretical cdf, etc.) may provide a qualitative answer. Classical statistical *goodness-of-fit tests* (chi-square, Kolmogorov, Cramer-Von Mises, Anderson-Darling, etc.) provide a less subjective and more quantitative answer (Shorack and Wellner, 1986).

- Each of these tests is implicitly based on the definition of a distance between the data and the candidate probability distribution; the type of distance is a first criterion for the choice of relevant test according to the goal of the uncertainty study: some distances are in fact averaged on the overall domain of variation of data (generally appropriate to the study of central dispersion), while others place the focus on distribution tails (generally more relevant for rare events).

- This distance can be used to reject the parametric models that are clearly not suited to describe the dataset, but also to rank the set of non-rejected probability distributions.

Nonetheless, note that goodness-of-fit tests are not absolutely fair. More precisely, on one hand, a model with many parameters will be more likely to offer the best fit (by minimizing the distance with data), and will therefore appear among the 'top parametric models'. But on the other hand, statistical uncertainty increases with the number of parameters. It may look more balanced to choose a simpler parametric model (with fewer parameters) which is less subject to statistical uncertainty, even if the goodness-of fit is not as good. *Model selection* techniques (Burnham and Anderson, 1989), such as the Akaike Information Criterion (AIC) and Bayesian Information Criterion (BIC), propose an answer: the parametric models are ranked via a criterion which depends on the goodness-of-fit (through the notion

---

[1]The term 'bootstrap' finds its origin in the famous 'Adventures of the Baron von Munchausen': the Baron, facing a critical danger, manages to reach safe heights by pulling his own 'bootstraps'. This is of course a magical trick and not a physically possible phenomenon. The principle of bootstrap is also a magical trick: fake new datasets are generated in order to evaluate the sensitivity of the estimates of θ. Executed correctly (i.e. abiding by the rigorous theoretical framework), this is completely admissible. Nevertheless, as for any statistical tool, a 'black-box' attitude could dramatically distort the results.

of likelihood) and which includes a penalization term depending on the number of parameters. The idea is thus to find the best compromise between goodness-of-fit and statistical robustness.

When such a sequence of statistical tools (estimation, confidence intervals, goodness-of-fit, model selection) has been applied, useful elements are in place to help choose the most relevant probability distribution for the uncertainty source X^i, and to assess the confidence that can be given to this distribution.

● **Building a univariate probability distribution via 'indirect' data**

Suppose now that no direct data are available for the uncertainty source X^i, but that there are indirect data $y_{m,1}, \ldots, y_{m,n}$, measurements of a vector $\underline{y}$ of variables linked to $\underline{x}$ by a pre-existing model:

$$\underline{y} = H(\underline{x}, \underline{d})$$

Measurements $\underline{y}_m$ differ from y because of the uncertainty in the model $H(.)$ and/or measurement uncertainty. Note:

$$\underline{y}_m = H(\underline{x}, \underline{d}) + \underline{\varepsilon}_m$$

An *inverse problem* has then to be solved: values of the outputs of a physical model are to be used to build an uncertainty model on some of the model inputs $\underline{X}$. A whole domain of techniques still undergoing development is associated with that purpose (Tarantola, 2005) and is only briefly introduced below, since it was described in *Radioprotection and Maintenance*. A famous simple algorithm is the Best Linear Unbiased Estimate (BLUE) method originating in data assimilation, which relies on a linear approximation of H but also on the hypothesis that $\underline{x}$ has a unique, albeit unknown, value such that the purely *epistemic* uncertainty in $\underline{X}$ would disappear if the number of data n tended to infinity. Linear assumptions can be relaxed when applying more refined data assimilation techniques (Talagrand and Courtier, 1987). The second assumption may not always be acceptable, especially if the uncertainty in $\underline{X}$ is due to natural variability (e.g. the annual maximum flow of a river in the future remains uncertain even if an exceedingly large historical dataset of maximum water level is available). In that latter situation, there are more relevant methods, such as the CIRCE algorithm (linked to the classical Expectation-Maximization (EM) statistical algorithm), if a linearization of H is acceptable (De Crecy, 1997). Methods adapted for strongly non-linear models are currently being developed in the literature, but they constitute a real numerical challenge when the physical model H is CPU-consuming. Note, finally, that methods have been developed in more specific cases in which, instead of an indirect dataset (y_{mj}), indirect expert judgment provides the complete pdf of $\underline{Y}$ (Kurowicka and Cooke, 2006).

16.2.3   The case of dependent uncertain model inputs

From this point on, it is supposed that the analysts who carry out the uncertainty study cannot ensure that the components of vector $\underline{X}$ are independent. A multidimensional or *multivariate* analysis is then necessary. As in the univariate case, the method to be used depends on the type of information available, with the same distinction between parametric and non-parametric approaches.

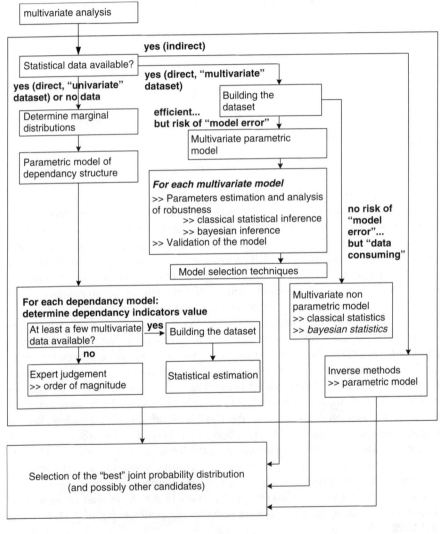

Figure 16.3 Diagram with possible dependencies taken into account.

• Building a multivariate probability distribution via 'direct' univariate data or with engineering/expert judgement

Suppose first that multivariate data (e.g. realizations $\underline{x}_1, \ldots, \underline{x}_n$ of $\underline{X}$, i.e. n *simultaneous* measurements of the uncertain model inputs in which dependencies are suspected) are *not* available – for instance, data have been collected separately for all the sources, or expert judgement is the only source of information. In such a case, clearly no multivariate statistical tool can be applied because of the absence of the required inputs. Yet, several solutions can be conceived for a multi-dimensional analysis.

A first possibility is to make assumptions on the *dependency structure* among uncertain model inputs. The approach is then divided into three steps:

- First, each source X^i is analysed separately (via expert judgement or statistical analysis, *cf.* previous paragraph); this makes it possible to determine the so-called *marginal probability distributions* of $\underline{X}$;

- Second, simple dependency indicators such as correlation coefficients (Pearson or Spearman) are determined;

- Finally, an arbitrary parametric model of the structure of dependencies (such as copulas (Nelsen, 2006), kriging models for spatial variability (Cressie, 1993), or autoregressive models for time variability (Box and Jenkins, 1970) is used to re-build the probability distribution of $\underline{X}$.

An approach which involves only correlation coefficients is attractive, since the complex problem of dependency is addressed through simple indicators; pragmatically, correlation coefficients are moreover the only indicators that can be correctly assessed in the absence of a large multivariate dataset. But such an approach cannot address *all types of dependencies*. Alternative methods (e.g. as previously mentioned, or a Bayesian approach) are thus of interest, even if they remain more complex.

At any rate, in a 'marginal plus dependency structure' approach, one has to determine correlation coefficients. This can be done statistically if some multivariate data are available; statistical tests (chi-square, Pearson, Spearman) can even be used to decide whether the observed correlation truly exists or is only an appearance caused by a natural statistical fluctuation. The other solution lies in expert judgement: experts may be able to provide correlation coefficients, whereas they have serious difficulties in estimating more complex dependency structures. In practice they are also able to provide the sign of the correlation coefficient (positive if the two uncertain model inputs tend to increase or decrease simultaneously, negative if one increases as the other decreases), and an *order of magnitude* of the relationship intensity (e.g. 0.2 if the relation exists but is not very strong, and 0.7 if the relation is strong, even though no convention exists on these 'typical' values).

- **Building a multivariate probability distribution via 'direct' multivariate data**

If a sample $\underline{x}_1, \ldots, \underline{x}_n$ is available, the problem is rather similar to the univariate case: the question 'parametric vs. non-parametric' and the sequence of statistical tools in the parametric case (estimation, confidence intervals, goodness-of-fit, model selection) remain the same. Nonetheless, the complexity of the statistical problem increases with the dimension – non-parametric models become even more data-consuming, and testing the goodness-of-fit of a multivariate parametric model is an issue on which a great deal of research is currently being carried out. Note that this situation is rare in an industrial study; yet it can be encountered when $\underline{X}$ is already the output of a physical model on which uncertainties have previously been propagated by Monte Carlo.

- **Building a multivariate probability distribution via 'indirect' multivariate data**

In the case of indirect data, the framework and the methods defined in the univariate case still apply. But an additional important issue arises: if the size of the dependent uncertain variables $\underline{X}$ is too important regarding the size of the observations $\underline{Y}_m$ and the size of the dataset, the information is simply not rich enough and the statistical problem may admit an infinite number of solutions: this problem is referred to as *non-identifiability*. Expert judgement has then to be solicited to choose arbitrarily the probability distribution of some components of $\underline{X}$, or even to turn them into deterministic variables.

16.3   Comments on level-2 probabilistic settings

When choosing level-2 settings, it is decided to model not only the uncertainty directly affecting $\underline{X}$, but also the uncertainty affecting the parameters $\underline{\theta}$ of an uncertainty model which is developed on $\underline{X}$. This is obviously a rather advanced option, for which the methods become more complex and diverse: as introduced in Chapter 15, varying level-2 settings may be built upon a level-1 probabilistic setting, either deterministic, probabilistic or even extra-probabilistic.

A first method, rather easy to grasp in theory, may be to involve expert judgment to assess plausible ranges, for instance, in the standard deviations of some Gaussian model inputs, and hence to build a deterministic level-2 setting with intervals for the corresponding $\underline{\theta}$ components (as exemplified in *CO$_2$ emissions*). In practice, eliciting this kind of information may already be quite a challenge for experts.

Beyond this, some of the statistical techniques already mentioned in relation to building the level-1 model may naturally generate a probabilistic uncertainty model for level 2. This is the case for the parametric maximal likelihood or method of moment, which both generate asymptotic approximations of the pdf characterizing the fluctuation of $\underline{\hat{\theta}}$: it can be shown to be Gaussian, with a covariance matrix closely linked to the Fisher information matrix which may be estimated on the sample (Saporta, 1990; Dixon and Massey, 1983). It is also the case for Bayesian statistics, whereby the posterior pdf of $\hat{\theta}$ represents the residual estimation uncertainty including both data and expertise: it may also be computed on the basis of the sample.

References

Beven, K. and Binley, A. (1992) The future of distributed models: model calibration and predictive uncertainty, *Hydrological Processes*, **6**(3), 279–298.

Box, G.E.P. and Jenkins, G. (1970) *Time Series Analysis: Forecasting and control*, San Francisco: Holden-Day.

Buishand, T.A. (1989) Statistics of extremes in climatology, *Statistica Neerlandica*, **43**, 1–30.

Burnham, K.P. and Anderson, D. R. (2004) *Model Selection and Multimodel Inference – A Practical Information-Theoretic Approach*, New York: Springer.

Castillo, E. (1988) *Extreme Value Theory in Engineering*, San Diego: Academic Press.

Coles, S.G. and Powell, E.A. (1996) Bayesian methods in extreme value modelling: a review and new developments, *International Statistical Review*, **64**(1), 119–136.

Coles, S.G. (2001) *An Introduction to Statistical Modeling of Extreme Values*, London: Springer.

Cover, T. M. and Thomas, J. A. (1991) *Elements of Information Theory*, New York: John Wiley & Sons Inc.

Cressie, N. (1993) *Statistics for Spatial Data*, Chichester: John Wiley & Sons, Ltd.

De Crécy, A. (1997) 'CIRCE: a tool for calculating the uncertainties of the constitutive relationships of Cathare2', in *Proceedings of the 8th International Topical Meeting on Nuclear reactor Thermo-Hydraulics (NURETH8)*, Kyoto.

Diebolt, J. and Girard, S. (2003) A note on the asymptotic normality of the ET method for extreme quantile estimation, *Statistics and Probability Letters*, **62**(4), 297–406.

Dixon, W.J. and Massey, F.J. (1983) *Introduction to Statistical Analysis*, London: McGraw Hill Book Company.

Efron, B. and Tibshirani, R. (1993) *An Introduction to Bootstrap*, London: Chapman & Hall.

Galambos, J. (1987) *The Asymptotic Theory of Extreme Order Statistics*, Malabar (Florida): R.E. Krieger Publishing Company.

Girard, S. (2004) A Hill type estimate of the Weibull tail-coefficient, *Communication in Statistics – Theory and Methods*, **33**(2), 205–234.

Gumbel, E.J. (1958) *Statistics of Extremes*, Colombia: Columbia University Press.

Kennedy, M. and O'Hagan, A. (2001) Bayesian calibration of computer models, *Journal of the Royal Statistical Society*, Series B **63**, 425–464.

Mengolini, A., Simola, K. and Bolado-Lavin, R. (2005) *Formal Expert Judgement: An overview*, European Commission – Joint Research Centre Report EUR 21772 EN.

Nelsen, R.B. (2006) *An Introduction to Copulas* (2nd edition), New York: Springer.

Nilsen, T. and Aven, T. (2003) Models and model uncertainty in the context of risk analysis, *Reliability Engineering and System Safety*, **79**(3), 309–317.

Olea, R.A. (1999) *Geostatistics for Engineers and Earth Scientists*, Dordrecht: Kluwer Academic Publisher.

Pickands, J. (1975) Statistical inference using extreme order statistics, *The Annals of Statistics*, **3**, 119–131.

Robert, C. P. (2001) *The Bayesian Choice*, New York: Springer.

Saporta, G. (2006) *Probabilités, Analyse de Données et Statistique* (2nd edition), Paris: Editions Technip.

Scott, D. W. (1992) *Multivariate Density Estimation Theory, Practice and Visualization*, Chichester: John Wiley & Sons, Ltd.

Shorack, G. R. and Wellner, J. A. (1986) *Empirical Processes with Applications to Statistics*, Chichester: John Wiley & Sons, Ltd.

Silvey, S. D. (1975) *Statistical Inference*, London: Chapman & Hall.

Tarantola, A. (1987) *Inverse Problem Theory and Methods for Data Fitting and Model Parameter Estimation*, London: Elsevier.

Talagrand, O. and Courtier, P. (1987) Variational assimilation of meteorological observations with the adjoint vorticity equation, *Q. J. R. Meteorological Society*, **113**, 1311–1328.

Tarantola, A. (2005) *Inverse Problem Theory and Methods for Model Parameter Estimation*, Philadelphia: Society for Industrial and Applied Mathematics.

Vanmarcke, E. (1983) *Random Fields: analysis and Synthesis*, Cambridge, MA: MIT Press.

Wallin, K. (1984) The scatter in K_{IC} results, *Engineering Fracture Mechanics*, **19**, 1085–1093.

Walter, E. and Pronzato, L. (1994) *Identification de Modèles Paramétriques à Partir de Données Expérimentales*, Paris: Collection MASC, Masson.

17

Uncertainty propagation methods

Uncertainty propagation in a pre-existing model consists in estimating the variability of the model outputs due to the uncertain model inputs. This variability could be characterized by various quantities of interest: statistics such as prediction intervals or moments, exceedance probability of a threshold, etc.[1] In some cases it may even be the full *measure of uncertainty*, represented by the probability distribution function. As explained in Section 13.2, the industrial uncertainty studies can be broadly categorized according to four major goals. The uncertainty propagation phase is at the core of studies concerned with Goal C(Comply), but this phase is also necessary in support of studies aiming for Goals U(Understanding), A(Accredit) and S(Select). Table 17.1 summarizes the characteristics of the main uncertainty propagation methods. Section 17.1 reviews the appropriate methods for each kind of quantity of interest and makes recommendations on their suitability based on the following 'key features':

- formulation of the decision criterion, if any – for example if a criterion requires a certain accuracy in the quantity of interest estimate, such as '$P(P(Z > z_s) < P_o) > \alpha$' or '$P(CoV(Z) < 3\%) > \alpha$' – this will have an influence on the choice of method;

- characteristics of the uncertainty model: discrete or continuous variables, dependencies, spatial field, time-dependent output, multi-dimensional output;

- characteristics of the pre-existing model: the CPU time constraint (or other constraint, such as no automatic meshing or no input/output file for linkage with an external code, which could influence the maximum number

---

[1] In this definition, the term 'uncertainty propagation' includes the field of the structural reliability.

Uncertainty in Industrial Practice Edited by E. de Rocquigny, N. Devictor and S. Tarantola,
© 2008 John Wiley & Sons, Ltd

of runs affordable), the degree of regularity (C^2, C^1, C^0, discontinuity), the availability of a gradient procedure, etc.

As the application of uncertainty and sensitivity methods often requires a large number of system model runs, this becomes a constraint if the system model is computationally intensive. An alternative is to search for *meta-models*, i.e. good predictors that are computationally cheaper, and hence can be run instead of the pre-existing model, with, for instance, Monte Carlo under large sample size. Section 17.2 describes the methodology and its use in support of uncertainty propagation; its use in support of sensitivity analysis is explained in Chapter 18.

Note: It is assumed that the reader is familiar with classical Monte Carlo methods and variance reduction techniques, but less so with FORM and SORM methods. For this reason more information on these latter methods will be given in Section 17.1.4.

17.1   Recommendations per quantity of interest

17.1.1   Variance, moments

This section is concerned with statistics based on moments. It also includes the expectation and the coefficient of variation defined by the relative standard deviation compared to the expectation. The methods suited to such quantities of interest are among families B (deterministic methods for numerical integration, Cacuci, 2003); C (Taylor approximation, ISO GUM, 1995); D (Monte Carlo simulation and some variants – see Robert *et al.*, 1999; Law *et al.*, 1991; Hamersley, 1979; Rubinstein, 1981); and F (methods based on a stochastic development – see Ghanem1. and Spanos, 1991; Sudret *et al.*, 2002).

When the decision criterion stipulates a certain level of accuracy in the estimate of the variance or another moment (for example, by a confidence interval), deterministic methods, such as those based on numerical integration or a Taylor development, are not appropriate. Methods based on a stochastic development could provide a variance matrix of the results, according to some recent research results. If the criterion is based on the expectation or the variance, then such a method may be adequate. Simulation methods, in particular SRS (Simple Random Sampling) and LHS (Latin Hypercube Sampling), are generally the most appropriate, even if the computed confidence intervals are only asymptotically valid and thus approximated.

Certain characteristics of the pre-existing model can restrict the choice of the propagation method. The first constraint is generally the computational time for one run. Depending on the power capacity of the available computers, the funds or the time allotted to the study, the affordable number of runs might be limited. Methods based on a numerical integration sometimes cannot be used, due to the large number of runs required when the number of uncertain model inputs is large ($p > 6$ *is often already unaffordable*). For the same reason, Monte Carlo methods like SRS or LHS might be unfeasible, depending on the accuracy desired in the

Table 17.1 Summary of the characteristics of the main propagation methods.

Family of methods	Relevant criterion (or quantity of interest)	Short description of method	Control of uncertainty in the results
A Deterministic method (or minimum/maximum)	Range	Determination of all extrema by an optimization method, using gradient if possible.	No
B Numerical integration	Mainly moments, possibly pdf and exceedance probability	Integration by quadrature to estimate moments, then fitting of an analytical pdf.	Error of the numerical integration (seldom convertible into a confidence interval).
C Taylor approximation (called 'quadratic approximation' for the variance)	Moments, but in practice mainly the variance	Development of Var[Z] or another moments using a low-degree differential approximation.	No
D Monte Carlo simulation and variance-reduction techniques (V.R.T.)	All probabilistic criteria	Simple Random Sampling, stratified sampling, LHS, control variates, antithetic variates.	Yes

(continued overleaf)

Table 17.1 (*continued*)

	Family of methods	Relevant criterion (or quantity of interest)	Short description of method	Control of uncertainty in the results
E	Variance-reduction Techniques (V.R.T) dedicated to exceedance probability	Exceedance probability	Conditional MC, importance sampling around limit state surface, Directional simulation...	Yes, to a certain extent
F	Methods based on a stochastic development (includes Stochastic Finite Elements, intrusive or not)	Mainly moments, possibly pdf and exceedance probability	Development of the v.i. Z onto a functional basis	Limited results, but research underway
G	FORM/SORM	Exceedance probability	Approximation by an optimization problem in a transformed space + approximated integration around the design point	No
H	Hybrid methods (FORM/SORM + V.R.T.)	Exceedance probability	Mainly importance sampling with a distribution centred on the design point	Yes, to a certain extent

project – it may alternatively be decided to reduce the required level of accuracy. The Taylor approximation is one of the least demanding methods in terms of the number of runs involved.

In the case of a practical difficulty due to the maximum possible number of runs, an alternative is to use a meta-model (see Section 17.2). Methods belonging to families B, C and F assume a certain *degree of regularity* according to the moment to estimate. Monte Carlo methods are not sensitive to discontinuities. If a procedure for evaluating the *gradient or the Hessian* is available (from an automatic differentiation or an adjoint method), the efficiency of methods from families C and F could be significantly increased, while Monte Carlo methods cannot profit from such a procedure. If some variables of interest are *multi-dimensional* or *time dependent*, only Monte Carlo methods are feasible.

Certain characteristics of the uncertainty model can also restrict the choice of the propagation method. For example, if at least one uncertain model input is a discrete random variable, methods from families B, C and F are not appropriate. An alternative is to re-apply the selected method for each event of the discrete variable, and then to aggregate the results.

Figure 17.1 shows the simplified decision tree for the quantity of interest 'variance/moment'. Three case studies of Part I exemplify such recommendations when the variance or the coefficient of variation is one of the quantities of interest:

- in *CO_2 emissions*, a Taylor approximation is used and its results are validated by a comparison with a Monte Carlo method with 20000 runs;

- in *Nuclear waste repository*, samples from Simple Random Sampling are used;

- in *Airframe maintenance contracts*, the first method used is based on Simple Random Sampling combined with a meta-model.

17.1.2   Probability density function

The methods suitable for building a probability density function, or the associated cumulative density function, belong to family B (deterministic methods for numerical integration, combined with Pearson or Johnson methods or another moment method for fitting a parametric or a non-parametric probability density function – see Pearson *et al.*, 1965); to family D (Monte Carlo simulation such as Simple Random Sampling (SRS) or Latin Hypercube Sampling (LHS, McKay, 1979), combined with a statistical method for fitting a parametric or a non-parametric probability density function); and to family F (methods based on a stochastic development combined with a fitting method for a parametric or a non-parametric probability density function). The quality of the fitting is assessed through statistical tests like Anderson-Darling, Kolmogorov-Smirnov, etc. (Gentle *et al.*, 2004).

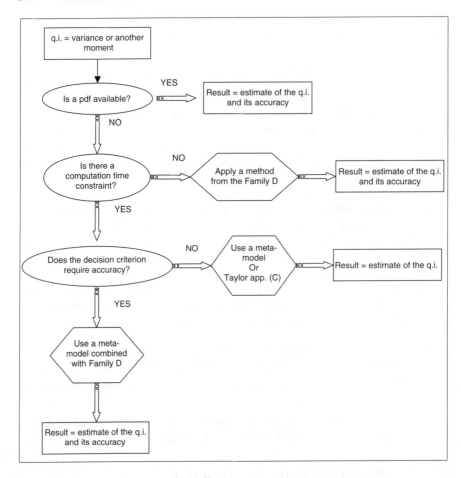

Figure 17.1 Decision tree for q.i. 'variance / moment' – time invariant case.

Note: FORM/SORM methods could be used to assess a set of probabilities associated with some percentiles. It involves repeating a full FORM/SORM calculation for each point. Such an approach may not be efficient in practice.

The first feature, namely the accuracy of the fitted probability density function in a decision criterion, is not relevant.

The use of the families B and F combined with a fitting method based on the moments is seldom relevant in practice when the number of uncertain model inputs is larger than 10, or if at least one of the uncertain model inputs is a discrete random variable. These methods are sensitive to the degree of regularity as with any method of integration. Even if the Pearson or Johnson method may be more efficient in small dimensions, a Monte Carlo-based strategy is used in the majority of applications, because this method is sensitive only to a CPU-time constraint.

Spent nuclear fuel behaviour exemplifies the use of Simple Random Sampling with a statistical fitting to compute the pdf of the minimal margin, one of the quantities of interest considered.

17.1.3   Quantiles

A quantile is easily derived from knowledge of a probability density function. The recommendations from Section 17.1.2 are relevant if this approach is chosen. Another option is to generate a random sample from any method of family D, and then to apply a post-treatment to estimate the quantile, typically an order statistic (David, 1981). Being based on a post-treatment of a random sample, this option is not sensitive to the presence of discrete variables, neither to multi-dimensional uncertain model inputs nor to the degree of regularity of the pre-existing model, etc.

The only constraint may be the number of affordable runs. The Wilks formula, a specific order statistic useful to determine the minimal number of runs necessary, could then be appropriate in some applications (Wilks, 1941; Nutt and Wallis, 2004). Wilks's formula refers to the computation of robust estimates at a given confidence level β of either:

- a quantile (i.e. one-sided prediction interval), the estimator $\hat{Z}^\alpha$ being defined so that $P(F_Z(\hat{Z}^\alpha) > \alpha) = \beta$ where F_Z denotes the cdf of Z; or
- a prediction interval (i.e. two-sided), the estimators $[\hat{Z}_d^\alpha, \hat{Z}_u^\alpha]$ being defined so that $P(F_Z(\hat{Z}_u^\alpha) - F_Z(\hat{Z}_d^\alpha) > \alpha) = \beta$.

In the case of a quantile i.e. one-sided prediction interval, such a relation means that it can be claimed, with at most $(1 - \beta)\%$ chance of error, that at least α percent of the variable of interest Z lies below $z_s = \hat{Z}^\alpha$. The first-order Wilks's formula determines the minimal size n of the sample $\{z_j, j = 1 \ldots n\}$ to be randomly generated according to the values of α and β; for example, the formula is $1 - \alpha^n > \beta$ for a one-sided statistical tolerance interval. Bound z_s of the prediction interval of Z is obtained by retaining the maximal values of the sample $\{z_j, j = 1 \ldots n\}$. Table 17.2 gives examples of sizes. One advantage of using Wilks's formula is that the number of runs needed is independent of the number of uncertain model inputs, but no information is provided on the difference between the quantile estimate and the true quantile. Higher order Wilks's formulae lead to more precise estimators (de Rocquigny, 2005).

Another alternative may be to use a meta-model (see Section 17.2), but its validity should be checked, particularly around the computed quantile.

Note: Indirectly, a multiple use of FORM/SORM or numerical integration, or perhaps even stochastic developments, could be used to approximate quantiles, as both can generate exceedance probabilities. Thus, through repeated use of those methods with a change of the thresholds, an approximate bounding may be produced. Such an approach may not be efficient in practice.

Table 17.2 Sample sizes for the 1st order Wilks formula.

	One-sided statistical tolerance limit			Two-sided statistical tolerance limit		
$\beta \neg \alpha \rightarrow$	0.90	0.95	0.99	0.90	0.95	0.99
0.90	22	45	230	38	77	388
0.95	29	59	299	46	93	473
0.99	44	90	459	64	130	662

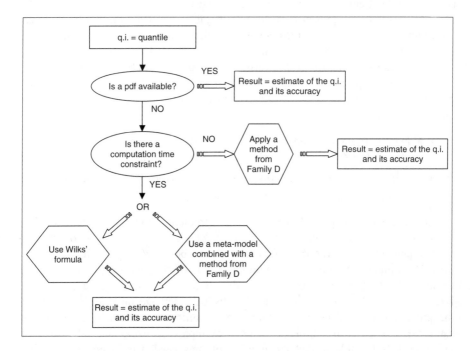

Figure 17.2 Decision tree for the q.i. 'quantile'.

Figure 17.2 shows the decision tree for the quantity of interest 'quantile'. The following case studies illustrate this situation:

- in *Nuclear waste repository*, the quantile and its prediction interval are obtained from samples obtained through Simple Random Sampling;

- in *Aircraft fatigue modelling*, the quantile is obtained from samples obtained through Simple Random Sampling on a meta-model, due to strong constraints on the affordable runs.

17.1.4   Exceedance probability

Apart from the Taylor approximation (C) and the deterministic min-max approach (A), all other methods mentioned above could in principle be used to compute an exceedance probability.

Numerical integration methods or stochastic developments can be used to assess the first four moments, from which a probability density function is fitted, for example, by using a Pearson or Johnson method. From this probability density function it is easy to estimate the probability that a variable of interest exceeds a fixed value. This method is seldom used when the pdf of the variable of interest is not presumed to be Gaussian, because of two main drawbacks:

- it then becomes very sensitive to the accuracy of the estimation of the moment and is not accurate for low exceedance probabilities;

- it can only with difficulty provide the accuracy of the estimated probability.

The most popular methods are based on Monte Carlo simulation, and mainly on Simple Random Sampling. This method consists in sampling the uncertain model inputs according to their pdf and then running the pre-existing model. Following this, the values of the samples of the variable of interest are compared to the threshold, and an estimate of the exceedance probability is provided by $\hat{P}_f = N_f/N$, where N_f is the number of runs for which $z^j \geq z_s$, and N is the size of the sample. The accuracy of this estimation can be evaluated in terms of its relative standard deviation (or coefficient of variation), using the Central Limit Theorem to express a Gaussian confidence interval:

$$CV(\hat{P}_f) \approx \sqrt{(1 - \hat{P}_f)/\hat{P}_f N} \qquad (17.1)$$

The smaller the coefficient of variation, the better the accuracy of the estimated probability of failure. Convergence of $\hat{P}_f$ to the true exceedance probability when N tends towards the infinite is both intuitive and guaranteed by the Law of Large Numbers.

Consequently, the number of runs needed increases inversely with the magnitude of the probability. This characteristic implies, in some applications, a strong constraint on the affordable number of runs. Variance reduction techniques have consequently been developed; for the same size of the random sample, such methods give a more precise estimate, i.e. a lower coefficient of variation when some conditions are fulfilled. The best known techniques of variance reduction are: stratified sampling, LHS, importance sampling, conditional Monte Carlo sampling, control variates, antithetic variates, directional simulation or subset simulation. (It is not in the scope of this book to describe these methods – see Hamersley, 1979; Rubinstein, 1981; Robert *et al.*, 1999; Law *et al.*, 1991; Rackwitz, 2001; Au, 2007). These methods from families D and E are often used because they are suitable if the decision criteria specify a level of 'accuracy' in the probability estimation, and

they may be applied whatever the characteristics of the pre-existing models and the uncertainty models.
Experience shows that:

- for low probabilities (lower than 10^{-3}), stratified sampling and LHS are seldom efficient, but they are easy to implement.

- importance sampling is very efficient if a suitable importance distribution has been found, but this search could be difficult.

- conditional Monte Carlo or the dimension reduction method is the most efficient if the selected model input is the most influential input on the probability.

Due to the intrinsic constraint on the affordable number of runs for the simulation methods in industrial applications, an alternative type of method was proposed a few decades ago, based on a deterministic algorithm and named First- and Second-Order Reliability Methods (FORM/SORM). These methods consist of three steps (see Ang *et al.*, 1990; Augusti *et al.*, 1984; Madsen *et al.*, 1986; Ditlevsen *et al.*, 1996; Melchers, 1999; Rackwitz, 2001; Lemaire, 2005):

1. Transform the uncertain model input space $X^1, X^2, \ldots, X^p$ into a space of standardized Gaussian variables $U^1, U^2, \ldots, U^p$. Rosenblatt and Nataf's transformations are generally used, and they assume that all the random inputs are distributed according to a continuous probability density function.

2. Search, in this transformed space, for the point on the transformed pre-existing model lying closest to the space's origin: this point is called the design point or the most probable failure point in the standard space, and is marked $\underline{u}^*$ in Figure 17.3. The Euclidian distance between the design point and the origin is called the Hasofer-Lind index and denoted β_{HL}. The calculation of β_{HL} consists in solving the following problem of optimization under the constraint $\min_{G(\underline{u})=0} \sqrt{\underline{u}^t \underline{u}}$, where $\underline{u}$ is the random vector in the

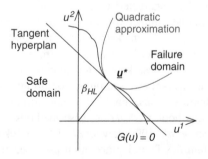

Figure 17.3 Principles of FORM-SORM methods.

standard space and $G(\underline{u})$ the transformed pre-existing model in the standard space. The characteristics of the pre-existing model (regularity degree, smoothness, etc.) and the size of the vector of uncertain model inputs both influence the choice of the optimization algorithm. Practitioners generally use the gradient-based algorithm as they need fewer runs of the pre-existing model to reach the design point than stochastic algorithms. This latter type of algorithm has been applied, but seldom in industrial applications because it generally requires thousands of runs. However, FORM/SORM are mainly used to circumvent the intrinsic constraint on the affordable number of runs for simulation methods. Note that the availability of a gradient procedure can help in the search for the design point.

3. Compute an exceedance probability for the approximated pre-existing model: in the case of an approximation by a tangent hyperplane, the FORM approximation of the exceedance probability is calculated by: $P_f = \Phi(-\beta_{HL})$, where Φ is the cumulative distribution function of a standardized Gaussian distribution. In the case of an approximation by a quadratic surface, several formulas exist for the second-order approximation SORM.

If there are several minima for the optimization problem, it is necessary to find all these minima in order to obtain a good approximation of the probability. This is done through extensions of FORM/SORM called multi-FORM and multi-SORM. In practice, for a low probability, FORM/SORM are extremely efficient compared to simulation methods, because the number of runs for FORM/SORM depends essentially on the size of the uncertain model input vector and characteristics of the pre-existing model (degree of regularity, monotonicity, convexity, number of local extrema, etc.), but not on the magnitude of the probability.

Since FORM and SORM approximations are the results of a deterministic method, it is not possible to obtain an estimate of their accuracy. This is the main drawback of FORM/SORM, meaning that it is often not sufficient to justify a decision based only on a FORM/SORM result, other than in certain cases enjoying quite strong assumptions of regularity (such as, for example, a demonstration of convexity in the transformed space). A validation of FORM/SORM could be obtained by a comparison with a confidence interval of the exceedance probability provided by a Monte Carlo method or a method of variance reduction. As this type of validation may not always be possible, it is thus advisable to use a hybrid method combining FORM/SORM with a simulation method (family H). The method most often used is Importance Sampling around the design point (DP), whereby the importance distribution is a multi-normal distribution centred on the design point (see Figure 17.4, Melchers, 1999). Such methods provide an estimate of the exceedance probability along with a measure of its accuracy. Alternative methods have been proposed in the literature, such as Axis-Orthogonal Simulation (Rackwitz, 2001; Tvedt, 2006).

When the variable of interest is time variant, families D and E are always relevant. But FORM/SORM present limitations, except when the evolution of the value of the pre-existing model is constantly decreasing, i.e. when it is enough to calculate the probability at the last moment to the transient. Research is being

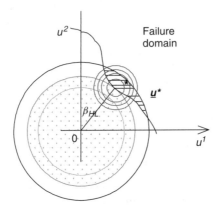

Figure 17.4 Importance sampling.

carried out to adapt FORM/SORM to this situation (Rackwitz, 2001; Andrieu, *et al.*, 2002).

Figure 17.5 shows the simplified decision tree for the quantity of interest *'exceedance probability'* for the time invariant case. The following case studies exemplify the situation in which an exceedance probability is one of the quantities of interest:

- in *Electromagnetic Interferences in Aircraft* and in *Dyke Reliability*, the FORM result was validated by a Monte Carlo result, and also by using a curvature argument in *Electromagnetic Interferences in Aircraft*;

- in *Radioprotection and Maintenance*, a Simple Random Sampling method was used;

- in *Spent Nuclear Fuel Behaviour*, a hybrid method was used with an importance distribution centred on the design point obtained through FORM. The comparison of the probabilities obtained by FORM/SORM and Importance Sampling shows that the curvatures have a non-negligible influence, and SORM gives a good approximation of the probability.

17.2   Meta-models

The meta-model methodology consists in building a mathematical function, which is cheaper from the point of view of computation time, and which approximates the behaviour of the pre-existing model over the domain of variation of its inputs, starting from a set of selected simulations of the pre-existing model in the uncertain space (Sacks, 1989; Fang *et al.*, 2006; Kleijnen, 2007). This set is called the *experimental design*.

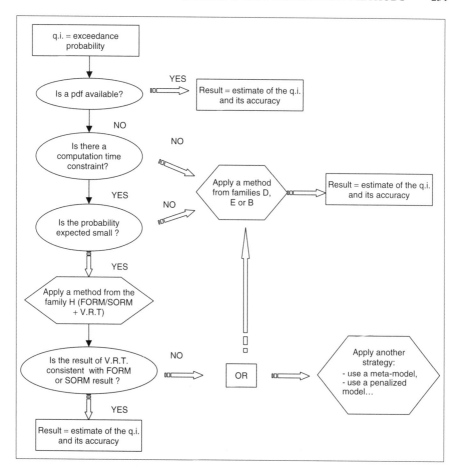

Figure 17.5 Decision tree for the q.i. 'exceedance probability' – time invariant case.

Note: Common synonyms of 'meta-model' are 'surrogate models' or 'response surface'. This last term is often linked to polynomial models for historical reasons (Box *et al.*, 1987); hence the terms 'meta-model' or 'surrogate models' are preferred.

17.2.1   Building a meta-model

In general, the following inputs are necessary to build a meta-model:

1. An experimental design D of points $(\underline{x}_j, z_j)_{j=1..n}$, where x is the vector $(x^1, \ldots, x^p)$, and $z_j = g(\underline{x}_j)$ denotes the result of the pre-existing model for each of those points $\underline{x}_j$;

2. A family F of functions $f(\underline{x}, \underline{c})$, where $\underline{c}$ is either a parameter vector or an index vector that identifies the different elements of F.

Many meta-model families F may be considered, such as: polynomials, generalized linear models (GLM), splines, interpolating radial functions, kriging, local polynomial kernel estimation, support vector machines, stochastic response surface methods using polynomial chaos expansions, partial least squares, neural networks, regression trees, etc. (McCullagh *et al.*, 1989; Wahba, 1990; Ghanem and Spanos, 1991; Antoniadis *et al.*, 1992; Fan *et al.*, 1996; Vapnik, 1998; Chilès *et al.*, 1999; Breiman, 2001; Hastie *et al.*, 2001; Santner *et al.*, 2003; Smola and Schölkopf, 2004; Fang *et al.*, 2006).

The process for building a meta-model consists in determining the function f_0 of F which best supports the following objectives:

- The meta-model should approximate as well as possible the training set (points of the experimental design D used to build the surface).

- The meta-model should predict as well as possible the value of the pre-existing model on a point not belonging to the training set.

It is then the function f_0 that minimizes a risk function $R(f) = \int L(z, f(\underline{x}, \underline{c}))dP(\underline{x}, z)$, where $P(\underline{X}, Z)$ is the probability distribution of the random vector $(\underline{X}, Z)$, which is unknown in practice. For example, the empirical risk function $R_E(f) = \frac{1}{n} \sum_{j=1}^{n} [z_j - f(\underline{x}_j, \underline{c})]^2$ is often used under certain assumptions, such as a constant standard deviation throughout D. Many alternatives are available to estimate the parameters of the meta-models; most techniques are based on a parametric or a non-parametric regression of a set of univariate (or multivariate) functions such as $E(Y|X^i)$(or$E(Y|X^i, \ldots, X^j)$ (see the references above).

In order to build the set of points D, results of the classical experimental design (Box *et al.*, 1987) could be considered, but bearing in mind their underlying assumptions. For example, many classical experimental designs (composite designs, etc.) assume a polynomial behaviour for the pre-existing model. Without good knowledge or intuition of the monotonicity or the smoothness of the model, it is often efficient to use a random experimental design based on Simple Random Sampling or LHS, because this choice does not make any assumptions about regularity or monotonicity of the pre-existing model. The construction of an efficient experimental design for computer experiments is a very active research field. Recent directions include the use of quasi-Monte Carlo samples or of iterative build-up of an experimental design optimized to limit the number of runs of the pre-existing model and to improve the quality of approximation and prediction of the meta-model (Fang *et al.*, 2006).

17.2.2   Validation of a meta-model

The most important issue is the validation of the meta-model for the purpose of the application. It should not be forgotten that the use of a meta-model instead of the pre-existing model means that an estimate of the quantity of interest is obtained for the meta-model rather than for the pre-existing model. Thus, even if the meta-model permits a very accurate estimate of the quantity of interest (variance, exceedance probability, etc.), through, for example, the use of highly robust propagation methods such as Monte Carlo with a high number of runs, it may still be a bad estimate of the true quantity, i.e. that of the pre-existing model. This fact could be important for a quantile or an exceedance probability, because the value of such quantities is strongly influenced by restricted areas of random space. If the residuals in that area are high, it is then extremely probable that the estimate of the meta-model will be poor.

This validation phase retains an element of subjectivity. A 'good approximation' is a subjective feature that depends on the use of the response surface, for example:

- the case considered could include additional constraints, such as a maximal acceptable residual in a given area, or, in the case of a safety application, assurance that the meta-model output overestimates (or underestimates) the pre-existing model output in the area of interest;

- in an exceedance probability study, a good quality of prediction in the domain where this exceedance is most probable is sufficient, and it is not necessary to insist on a good quality of prediction in the entire field of variation of the uncertain input;

- if the uncertain model inputs have been poorly established, it is not necessary to identify a meta-model explaining 99.9% of the variability.

In order to secure explicit confidence in the meta-model, many recent works have proposed either statistical criteria (based on cross-validation or bootstrap on the residue analysis, etc.), or graphical tools (such as a linear regression for comparing the outputs of the meta-model and the pre-existing model – see, for example, Figure 17.6) (Fang et al., 2006; Iooss et al., 2006; Kleijnen, 2007). Confidence could be increased by a comparison of statistical quantities (average, standard deviation, minimum and maximum, etc.) obtained from the meta-model and from the pre-existing model on an identical sample. For discontinuous pre-existing models, no usual response surface family is suitable. In practice, discontinuous behaviour generally means that more than one physical phenomenon has been implemented in the pre-existing model G. In this case, to avoid a misleading interpretation of results of uncertainty and sensitivity analysis, discriminant analysis could be employed to define subsets where the function is continuous, before performing statistical analyses on each continuous subset.

17.3   Summary

The following table summarizes the characteristics and recommendations for the methods to be used in uncertainty propagation, except for the meta-modelling method described above. This latter is an alternative to the direct use of the system model rather than a dedicated propagation method.

Table 17.3 Summary of the characteristics of the methods.

Family of methods	Relevant q.i.	Number of runs	Accuracy control of the q.i.*	Number of unc. model inputs	Discrete unc. model inputs	Dependent unc. model inputs	Maximum number of v.i.	Functional v.i.*	Degree of regularity	Gradient procedure
B Numerical integration	Mainly moments, possibly pdf and exceedance probability	Exponential with the number of model inputs.	Dependent of the numerical scheme	Low (<10)	No	Yes	1	No	Requirement increasing with the order of the moment	No influence
C Taylor approximation	Variance, moments	Linear with number of model inputs	No	No constraint	No	Yes (but restricted to the use of the moments)	>1	No	Increase with the order of the moment	Benefit
D Monte Carlo simulation and variance-reduction techniques (V.R.T.)	All probabilistic q.i.	Independent of the number of unc. model inputs. Depends on the desired accuracy.	Yes	No constraint	Yes	Yes	>1	Yes	No constraint	No influence
E Variance-reduction Techniques (V.R.T) dedicated to exceedance probability	Exceedance probability	Depends on the exceedance probability and accuracy expected.	Yes	No constraint	Yes	Yes	1		No constraint	No influence

Table 17.3 (*continued*)

Family of methods	Relevant q.i.	Number of runs	Accuracy control of the q.i.*	Number of unc. model inputs	Discrete unc. model inputs	Dependent unc. model inputs	Maximum number of v.i.	Functional v.i.*	Degree of regularity	Gradient procedure
F Methods based on a stochastic development	Mainly moments, possibly pdf and exceedance probability	Efficient in many cases but increases quickly with number of model inputs.	No	Low (<10)	No	Yes	> 1*		Regularity needed to converge quickly	Benefit*
G FORM/SORM	Exceedance probability	Depends on the number of model inputs.	No	Depends on the optimization algorithm	No	Yes	> 1*		Requirement depends on the optimization algorithm	Benefit
H Hybrid methods (FORM/SORM + V.R.T.)	Exceedance probability	Depends on the number of model inputs.	No	Depends on the optimization algorithm	No	Yes	> 1*		Requirement depends on the optimization algorithm	Benefit

*except for the probability density function.

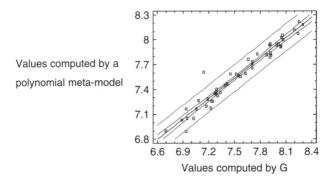

Values computed by a polynomial meta-model

Values computed by G

Figure 17.6 Example of a regression between the outputs of the meta-model and the pre-existing model (G).

References

Andrieu, C., Lemaire, M. and Sudret, B. (2002) 'The PHI2 method: a way to assess time-variant reliability using time-invariant reliability tools', in *Proceedings of European Safety and Reliability Conference ESREL'02*, 472–479.

Ang, A.H.S. and Tang, W.H. (1984) *Probability Concepts in Engineering, Planning and Design* (2 volumes), Chichester: John Wiley & Sons.

Antoniadis, A., Berruyer, J. and Carmona, R. (1992) *Régression Non Linéaire et Applications*, Collection Economie et Statistiques Avancées, Paris: Economica.

Au, S.K., Ching, J. and Beck, J.L. (2007) Application of subset simulation methods to reliability benchmark problems, *Structural Safety*, **29**, 183–93.

Augusti, G., Baratta, A. and Casciati, F. (1984) *Probabilistic Methods in Structural Engineering*, London: Chapman & Hall.

Bjerager, P. (1990) On computation methods for structural reliability analysis, *Structural Safety*, **9** (2), 79–96.

Box, G.E.P and Draper, N.R. (1987) *Empirical Model Building and Response Surface*, Chichester: John Wiley & Sons, Ltd.

Breiman, L. (2001) Random Forest, *Machine Learning*, **45** (1), 5–32.

Breiman, L., Friedman, J., Olshen, R. and Stone, C. (1984) *Classification and Regression Trees*, Wadsworth and Brooks/Cole.

Chilès, J.-P. and Delfiner, P. (1999) *Geostatistics: Modeling Spatial Uncertainty*, Chichester: John Wiley & Sons, Ltd.

David, H.A. and Nagaraja, H.N. (2003) *Order Statistic* (3rd edition), Chichester: John Wiley & Sons, Ltd.

Deming, W. E. (1966) *Some Theory of Sampling*, New York: Dover Publications.

Ditlevsen, O. and Madsen, H.O. (1996) *Structural Reliability Methods*, Chichester: John Wiley & Sons, Ltd.

Fan, J. and Gijbels, I. (1996) *Local Polynomial Modelling and Its Applications*, London: Chapman & Hall.

Fang, K.-T., Li, R. and Sudjianto, A. (2006) *Design and Modeling for Computer Experiments*, London: Chapman & Hall/CRC.*

---

*Also suitable for beginners.

Frangopol, D.M. (1985) Structural optimization using reliability concepts, *Journal of Structural Engineering*, **11**, 2288–2301.

Freeman, P.R. and Smith, A.F.M. (1994) *Aspects in Uncertainty: A Tribute to D.V. Lindley*, Chichester: John Wiley & Sons.

Gentle, J.E., Härdle, W. and Mori, Y. (2004) (Eds) *Handbook of Computational Statistics: Concepts and Methods*, New York: Springer-Verlag.

Gentle, J.E. (2003) *Random Number, Generation and Monte Carlo Methods*, New York: Springer.

Ghanem, R.G. and Spanos, P.D. (1991) *Stochastic Finite Elements: A Spectral Approach*, New York: Springer-Verlag.

Goupy, J. (2006) *Introduction aux Plans d'Expériences* (3rd edition), Paris: Dunod.

Hammersley, J. and Handscomb, D. (1979) *Monte-Carlo Methods*, London: Chapman & Hall.

Hastie, T., Tibshirani, R. and Friedman, J. (2001) *The Elements of Statistical Learning*. New York: Springer.

Helton, J.C. and Davis F.J. (2003) Latin hypercube sampling and the propagation of uncertainty in analyses of complex systems, *Reliability Engineering & System Safety*, **81**, 23–69.

Iooss, B., Van Dorpe, F. and Devictor, N. (2006) Response surfaces and sensitivity analyses for an environmental model of dose calculations, *Reliability Engineering & System Safety*, **91** (10–11), 1241–1251.

Kleijnen, J.P.C. and Sargent, R.G. (2000) A methodology for fitting and validating metamodels in simulation, *European Journal of Operational Research*, **120**, 14–29.

Kleijnen, J.P.C. (2007) *Design and Analysis of Simulation Experiments*, New York: Springer.

Kotz, S., Read, C.B., Balakrishnan, N. and Vidakovic, B. (2006) (Eds) *Encyclopedia of Statistical Sciences* (2nd edition), Chichester: John Wiley & Sons, Ltd.

Kurowicka, D. and Cooke, R. (2006) *Uncertainty Analysis with High Dimensional Dependence Modelling*, Unpublished lecture notes, Delft University of Technology.

Lemaire, M. (2005) *Fiabilité des Structures – Couplage Mécano-fiabiliste Statique*, Paris: Lavoisier Hermès.

McCullagh, P. and J.A. Nelder J.A. (1989) *Generalized Linear Models* (2nd edition), London: Chapman & Hall.

Madsen, H.O., Krenk, S. and Lind, N.C. (1986) *Methods of Structural Safety*, New Jersey: Prentice Hall.

McKay, M.D., Conover, W.J. and Beckman, R.J. (1979) A comparison of three methods for selecting values of input variables in the analysis of output from a computer code, *Technometrics*, **21**, 239–245.

Melchers, R.E. (1999) *Structural Reliability Analysis and Prediction* (2nd edition), Chichester: John Wiley & Sons, Ltd.*

Montgomery, D.C. (2004) *Design and Analysis of Experiments* (6th edition), Chichester: John Wiley & Sons, Ltd.

Moses, F. (1998) Probabilistic-based structural specifications, *Risk Analysis*, **18** (4), 445–454.

Myers, R.H. (1995) *Response Surface Methodology*, Chichester: John Wiley & Sons, Ltd.

Nutt, W.T. and Wallis, G.B. (2004) Evaluation of nuclear safety from the outputs of computer codes in the presence of uncertainties, *Reliability Engineering & System Safety*, **83** (1), 57–77.

---

*Also suitable for beginners.

Pearson, E.S. and Tukey, M. (1965) Distributions whose fourth moments are known, ne*Biometrika*, **6**, 133–137.

Robert, C. and Casella, G. (1999) *Monte-Carlo Statistical Methods*, New York: Springer-Verlag.

De Rocquigny, E. (2005) 'A statistical approach to control conservatism of robust uncertainty propagation methods; application to accidental thermal hydraulics calculations', in *Proceedings of ESREL-05*, Tri City.

Rackwitz, R. (2001) Reliability analysis – a review and some perspectives, *Structural Safety*, **23**, 365–395.

Rubinstein, R.Y. (1981) *Simulation and the Monte-Carlo Method*, Chichester: John Wiley & Sons, Ltd.*

Sacks, J., Welch, W.J., Mitchell, T.J. and Wynn, H.P. (1989) Design and analysis of computer experiments, *Statistical Science*, **4**, 409–435.

Santner, T., Williams, B. and Notz, W. (2003) *The Design and Analysis of Computer Experiments*, New York: Springer.

Smola, A.J. and Schölkopf, B. (2004) A tutorial on support vector regression, *Statistics and Computing*, **14**, 199–222.

Sudret, B. and Der Kiureghian, A. (2002) *Stochastic Finite Elements and Reliability – A state-of-the-art report*, Report no. UCB/SEMM-2000/08 (Available online at: http://bruno.sudret.free.fr/).

Sudret, B. and Der Kiureghian, A. (2002) Comparison of finite element reliability methods, *Probabilistic Engineering Mechanics*, **17**, 337–48.

Tvedt, L. (2006) PROBAN probabilistic analysis, *Structural Safety*, **28**, 150–163.

Vapnik, V. (1998) *Statistical Learning Theory*, Chicester: John Wiley & Sons, Ltd.

Wahba, G. (1990) *Spline Models for Observational Data*, Philadelphia: Society for Industrial and Applied Mathematics.

Wilks, S. (1941) Determination of sample sizes for setting tolerance limits, *The Annals of Mathematical Statistics*, **12**(1), 91–96.

Woodbury, A.D. and Ulrych, T.J. (1993) Minimum relative entropy: forward probabilistic modeling, *Water Resources Research*, **29**(8), 2847–2860.

Zellner, A. (1986) Bayesian estimation and prediction using asymmetric loss functions, *Journal of the American Statistical Association*, **81**(394), 446–451.

---

*Also suitable for beginners.

18

Sensitivity analysis methods

The previous chapters discussed the set-up of the uncertainty model to describe current knowledge of the model inputs and the propagation of this uncertainty through the model representing the system. As an essential and critical step in the modelling process, sensitivity analysis (SA) is the study of how uncertainty in the output of the model can be apportioned to different sources of uncertainty in the model inputs (Saltelli *et al.*, 2004). The generic mathematical formulation of the pre-existing system is given by

$$G(\underline{x}, \underline{d}) = z$$

where G accounts for the model representing the system, $\underline{x}$ for the inputs that are allowed to vary, $\underline{d}$ the fixed inputs and $\underline{z}$ the set of variables of interest. The objective of **sensitivity analysis (SA) is to instruct the modeller as to the relative importance of the uncertain inputs in determining the variable of interest.** The decision of which input to keep fixed and which to vary can be very subjective. However, in some situations, decision-makers or regulators deliberately fix, for instance, hypothetical risk scenarios for comparative studies in which it would be presumptuous to describe their completely unknown variability (e.g. human settlements around nuclear waste storage after a million years). An obvious consequence is that SA cannot analyse the effect of those variables which have been kept fixed.

Recommendations for the allocation of $\underline{x}$ and $\underline{d}$ are provided in Section 15.9.

While in the deterministic framework the pre-existing system model is analysed at specified values for the model inputs $\underline{x}$ (Cacuci, 2003), the space of uncertain model inputs may be explored in statistical approaches (Saltelli, 2000). In the standard probabilistic framework, which will be the focus of this chapter, the set of possible inputs or outcomes is represented by probability distributions

Uncertainty in Industrial Practice   Edited by E. de Rocquigny, N. Devictor and S. Tarantola,
© 2008 John Wiley & Sons, Ltd

and therefore the p inputs and r variables of interest are considered as random variables denoted $\underline{X} = (\underline{X}^1, \ldots, \underline{X}^p)$ and $\underline{Z} = (\underline{Z}^1, \ldots, \underline{Z}^r)$. When there is limited information available on the characterization of uncertainty, rather than probability distributions, alternative representations may be of interest in practice (see, for instance, Appendix C, devoted to the Dempster-Shafer evidence theory, and *(Automotive) Reliability in Early Stages*). Although sensitivity analysis can be carried out in conjunction with evidence theory or probability bound analysis (see Appendix C), the associated literature is still quite sparse and rather recent.

As concerns the measure of uncertainty of the variable(s) of interest, the quantities of interest considered here are the probability density function of the output $f_z(Z)$ (or the cumulative distribution function $F_z(Z)$) or any derived quantity of $f_z(Z)$ (variance, standard deviation, coefficient of variation, exceedance probability, quantiles, ranges, etc.).

In the standard probabilistic framework, uncertainty analysis is conducted first through Monte Carlo simulations. Sensitivity indices are subsequently derived from the input samples and related model realizations. Although the reverse path is adopted in the deterministic approach, ideally uncertainty and sensitivity analysis should be run in tandem (iterative strategy). Sensitivity analysis can serve a number of useful purposes in the modelling process. It is a valuable and impartial step, carried out to understand and corroborate the model structure, to identify critical regions in the space of the inputs, to establish priorities for updating the uncertainty model (including model simplification) and to guard against falsifications of the analysis.

Without explicitly recommending a 'best' method for all circumstances, the objective of this chapter is to provide guidelines for the use of sensitivity measures according to the study settings and the quantities of interest. In the first section the role of sensitivity analysis in relation to the four main uncertainty assessment goals will be explained. In the second section the seven different families of methods will be reviewed in the light of their features and limitations. Depending on the quantity of interest selected for a given problem, the reader will be directed to one or several suitable families of methods. In the last section, for each family of methods, the available techniques are discussed and compared.

18.1 The role of sensitivity analysis in quantitative uncertainty assessment

This section will explain the role of sensitivity analysis in relation to the four main uncertainty assessment goals introduced in Part I. Industrial practice is generally characterized by an iterative and repeatable achievement (at least in part) of the four uncertainty assessment goals. The complementarity and interdependence of these assessment goals has already been highlighted. Sensitivity analysis can be seen as the cement linking these different stages in the industrial process. In fact, the feedback processes underlying the various quantitative uncertainty assessment goals refer directly or indirectly to sensitivity analysis. The four quantitative uncertainty goals are recalled below:

- **U** (Understand): to understand the influence or to rank the importance of uncertainties, thereby to guide any additional measurement, modelling or R&D efforts;

- **A** (Accredit): to give credit to a model or a method of measurement, i.e. for it to reach an acceptable level of quality for use (through calibration, simplification, validation);

- **S** (Select): to compare relative performance and optimize the choice of maintenance policy, operation or design of the system;

- **C** (Comply): to demonstrate compliance of the system with an explicit criterion or regulatory threshold.

Whatever the predominant uncertainty assessment goal, sensitivity analysis is closely related to 'understanding influence and ranking importance' (Goal U). Nonetheless, the anticipated feedback processes and associated methodologies may be very diverse. In this chapter the potential of sensitivity analysis to support the four uncertainty assessment goals is outlined with reference to the case studies of Part II. Correspondences with classic sensitivity assessment settings (input prioritization, input fixing, (output) variance cutting and input/output mapping) proposed by Saltelli *et al.* (2004) are also provided.

18.1.1   Understanding influence and ranking importance of uncertainties (goal U)

The use of numerical models is very often characterized by an incomplete understanding of the input-output relationship. Most studies account for a very limited assessment of model structure and therefore poorly corroborate the hypothesis encoded into the model formulation. Very often the inputs of a model follow very asymmetric distributions of importance, with few inputs accounting for most of the output uncertainty and most inputs playing little or no role.

The **prioritization** setting, aiming to rank the model inputs in order of importance, is indisputably the most common function of sensitivity analysis. This assessment is of prime importance, especially when the system is not well known and the model-building is at an early stage of development.

Depending on the scope, stage and context of the study, the understanding of how changes in the inputs influence the results of the model will constitute the main objective (*Nuclear Waste Repository*) or represent a supporting goal of the analysis (*Hydrocarbon Exploration*). As the ordering of model inputs by importance may be an issue of great significance when the model is used, for example, in risk analysis (*Dyke Reliability*) or decision-making, a rigorous definition of 'importance' is necessary, especially in relation to the quantity of interest it refers to.

The outcomes of the analysis, possibly accounting for a spatial and/or temporal dimension, provide a real insight into the model's behaviour. The results usually provide guidance for the allocation of resources to future data collection campaigns and R&D efforts (*Electromagnetic Interferences in Aircraft* and *Spent Nuclear Fuel Behaviour*).

18.1.2   Calibrating, simplifying and validating a numerical model (goal A)

However complex the mathematical representation of natural or industrial systems, it is inevitably a simplification of real processes. The warranted complexity of any model depends mainly on the intended use of the model (diagnostic or prognostic), on the model type (data driven or law driven) and on the quantity and quality of data available for calibration and validation.

Whatever the framework adopted for model calibration, it requires the use of one of several objectives, cost, performance or likelihood functions in order to condition the inputs to be estimated on the available observations. Given the irreducible (or not yet reduced) sources of uncertainty, and since the complexity of the model is not necessarily tailored to the information derived from observations, the estimation of model inputs can lead to ill-posed inverse problems. Sensitivity analysis can be a valuable tool to identify the inputs really affecting the calibration criteria (i.e. **input fixing** setting) and observations truly constraining the inputs (i.e. useful observations). Given its close relation with identifiability, sensitivity analysis will play a key role in assessing and enhancing the identifiability of model inputs. Usually only a minority of inputs has a chance of being estimated – the majority does not have any appreciable influence on the variable of interest.

Since the calibration and validation of numerical models is very often based on a very partial appraisal of model performance (due to scarce observations), different models can lead to the same results. **Equifinality** (Beven and Binley, 1992) or **model indeterminacy** are the most popular designations for this phenomenon. Competing model structures (with different constitutive equations, considering different types of process, spatial/temporal resolution, etc.) may all be compatible with the same empirical evidence.

Once again, the construction of a generalized model can greatly extend the scope of the analysis. The objective of sensitivity analysis is not only to quantify and rank the importance of the sources of prediction uncertainty, but also (and which is much more relevant to calibration) to identify the elements (inputs, assumptions, structures, etc.) which are most responsible for the model realizations in the acceptable range (**i/o mapping** setting). Although performance in calibration is usually different from that achieved in validation or forecasting, if the various response modes of the system are explored, the redundant complexity of some model components can be revealed.

18.1.3   Comparing relative performances and decision support (goal S)

Industrial engineering usually involves complex decision-making situations calling for efficient, powerful and robust decision-support tools. Given the uncertainties underlying any real engineering assessment, the combined use of uncertainty and sensitivity analysis is now gaining widespread acceptance. However, although it seems to have been widely acknowledged that meaningful comparisons require estimates of uncertainties in prediction, the role of sensitivity analysis is generally still neglected.

The various options (model structures, designs, organizational changes, etc.) can be understood as different encodings of the system under study. The construction of a generalized model enables the confrontation of internally consistent but mutually exclusive representations of the system (i.e. the different options). Formally, variations in the function representing the system model (i.e. G), in the fixed or/and uncertain inputs (i.e. $\underline{x}$ and/or $\underline{d}$) will stand for the different options. In most cases the different options will be activated using a scenario trigger, i.e. a discrete random variable such as that used in CO_2 *Emissions* or *Radioprotection and Maintenance*. However, the performances to be compared can also refer to variables of interest at different spatial locations of the same computational domain (e.g. oil/gas traps in *Hydrocarbon Exploration*).

While uncertainty analysis should lead to an informed choice, sensitivity analysis can play a key role in understanding and controlling the drivers of the decision process. In fact, if sensitivity analysis is essential to assess the robustness of the final ranking of the options (classical **prioritization** setting), the lessons learned from the outcomes can be further extended. For instance, when two options have to be distinguished, aggregated responses on the performance of both options (difference, ratio, log[ratio], etc.) can be formulated (see *Radioprotection and Maintenance*) and may lead to a profitable re-definition of the variable of interest. In such cases the relative position of the probability density function of the re-defined variable of interest, with respect to a reference axis representing equal performances for both options, already highlights the merits of both. As mentioned above, the realizations of the output can be classified, and a detailed analysis of the mapping can identify the model inputs most responsible for producing realizations of the aggregate response favouring a given option (i.e. **i/o mapping** setting). Although this is not strictly a sensitivity analysis outcome, when the model inputs are controllable, the localization of the regions in the space of the inputs producing a targeted output (advantage over the other option) can guide constructive action on the system, and lead to an updating of the uncertainty model.

18.1.4   Demonstrating compliance with a criterion or a regulatory threshold (goal C)

In order to demonstrate compliance with a criterion or regulatory threshold, it is necessary to show that, given the uncertainty in the inputs, all realizations of the variable of interest fall into the acceptable region for the output. In some cases, such as CO_2 *Emissions* or *Dyke Reliability*, there is a very clear official uncertainty criterion (i.e. a regulatory environment). In others, such as *Electromagnetic Interferences in Aircraft*, the quantitative decision criterion has not yet been defined. If this assessment is mainly perceived as an uncertainty analysis appraisal, sensitivity analysis is essential at an early stage of the study in order to guide the updating of the uncertainty model towards compliance of the output.

When the quantity of interest is linked to variance (i.e. central dispersion q.i., etc.), this assessment is intimately related to the **variance cutting** setting for sensitivity analysis. The objective is to support an informed choice, aiming to reduce the variance of the variable of interest by fixing the smallest number of model inputs. More generally, whatever the quantity of interest, the feedback process consists in acting on the system, which may lead to an update of the cumulative probability distribution function (cdf) of some of the inputs or modification of the system model. For instance, in CO_2 *Emissions*, when the system performs beyond the official uncertainty criteria, sensitivity measures lead to calibration and intensification of measurements for the corresponding metrological chains. In *Dyke Reliability*, non-compliance leads to a change in the design of the system.

According to the criterion or regulatory threshold, the realizations of the variable of interest can be classified as either acceptable or non-acceptable. With the **i/o mapping** setting for sensitivity analysis, all details of the transformation can be analysed, and the model inputs that are most responsible for producing realizations of the variable of interest (v.i.) in the region of interest, identified. Lastly, once the criteria or regulatory threshold have been reached, sensitivity analysis may also play a key role in testing the robustness of the attained compliance. As underlined in *Radioprotection and Maintenance*, when the model inputs controlling the outcomes have been identified, the choice of the corresponding probability distributions must be rigorously validated.

18.2   Towards the choice of an appropriate Sensitivity Analysis framework

In this section a review of the different families of sensitivity analysis methods is offered and guidelines for the choice of the appropriate family of methods provided. When different families of methods are suitable for a given quantity of interest (e.g. variance or exceedance probability), other aspects of the problem should be taken into account in the choice of the appropriate family of methods.

Sensitivity analysis has an increasingly important role in quantitative modeling. In latter decades methodological advances have been achieved in a number of scientific disciplines. A diversity of sensitivity analysis approaches can now be found in the literature and an effort will be made here to group these methods into seven distinct families (see Table 18.1). While the scope of some local techniques, such as differential methods, is limited to a base point (nominal values for the uncertain model inputs), others aim to explore the full space of model inputs (global techniques).

Sometimes a variety of methods may be applied to a problem with no proper definition of the objectives of the sensitivity analysis nor a careful consideration of the question that the model is supposed to answer. This can lead to confused or inconclusive results 'with any clue as to which one should believe' (Saltelli and Tarantola, 2002).

For the choice of the appropriate family of sensitivity analysis methods, some attributes, such as the features of the pre-existing system model (dimension of the space of inputs, computational cost of the model, etc.), play an important role. However, among the characteristics to be specified for the selection of the most appropriate approach (see Part I), the final goal of the study and the quantity of interest are the most critical.

In the previous section it was shown that the different goals lead to certain expected feedback processes, and therefore to specific sensitivity analysis settings. The final goal of the study and the quantity of interest are not completely independent – as emphasized above, a given quantity of interest is not always meaningful for all the goals, but rather should be suited to answer the particular goals of the study. The relation between the quantity of interest and the final goal was discussed in Section 15.3.

Since variance or related quantities (standard deviation, coefficient of variation) are widely used to quantify the uncertainty in model predictions, many sensitivity analysis methods rely on the decomposition of the variance of the output. However, **variance is not the only nor the best measure of uncertainty** and other quantities of interest may be of prime importance in many industrial quantitative uncertainty assessments.

For each class of methods several quantities of interest exist and could be applied. However, with a given problem, the first need is to identify the proper quantity of interest, and on this basis select the most appropriate sensitivity analysis method. It is therefore necessary to establish a correspondence between the main quantities of interest and the available families of sensitivity analysis methods. This correspondence is given in Table 18.2.

Note that graphical methods can be used for any quantity of interest, so it is always possible to choose between graphical methods and some other family. However, depending on the goal of the study, a decision is needed on whether a qualitative or quantitative measure of sensitivity should be used.

The most widely used quantities of interest are related to central dispersion quantities. For instance, confidence intervals, when not too wide (e.g. <95%) may be approximated on the basis of the calculation of the mean and standard deviation

Table 18.1   Review of the different approaches to sensitivity analysis.

	Features	Caveats
Differential methods	Appropriate for computationally intensive (CPU time > 1 h per run) and/or high dimensional models (number of inputs >20). Algebraic operations on the continuous/discrete model or directly on the source code.	Local Analysis. Skills and efforts for accurate and efficient estimation of the derivatives. Interactions characterized only by expensive second- or higher order analysis. Work only with continuous inputs.
Approximate reliability algorithms	Focus on a particular mode of the system (failure), dedicated to exceedance probability. Deterministic measures combined with statistical properties of inputs.	Search for most probable failure point tailored to I/O complexity (i.e. regularity hypotheses for FORM/SORM approximations). Validation with Monte Carlo simulation recommended.
Regression/ correlation	Global methods. Appropriate for quick assessment. Simple to code, no sampling design required.	Model assumptions (linearity, monotonicity). Provides only main effects.
Screening methods	Global methods. Appropriate for the analysis of computationally intensive (but < 1 h per run) and/or high-dimensional models (20 < number of inputs <100), analysis by groups of inputs.	Require special designs for the sample. The importance of interactions is qualified but these are not quantified.
Variance analysis of Monte Carlo simulations	Global methods. Can capture interactions of any order. Available shortcuts for total indices and analysis by groups.	Cannot cope with high-dimensional (number of inputs limited to 20) and/or expensive models (CPU time limited to 1 min per run). Relies on the second-order moment of the distribution.

Table 18.1   (*continued*)

	Features	Caveats
Non-variance analysis of Monte Carlo Simulations	Global methods. No specific sampling design. Exploration of the I/O mapping. Flexibility in the treatment of model output realizations.	For statistical testing methods: qualitative measures, subjective choice of significance levels, reliability of inferences depend on robustness of the test results.
Graphical methods	Global/local methods. Intuitive and model free (no sampling design). Exploration of the I/O mapping. Flexibility in the treatment of model output realizations.	Qualitative assessment limited to first-order effects. Impractical when more then 10 inputs have to be analysed.

Table 18.2   Correspondence between the quantities of interest and the families of sensitivity analysis methods.

Type of quantity of interest	Suitable classes of sensitivity analysis methods
Central dispersion quantities (variance, standard deviation, coefficient of variation, confidence interval)	Variance analysis of Monte Carlo simulations, screening methods, regression/correlation methods, differential methods, graphical methods.
Exceedance probabilities	Approximate reliability algorithms, non-variance analysis of Monte Carlo simulations, graphical methods.
Quantiles or complete probability density function	Non-variance analysis of Monte Carlo simulations, graphical methods.

of the output realizations. Variables of interest like the maximum or the minimum of the model prediction over time (and/or over space) or specific values (e.g. at a particular time step and/or spatial location) can be treated with the classical variance analysis of Monte Carlo simulations. For example, in *Nuclear Waste Repository*, the variance of the peak dose (i.e. the maximum radiological dose over time) and the variance of the dose at a particular time are decomposed with a variance-based method. However, when the variable of interest is deterministic (e.g. maximum over all realizations, worst/best case scenario), a local sensitivity analysis is usually sufficient to investigate the corresponding behaviour of the system.

Although for the analysis of less common quantities of interest (e.g. complete probability density function or quantiles) only one family of methods can be used in addition to graphical methods, this is not the case for the most widely used quantities (i.e. central dispersion and exceedance probabilities). In this case, other criteria intervene in the choice of the appropriate sensitivity analysis method. They mainly concern:

- properties of the pre-existing system model: linear or non-linear, monotonic or non-monotonic (this may or may not be known *a priori*), CPU time;

- features underlying the model inputs: number, independent or correlated, presence of discrete model inputs;

- the goal of the study, the characteristics of the candidate method: qualitative or quantitative, local or global, ability to capture interactions, specific sampling strategy required.

The following provides guidelines to direct the reader towards the most suitable class of methods. Additional information will be available in the description of the family of methods for the selection of a particular technique (see Section 18.3).

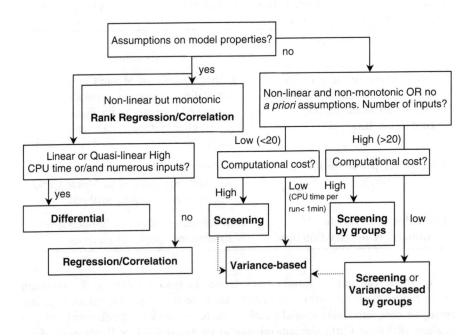

Figure 18.1 Decision tree for the choice of the appropriate framework for central dispersion quantities of interest (disregarding meta-models, which could be used with any method in the case of high computational cost).

When the quantity of interest refers to central dispersion quantities, the decision tree proposed in Figure 18.1 provides advanced guidance to the analyst. When the number of model inputs is large, screening methods may constitute a very good preliminary analysis for the reduction of the number of uncertain inputs prior to any other advanced and computationally intensive analysis.

For the 'probability of exceedance' quantity of interest, the choice between approximate reliability algorithms and non-variance analysis of Monte Carlo simulations depends mainly on the regularity of the model and on the rareness of the failure mode. The difficulties related to the estimation of low probabilities when using sampling-based methods and the problems related to the convergence of approximate reliability algorithms have been discussed in the previous chapter on uncertainty propagation. Approximate reliability methods provide a computational advantage over Monte Carlo simulation for low-probability events, and provide, at no extra computational cost, factors quantifying approximate contribution of each uncertain model input to the probability of failure.

18.3   Scope, potential and limitations of the various techniques

The previous section gave a summary of the various families of sensitivity analysis methods and some guidelines for the choice of the appropriate family, according to the relevant quantity of interest. This section reviews the main techniques belonging to each family of methods. Their features, advantages and limitations are examined in order to provide the analyst with recommendations for the choice of the optimal sensitivity analysis technique for a given problem.

18.3.1   Differential methods

The objective of differential methods is to quantify the rate of change in the variable of interest due to small variations in the uncertain model inputs (note that only continuous inputs can be considered). Although the final sensitivity measure should include some statistical properties of the inputs, importance measures rely on derivatives which are approximated or calculated analytically. The singular value decomposition of matrices of partial derivatives (typically Jacobians or Hessians) can provide additional insights into the I/O mapping. However, this approach is intrinsically local and refers to a particular point in the space of the uncertain inputs which is defined by the nominal values for the inputs. The choice of the base-case can greatly influence the sensitivity analysis outcomes. A valuable review of deterministic sensitivity analysis methods is proposed by Turanyi and Rabitz (in Saltelli *et al.*, 2000).

The method of finite differences (sometimes referred to as the *brute force method*), involving recalculations of the variable(s) of interest, is the simplest for the approximation of derivatives. In order to avoid the critical choice of a perturbation parameter (involving a trade-off between approximation errors and round-off errors), the calculation of exact derivatives requires some form of differentiation of the mathematical/numerical/algorithmic representation of the system under study. The most general formalism, based on the concept of the Gâteaux differential, is proposed by Cacuci (1981a, 1981b). While the computational cost of the **forward sensitivity analysis** method (suitable when $p < r$) is still dependent on the number of model inputs, the **adjoint sensitivity analysis** method (appropriate when $p \gg r$) uses the concept of the adjoint of a linear operator in order to make it independent from the number of model inputs p (Cacuci, 1981a,b). There is no general rule for the implementation of the approach (operations can be performed on the continuous or discrete representation of the model), which may be rather time-consuming. In the direct and reverse modes of **algorithmic differentiation** the algebraic operations are carried out directly on the source code implementing the model (see Griewank, 2000, for a recent and comprehensive description).

Apart from the **global adjoint sensitivity analysis** procedure (GASAP, see Cacuci, 2003), which aims to identify all the critical points (using forward/adjoint sensitivity analysis) of the I/O mapping for subsequent analysis, all differential sensitivity analysis methods are restricted to a point of the input space. This constitutes a very limited assessment for non-linear models. However, the adjoint method can cope with very high-dimensional systems, and the ability to compute exact derivatives (forward and adjoint) can benefit approximate reliability algorithms (in the search for the failure point using gradient-based methods) and sampling-based techniques (Isukapalli *et al.*, 2000; Cao *et al.*, 2006).

It is crucial to take into account the relative uncertainty of the model inputs when estimating sensitivity measures from derivatives. As illustrated by Helton (1993), the contribution of a particular model input to the variance of the variable of interest can be estimated with differential methods (see *CO_2 Emissions*). While the characterization of local interactions requires the calculation of higher order derivatives, usually only first-order derivatives are calculated. The widespread use of differential methods is mainly due to the cultural/mathematical traditions of modelling communities, which generally use models based on partial differential equations. This constitutes a strong bias in the choice of a sensitivity analysis method and the critical limitations of differential methods are not always acknowledged.

18.3.2   Approximate reliability methods

When the modeller's interest is not in the entire range of probabilistic outcomes, but rather in a particular mode of failure, approximate reliability algorithms may be very efficient for the calculation of the probability of failure (Madsen *et al.*, 1986). In this case the quantity of interest is generally the probability of exceeding a threshold. Illustrative examples are provided by three case studies of this book: *Spent Nuclear Fuel Behaviour*, in which the probability of the cladding losing its integrity is assessed; *Dyke Reliability*, in which the probability of failure is analysed; and *Electromagnetic Interferences in Aircraft*, in which the probability of interference is investigated. When extremely low-probability events are of interest (of the order of 10^{-6} for *Spent Nuclear Fuel Behaviour* and *Electromagnetic Interferences in Aircraft*), approximate reliability methods provide a computational advantage over Monte Carlo simulation. Moreover, taking into account the statistical properties characterizing the model inputs, they also provide sensitivity measures identifying the inputs which have the greatest impact on the failure probability. Cawlfield (in Saltelli, 2000) offers a good overview of reliability approaches to sensitivity analysis.

A description of the **First- and Second-Order Reliability Algorithms** was given in the previous chapter on uncertainty propagation. Regardless of which method is used for their estimation, partial derivatives are available at the failure point. They are therefore usually used to assess the sensitivity of the probability of failure to changes in the model inputs' distribution parameters. Although it should be emphasized that importance measures are mainly a function of the failure point and not really related to the quality of the approximation of the failure surface, a rigorous interpretation is much more straightforward for FORM estimates. The most widely used probabilistic sensitivity measures, like the Gamma sensitivities, refer to FORM and take the statistical properties of the inputs (mean, standard deviations and correlations) into account.

As shown in *Spent Nuclear Fuel Behaviour*, *Dyke Reliability* and *Electromagnetic Interferences in Aircraft*, validation of FORM/SORM approximations with Monte Carlo simulation is preferable for the calculation of the probability of exceedance. Given that the analysis usually concerns rare events, the use of variance reduction techniques (e.g. use of importance sampling in *Spent Nuclear Fuel Behaviour*) is generally advantageous. Although some techniques can provide comparable results for a reasonable computational cost, Monte Carlo reliability analysis does not directly provide importance measures.

18.3.3   Regression/correlation

Regression and correlation approaches to global sensitivity analysis represent a very simple and intuitive assessment for the (approximate) decomposition of the variance of the variable of interest for linear (or quasi-linear) models. McKay (1997) shows that regression-based methods are special cases of variance-based methods. Although specific sampling techniques (ensuring space filling) usually enhance the quality of the results, no particular sampling design is required for such techniques and their implementation is straightforward.

One of the most widely used measures to assess the strength of the relation between variables of interest and model inputs is **Pearson's Correlation Coefficient** (CC). The calculated values for a given pair (variable of interest, input) range between -1 and 1 and the interpretation of the sign and magnitude provides useful information. When both inputs and variables of interest are standardized (i.e. 0 mean and 1 standard deviation), the sample correlation coefficient between the variable of interest and a given input can be viewed as the regression coefficient in a linear regression in which the independent variables are the model inputs and the dependent variable is one of the variables of interest. However, the **Standard**

Regression Coefficient (SRC) might be preferred because it directly indicates the fraction of the original model variance explained by the linear regression model. Results can be improved by applying suitable transformations both to the inputs and to the variables of interest (logarithmic, square root, etc.) when these have peculiar uncertainty distributions. However, effective transformations can be difficult to identify.

The Partial Correlation Coefficient (PCC) gives the strength of the correlation between the variable of interest and a given input. This latter is cleaned of any effect due to correlation with any other input. In other words, PCC provides a measure of input importance that tends to exclude the effects of other inputs. In the particular case in which the inputs are not correlated, the importance based on any of CC, SRC or PCC (in their absolute values) is exactly the same.

When a small portion of the variance of the variable of interest can be explained by a linear regression, the non-linearity of the model can be taken into account by taking a rank transformation of both the inputs and the variable of interest before application of regression/correlation techniques. This usually leads to a substantial improvement in the sensitivity results, provided that the relationships between the inputs and the variable of interest are monotonic (i.e. either increasing or decreasing over the range of variation of each input). The resulting measures of importance are the (Spearman) **Rank Correlation Coefficients** (RCC), the **Standardized Rank Regression Coefficients** (SRRC) and the **Partial Rank Correlation Coefficients** (PRCC). Note that the rank transformation modifies the original model, so the sensitivity results have to be interpreted with respect to the transformed model, and cannot be generalized to the original.

Moreover, instead of constructing a regression model with all inputs simultaneously (original or rank-transformed), a sequence of regression models can be constructed. The inputs are introduced one-by-one via a stepwise procedure (i.e. **Stepwise Regression**). Their order of inclusion is determined by the square of the correlation coefficient between the variable of interest and each input. Only variables explaining statistically significant amounts of the variance are included in the regression (stopping criteria). As explained by Helton *et al.*, (2006), the importance of a given model input is then indicated by the changes in cumulative R^2 values as additional inputs are added to the regression model, and by the SRCs for the same inputs in the final regression model.

Note that in the case of small values of the coefficient of determination (either in original values or in rank-transformed data), the reliability of the corresponding sensitivity measures (original or ranked) is limited. In *Spent Nuclear Fuel Behaviour* the combined use of regression/correlation and rank regression correlation approaches can characterize the type of relation (linear or/and monotonic) between the variable of interest and the various model inputs. Since the computational cost for the calculation of those sensitivity measures is negligible, they are frequently used for the analysis of temporally-dependent responses (see *Nuclear Waste Repository*). The same case study demonstrates the limitations of these techniques in the analysis of complex models (non-linear and non-monotonic).

18.3.4   Screening methods

Screening techniques are well-designed and cheap sensitivity analysis methods which were conceived in order to assess efficiently models containing tens or hundreds of model inputs. In fact, computationally intensive methods of sensitivity analysis can be precluded if a model contains many inputs and/or is computationally expensive to execute. In such cases, a possible alternative is the use of screening methods, which enable the analyst to identify the most important among many model inputs. The underlying assumption is that the model inputs follow Pareto's law: the number of important inputs is small compared to the total number of inputs. Therefore, the purpose of screening is to eliminate negligible inputs in order to concentrate efforts on the most important.

The set-up of numerical experiments for sensitivity analysis can be quite similar to that for physical experimentation. Screening methods usually commence with a design of experiments in order to assign levels (i.e. specific values) to the various inputs. Outcomes are then analysed in order to make inferences about the relative importance of the inputs. A detailed description of the different strategies and guidelines for the choice of the appropriate approach for the design of experiments can be found in Montgomery (2004).

The simplest screening design consists in modifying only one input between two consecutive simulations, e.g. **One-At-a-Time** designs (OAT). Starting usually with the nominal value of each input, different strategies can be adopted for the path defined by successive input changes (amplitude and direction of the change), but the approach is inefficient when the proportion of important inputs is small. When only two extreme values are selected (i.e. two-level analysis), the approach is comparable to the most common use of **Factorial Design** (FD), in which two discrete levels for each input (i.e. 2^p model runs) are used to explore all possible combinations of levels and inputs. When more than two levels are considered, the larger the number of levels, the better the space-filling properties of the design. Although the approach is able to identify interactions, it is inefficient and computationally intensive for an important number of inputs. Assuming that higher-order interactions are negligible, information on the main effects and low-order interactions can generally be characterized by running only a fraction of the complete factorial (i.e. 2^{p-q} model runs).

The ability to capture interactions will depend on the resolution of the **Fractional Factorial Design** (FFD). While a fractional factorial design of resolution III will identify only the important main effects, a resolution IV design will also capture a selected number of two-way interactions, while a resolution V will determine the most important two-way interaction effects. When the design aims to estimate more effects than the number of required model runs, it will fall into the class of **Supersaturated Designs** (SSD).

Examples of supersaturated designs are the **Iterated Fractional Factorial Design** (IFFD) proposed by Andres *et al.* (1993) and **Sequential Bifurcations** (SB) from Bettonvil and Kleijnen (1997). Both are group screening techniques, implementing the divide-and-conquer strategy in order to identify the important inputs. While both techniques provide main effects and two-way interactions, sequential bifurcations provide reliable estimates only when the I/O relation is monotonic and the sign of each input effect is known.

Among the screening methods, the global and model-free **Elementary Effect** method proposed by Morris (1991) is receiving an increasing amount of interest. In this method each input is allowed to vary over l levels and t trajectories (the latter are randomly generated). Each trajectory is built in such a way that inputs are varied one-at-a-time across their levels. Along each trajectory, the so-called elementary effects are evaluated. For each input, the mean and standard deviations of all elementary effects are used to infer main and interaction effects. In a recent paper Campolongo *et al.* (2007) revise the definition of the sensitivity measure and sampling strategy, extend the original method to groups of inputs and provide a link with the sensitivity indices calculated with the widely accepted variance-based methods.

18.3.5   Variance analysis of Monte Carlo simulations

When the quantity of interest is related to the central dispersion of the variable(s) of interest, the widely accepted variance-based methods yield robust and accurate global sensitivity measures without relying on any assumption on the nature of the I/O relation. These techniques are recommended for models which require a modest amount of CPU time (i.e. up to the order of 1 min per run), and with a limited number of inputs (i.e. not exceeding 20, for example). Otherwise, rather than a complete decomposition of the variance of the output, the screening methods described in the previous section are preferred for a less accurate but computationally more amenable assessment.

The fraction of the variance of the output Y due to the uncertainty of a given model input X^i when the function is averaged over all other uncertainties (i.e. $V(E(Y|X^i))/V(Y)$) was employed as a sensitivity measure in the **Fourier Amplitude Sensitivity Test** (FAST) of Cukier *et al.* (1978). However, the calculated partial variances represent only the first-order effects and the reliability of the analysis is compromised when their sum does not approach unity (for example, below 0.6–0.7 in practice). A general formalization for variance decomposition methods is proposed by *Sobol'* (1993). This formalization enables the characterization of interactions at any order. The approach is based on a Monte Carlo exploration of the input space with the recommended quasi-random sampling strategy to generate uniformly distributed sample points (Sobol', 1967). In order to avoid the computational burden related to the evaluation of the $2^p - 1$ terms from the Sobol' decomposition (p first-order effects plus all the interactions terms), Homma and Saltelli (1996) introduced the so-called total effect S_{Ti} which includes the fraction of variance accounted for by any combination of X^i with the remaining inputs. With S accounting for the sample size, first-order and total effect indices can be computed very efficiently using $(p + 2)S$ model evaluations with the **extended Sobol'** method proposed by Saltelli (2002) or the method described in Helton *et al.* (2006). Using the FAST transformation in the frequency space, only pS model evaluations are required with the **extended FAST** method (Saltelli *et al.*, 1999) for the calculation of first-order and total effect indices. The extended Sobol' method was used in *Hydrocarbon Exploration* and *Airframe Maintenance Contracts* while FAST was adopted in *Spent Nuclear Fuel Behaviour* and *Nuclear Waste Repository*.

Although they are usually limited when used for the quantification of interactions, other techniques have recently been proposed for the decomposition of the variance of the variable of interest. While the method proposed by McKay (1997) is computationally quite intensive, other techniques are very efficient when compared to the extended Sobol' or the extended FAST. By combining the classic FAST with **Random Balance Designs** (Tarantola *et al.*, 2006), first-order indices can be computed for a computational cost independent of the number of inputs (i.e. S).

Besides providing importance measures, some approaches produce a surrogate model. For instance, the **Gaussian Process Emulator** with Bayesian analysis of Oakley and O'Hagan (2004) and the **State Dependent Regression** models based on the recursive filtering and smoothing estimation of Ratto *et al.* (2004) are representative examples among the increasing number of meta-modelling approaches to sensitivity analysis. While the mapping should be sufficiently smooth for the Gaussian process emulator, **State Dependent Parameter** (SDP) modelling works for non-smooth and even discontinuous mappings and allows for the characterization of low-order (2nd or 3rd order) interactions. This technique was used for the estimation of the first-order effects in *Hydrocarbon Exploration* and compared to the results obtained with the Sobol' method.

Although the use of variance-based methods can be computationally difficult with large numbers of inputs, the analysis can be also performed by partitioning the model inputs into groups of different logical meaning for the analyst (Saltelli *et al.*, 2005). The analysis by group provides sensitivity measures for the groups rather than for the single inputs – it is cheaper than the classic analysis and can provide very useful insights. Provided that it is possible to design the groups so that model inputs in different groups are independent (although the inputs within the same group could be dependent), this strategy can be employed to tackle the difficult problem of dependent model inputs (Jacques *et al.*, 2006).

When the inputs are not independent, the computational shortcuts available for orthogonal inputs are no longer applicable. In this case, the sensitivity estimators require larger sample sizes to attain a given precision of the estimates, and the convergence of the sensitivity estimates with increasing sample size is much slower with respect to the case of independent inputs. In addition, dependent input samples are more laborious to generate than independent; this difficulty often arises when a specific design is required by the sensitivity estimator.

In the case of dependent inputs, the estimation of the first- and higher-order conditional variances, $V[Z|X^i]$ and $V[Z|X^i, X^j, \dots]$, respectively, is much more computationally expensive. Saltelli and Tarantola (2002) proposed a stepwise approach to keep their number to a minimum. An alternative approach is to approximate the conditional expectations or conditional variances with certain meta-models (such as local polynomial smoothers, etc.). In sum, as emphasized by Saltelli *et al.* (2004), recourse to dependent inputs is recommended only when essential.

The most commonly used approach for dependent inputs is currently the extension of the correlation ratios of McKay (1997) employed in Saltelli and Tarantola (2002). This approach uses the methods of Iman and Conover (1982) and Stein (1987) to generate replicated Latin hypercube-correlated samples. A promising strategy recently proposed by Xu and Gertner (2007) uses the FAST sampling design coupled with Iman and Conover's method to induce correlations and the standard FAST formula to estimate the main effect indices.

18.3.6 Non-variance analysis of Monte Carlo simulations

This class of techniques is also based on a Monte Carlo exploration of the space of uncertain inputs but permits differentiated flexibility in the treatment of realizations of the variable of interest. Although the outcomes are not always as robust and precise as those obtained with variance-based methods, the advantages of this family of methods are that no specific sampling design is required, the necessary number of model evaluations is likely to be smaller, and various quantities of interest can be treated. This section will review the main techniques based on statistical testing and on classification/regression trees.

Statistical testing approaches

Although the visual analysis of scatterplots can already provide important insights (see Section 18.4 on graphical methods), some techniques referred to as **grid-based methods** have been developed in order to assess the randomness of the distribution of points across grid cells. The appearance of a non-random pattern in the scatterplots (X^i, Z) indicates that X^i has a significant effect on Z. Various statistical tests have been developed in order assess **common means** (CMNs), **common distributions or locations** (CLs) (Kruskal and Wallis, 1952), **common medians** (CMDs) or **statistical independence** (SI). Using the same grid-based approach, **entropy-based measures** can also be used to assess the strength of the relation between X^i and Z. When the mapping is linear, all tests produce the ranking obtained with regression/correlation approaches. However, when the relation is non-linear and non-monotonic, the statistical tests based on grids perform better. However, the violation of the tests' assumptions can lead to misleading rankings of the inputs. Note that the points in the sample have to be chosen independently. See Helton *et al.* (2006) for a comparison and detailed description of the approaches.

Grid-based measures are based on a segmentation of the inputs ranges without regard to the values of the variable of interest. Other sensitivity analysis techniques relying on **non-parametric statistical tests** conduct a segmentation based on an associated partitioning of the realizations for the variable of interest (Monte Carlo Filtering). The partitioning of the output can occur at any value or percentile, thus providing interesting outcomes for a large range of quantities of interest. For instance, the **Regional Sensitivity Analysis** (RSA) of Hornberger and Spear (1981) relies on the **Kolmogorov-Smirnov test** of the maximum distance separating empirical distributions resulting from the partitioning. The test statistics produce a qualitative sensitivity measure. The **Cramer-Von Mises test** adopts the same strategy but uses the sum of all squared distances, while the **Mann-Whitney test** (also known as Wilcoxon test) is based on the sum of the ranks. In order to give more statistical power to the test, the ranks can be squared (squared rank test). All theses techniques share the advantages and limitations of grid-based approaches but allow for the identification of the model inputs which drive model realizations in any defined subset for the variable of interest. The outcomes from the analysis are mostly qualitative and interactions cannot be characterized.

Strategies based on classification and regression trees (CART)

For the use of **classification and regression trees**, a set of rules are provided for the partitioning of realizations for the variable of interest. An appropriate splitting criterion has to be specified, and various quantities of interest can be addressed. Using a strategy presenting some analogies with stepwise regression, a tree is grown until the satisfaction of a termination criterion. Each node refers to a model input on which the split is based. The importance of an input is therefore determined by its level (precedence in the tree structure) and frequency of appearance. A comprehensive description of CART principles can be found in Breiman *et al.* (1984), and typical applications for the sensitivity analysis of rather complex numerical models are proposed by Mishra *et al.* (2003) and Mokhtari *et al.* (2006). Generally this kind of analysis identifies interactions and provides extensive understanding of the mapping. Although there is no clear summary measure of importance, the model inputs contributing most significantly to pre-defined subsets of the realizations of the variable of interest can be identified. This is the primary reason that CARTs are used to extend the ability of the RSA approach to highlight high-order interactions between inputs. This method, called **Tree Structured Density Estimation** (TDSE), was proposed by Spear *et al.* (1994). Relying on the assumption that non-random density patterns indicate an influence of inputs on the model output, TSDE can be linked to the previously mentioned **grid-based methods**. However, the tree structure can be relatively sensitive to the data set and the adopted splitting criteria. Using the concept of **Random Forest** (Breiman, 2001), which essentially involves growing a set of regression trees, Pappenberger *et al.* (2006) propose a sensitivity analysis method which seems particularly suitable for the identification of interactions and the investigation of sub-regions from the space of uncertain model inputs.

Moment-independent sensitivity analysis methods

Most sensitivity analysis techniques presented in this chapter refer, explicitly or implicitly, to a specific moment or targeted proportion of the distribution of the variable of interest. However, because of the shape of this distribution, or because of specific needs related to a particular domain of application, the focus on a particular moment of the distribution (such as variance) can lead to non-informative conclusions (Borgonovo, 2007). Although the **bivariate importance measure** proposed by Iman and Hora (1990), using the α and $1 - \alpha$ quantiles of the distribution, partly addresses this limitation, it still does not account for the whole distribution (Park and Ahn, 1994). When the input X^i is fixed at one of its possible values, both the conditional probability density function (i.e. $f(Z|X^i)$) and the conditional probability distribution function (i.e. $F(Z|X^i)$) are altered and can significantly differ from the unconditional distribution function of the variable of interest (i.e. $F(Z)$). The specification of a difference measure (distance- or area-based) and the exploration of all possible values of X^i (by calculating the expectation of the previously mentioned difference measure) are the key elements of a **moment-independent global sensitivity analysis** method (Chun *et al.*, 2002; Borgonovo, 2007). As emphasized by Borgonovo (2007), this type of approach can be extended to any group of inputs and requires the specification of the joint distribution function for the inputs without requiring their independence. Finally, an example of sensitivity analysis based on the cumulative distribution function of the response to the uncertain input distributions is proposed by Mohanty and Wu (2001). In this method the response *cdf* is defined as the integral of the joint probability density function of the model inputs, with a domain of integration defined by a subset of the samples which satisfies a constraint on the model values. The sensitivities are then calculated from the derivatives of the probability integral.

18.3.7   Graphical methods

The proper use of graphics constitutes a key element of any quantitative assessment. While bars, tornado graphs or radar charts can be particularly useful to communicate importance measures, box-and-whisker plots[1] are more suitable for the representation of uncertainty analysis results. Without describing every type of graphic which can be used for the visualization of uncertainty and sensitivity analysis results, this section will review some techniques which can support the extraction of relevant information on the I/O mapping.

Since the use of these techniques requires the availability of a data set characterizing this I/O mapping, they are particularly suitable in the probabilistic framework. In this case, and similarly to sampling-based approaches to sensitivity analysis (variance and non-variance analysis on Monte Carlo simulations), the basic information required is one or several samples (replicates) characterizing the relation between the uncertain model inputs $X = (X^1, \ldots, X^p)$ and the uncertain variables of interest $Z = (Z^1, \ldots, Z^r)$. Specific sampling strategies can always enhance the quality of the results, but no particular sampling design is required. Given the limitations related to the display of multi-dimensional information, the approach is also affected by the curse of dimensionality and becomes impractical for a

---

[1] A graphical approach to summarizing a probabilistic distribution function.

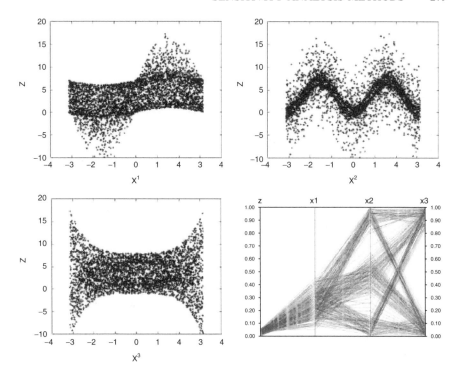

Figure 18.2 Scatterplots and 5th percentile cobweb plot for the Ishigami function[2].

large number of inputs or variables of interest (i.e. p and/or $r > 10$). In order to simplify the argument developed in the following paragraphs, a scalar variable of interest Z is considered. When compared to the previously described sampling-based quantitative sensitivity analysis techniques, although numerical operations might be required for the transformation of the raw data (logarithmic, ranks, etc.), the outcome is mainly graphical and therefore intrinsically qualitative. Depending on the complexity and quality (i.e. accessibility of information content, clarity) of the graph, the analyst may require some training.

The simplest and most widely used visualization of the I/O mapping is provided by so-called **scatterplots**. For a given model input X^i, a scatterplot corresponds to a projection of the sample points defining the (X, Z) hyper-surface in the (X^i, Z) plane (Figure 18.2). It is then possible to visualize the values taken by the variable of interest Z across the range of X^i. When a pattern can be observed in the scatterplot, the stronger the pattern, the more important the influence of the corresponding input on the variable of interest. Although a visual inspection can be seen as an empirical and somehow subjective appraisal of pattern randomness,

---

[2]An analytical function often used in benchmarks: $Z = \sin(X^1) + 7*(\sin(X^2))^2 + 0.1*(X^3)^4*$ $\sin(X^1)$

scatterplots provide rich information on the I/O mapping, which the other families of sensitivity analysis methods tend to condense into a few sensitivity indices.

Using scatterplots, the relatively efficient and intuitive processing of the graph carried out by the analyst presents some analogies to the grid-based methods presented in Section 18.1. The same applies to parametric and non-parametric regression approaches to sensitivity analysis. As shown for this particular type of method, the examination of scatterplots usually represents a very fruitful preliminary analysis for the choice of the most suitable class of function (or smoothing strategy), and more broadly for the selection of the appropriate sensitivity analysis method. The detection of important features like linearity, monotonicity or 'jumps' can speak in favour of (or against) the use of certain techniques.

The use of scatterplots can be greatly extended in order to identify which region in the space of uncertain inputs corresponds to a given subset of the realizations of the variable of interest. Model inputs can be plotted against each other with an intensity ramp corresponding to the values of the variable of interest (**matrix of scatterplots**), and different colours corresponding to different subsets can be used on a single graph (**overlayed scatterplots**).

A very interesting but qualitative assessment which can be seen as a one-dimensional summary of all scatterplots is provided by the **contribution to the mean plot** exploited in *Nuclear waste repository*. For each model input X^i, the realizations of the variable of interest are sorted with respect to increasing values of x^i; the cumulative sum of z_j (i.e. cumulative contribution to the mean) is then plotted against the distribution function of X^i. The realizations z_j which differ significantly from $E(Z)$ produce local deviations from the diagonal, and show qualitatively the influence of X^i on Z. A single plot facilitates the simultaneous appraisal of contributions, for all model inputs, across their range.

Moreover, valuable information is also presented in condensed for in so-called **cobweb plots** (Cooke and Noortwijk, in Saltelli *et al.*, 2000), which are able to represent graphically multi-dimensional distributions with a two-dimensional plot (see Figure 18.2 for instance). The possible values for the variable of interest and the model inputs correspond to locations on the vertical axis (i.e. $p + 1$ axis when $r = 1$) and a jagged line represents each sample realization (Wegman, 1990). Coloured lines can be used to display the different percentiles of the variable of interest distribution. Moreover, flexible conditioning capabilities facilitate an extensive insight into particular regions of the mapping. Cobweb plots are usually provided with 'cross densities' showing the density of line crossings midway between the vertical axes. Therefore, an informed and careful analysis of cobweb plots facilitates the characterization of dependence and conditional dependence. In sum, graphical methods can be characterized mainly by qualitative outcomes and an advantageous flexibility in the choice of the quantity of interest.

18.4 Conclusions

Given the tremendous diversity of approaches to sensitivity analysis, the variety of outcomes and applicability conditions, special care should be taken in the framing of the problem at hand (in order to minimize type-III error). All sensitivity analysis methods may be vulnerable to misclassifications of model inputs, such as treating important inputs as non-influential (type-II error) or (although less risky for the analysis) non-influential inputs as important (type-I error). For quantitative techniques, an appropriate characterization of interactions generally constitutes an effective protection against type II error.

The complementary nature of the various approaches associated with the increasing availability of efficient and user-friendly software and libraries is an argument for the adoption of several approaches, possibly in a stepwise way (Kleijnen and Helton, 1999; Frey and Patil, 2002). In fact, quite apart from the feedback process related to action on the system and eventual updating of the uncertainty model, the results of an analysis can benefit from a subsequent and more sophisticated assessment. For instance, concerning variance in the quantity of interest, the iterative use of graphical, regression/correlation (or screening) and variance-decomposition methods can lead to a very robust and informative analysis (as evidenced in *Spent nuclear fuel behaviour* and *Nuclear waste repository*). Similarly, the parallel application of techniques providing comparable results reinforces the robustness of the sensitivity analysis results.

For techniques involving random or quasi-random sampling of the space of model inputs, the sample on which the inference is based only partly characterizes the resulting mapping. When possible, replicating sensitivity analysis experiments is strongly encouraged to confirm the robustness of the sensitivity analysis. Kendall's coefficient of concordance (KCC), top-down coefficient of concordance (TDCC) or Savage scores can be used to this end (Helton *et al.*, 2005). Moreover, given the difficulties related to the choice of the appropriate sample size and sampling strategy to obtain meaningful results, convergence of the estimates with increasing sample size is also recommended.

For a given problem setting, the concept of importance has to be defined rigorously, so that the sensitivity analysis results can be interpreted unambiguously and transparently (Saltelli and Tarantola, 2002). Although good practice in sensitivity analysis is gaining acceptance, the use of primitive sensitivity analysis methods is still widespread, mainly because of divisions between scientific disciplines and industrial sectors. However, among all the challenges and research needs which have been identified in the field of sensitivity analysis, techniques to help overcome the curse of dimensionality and to cope with dependent, spatially and temporally variable model inputs are the object of ongoing investigations.

References

Andres, T. (1997) Sampling methods and sensitivity analysis for large parameter sets, *Journal of Statistical Computation and Simulation*, **57**(1–4), 77–103.

Bettonvil, B. and Kleijnen, J. (1997) Searching for important factors in simulation models with many factors: sequential bifurcation, *European Journal of Operational Research*, **96**(1), 180–194.

Bolado, R., Castaings, W., Tarantola, S., Pagano, A. and Ratto, M. (2007) 'Estimation of the contribution to the sample mean and sample variance using random samples', in *Proceedings of International Symposium on Sensitivity Analysis of Model Output (SAMO 2007)*, Budapest.

Borgonovo, E. (2007) A new uncertainty importance measure, *Reliability Engineering & System Safety,* **92**(6), 771–784.

Breiman, L. (2001) Random Forest, *Machine Learning*, **45**(1), 5–32.

Breiman, L., Friedman, J., Olshen, R. and Stone, C. (1984) *Classification and Regression Trees*, California: Wadsworth and Brooks/Cole.

Cacuci, D.G. (1981) Sensitivity theory for nonlinear systems. I. Nonlinear functional analysis approach, *Journal Mathematical Physics*, **22**(12), 2794–2802.

Cacuci, D.G. (1981) Sensitivity theory for nonlinear systems. II. Extensions to additional classes of responses, *Journal Mathematical Physics*, **22**(12), 2803–2812.

Cacuci, D.G. (2003) *Sensitivity and Uncertainty Analysis, Volume I Theory*, London: Chapman & Hall/CRC.

Campolongo, F., Cariboni, J. and Saltelli, A. (2007) An effective screening design for sensitivity analysis of large models, *Environmental Modelling & Software*, **22**(10), 1509–1518.

Cao, Y., Hussaini, M.Y., Zang, T. and Zatezalo, A. (2006) A variance reduction method based on sensitivity derivatives, *Applied Numerical Mathematics*, **56**(6), 800–813.

Chun, M., Han, S. and Tak, N. (2000) An uncertainty importance measure using a distance metric for the change in a cumulative distribution function, *Reliability Engineering & System Safety*, **70**(3), 313–321.

Cukier, R., Levine, H. and Shuler, K. (1978) Nonlinear sensitivity analysis of multiparameter model systems, *Journal of Computational Physics,* **26**, 1–42.

Frey, H.C. and Patil, S.R. (2002) Identification and review of sensitivity analysis methods, *Risk Analysis*, **22** (3), 553–578.

Griewank, A. (2000) *Evaluating Derivatives: Principles and Techniques of Algorithmic Differentiation*, Philadelphia: Society for Industrial and Applied Mathematics.

Helton, J.C., Davis, F. and Johnson, J. (2005) A comparison of uncertainty and sensitivity analysis results obtained with random and Latin hypercube sampling, *Reliability Engineering & System Safety,* **89**(3), 305–330.

Helton, J., Johnson, J., Sallaberry, C. and Storlie, C. (2006) Survey of sampling-based methods for uncertainty and sensitivity analysis, *Reliability Engineering & System Safety,* **91**(10–11), 1175–1209.

Helton, J.C. (1993) Uncertainty and sensitivity analysis techniques for use in performance assessment for radioactive waste disposal, *Reliability Engineering & System Safety,* **42**(2–3), 327–367.

Homma, T. and Saltelli, A. (1996) Importance measures in global sensitivity analysis of nonlinear models, *Reliability Engineering & System Safety*, **52**(1), 1–17.

Hornberger, G. and Spear, R. (1981) An approach to the preliminary analysis of environmental systems, *Journal of Environmental Management*, **7**, 7–18.

Iman, R.L and Hora, S. (1990) A robust measure of uncertainty importance for use in fault tree system analysis, *Risk Analysis*, **10**(3), 401–406.

Iman, R.L. and Conover, W.J. (1982) A distribution free approach to inducing rank correlation among input variables, *Communications in Statistics*, **B11**(3), 311–334.

Isukapalli, S.S., Roy, A. and Georgopoulos, P.G. (2000) Efficient sensitivity/uncertainty analysis using the combined stochastic response surface method and automated differentiation: application to environmental and biological systems, *Risk Analysis*, **20**(5), 591–602.

Jacques, J., Lavergne, C. and Devictor, N. (2006) Sensitivity analysis in presence of model uncertainty and correlated inputs, *Reliability Engineering & System Safety*, **91**(10–11), 1126–1134.

Kleijnen, J.P.C. and Helton, J.C. (1999) Statistical analyses of scatterplots to identify important factors in large-scale simulations, 1: Review and comparison of techniques, *Reliability Engineering & System Safety*, **65**(2), 147–185.

Kleijnen, J.P.C. and Helton, J.C. (1999) Statistical analyses of scatterplots to identify important factors in large-scale simulations, 2: Robustness of techniques, *Reliability Engineering & System Safety*, **65**(2), 187–197.

Kleijnen, J.P.C. (2005) An overview of the design and analysis of simulation experiments for sensitivity analysis, *European Journal of Operational Research*, **164**, 287–300.

Kleijnen, J.P.C. (2007) *Design and Analysis of Simulation Experiments*, New York: Springer.

Kruskal, W.H. and Wallis, W.A. (1952) Use of ranks in one-criterion variance analysis, *Journal of the American Statistical Association*, **47**(260), 583–621.

McKay, M.D. (1997) Nonparametric variance-based methods of assessing uncertainty importance, *Reliability Engineering & System Safety*, **57**(3), 267–279.

Mishra, S., Deeds, N.E. and RamaRao, B.S. (2003) Application of classification trees in the sensitivity analysis of probabilistic model results, *Reliability Engineering & System Safety*, **79**(2), 123–129.

Mohanty, S. and Wu, Y. (2001) CDF sensitivity analysis technique for ranking influential parameters in the performance assessment of the proposed high-level waste repository at Yucca Mountain, Nevada, USA, *Reliability Engineering & System Safety*, **73**(2), 167–176.

Montgomery, D.C. (2004) *Design and Analysis of Experiments* (6th edition), Chichester: John Wiley & Sons, Ltd.

Morris, M.D. (1991) Factorial sampling plans for preliminary computational experiments, *Technometrics*, **33**, 161–174.

Oakley, J. and O'Hagan, A. (2004) Probabilistic sensitivity analysis of complex models: a Bayesian approach, *Journal of the Royal Statistical Society*, **B66**, 751–769.

Pappenberger, F., Iorgulescu, I. and Beven, K.J. (2006) Sensitivity analysis based on regional splits and regression trees (SARS-RT) *Environmental Modelling & Software*, **21**(7), 976–990.

Park, C.K. and Ahn, K. (1994) New approach for measuring uncertainty importance and distributional sensitivity in probabilistic safety assessment, *Reliability Engineering & System Safety*, **46**(3), 253–261.

Ratto, M., Tarantola, S., Saltelli, A. and Young, P.C. (2004) 'Accelerated estimation of sensitivity indices using State Dependent Parameter models', in *Proceedings of the 4th International Symposium on Sensitivity Analysis of Model Output (SAMO) 2004*, Santa Fe.

Saltelli, A. (2002) Making best use of model valuations to compute sensitivity indices, *Computer Physics Communications*, **145**, 280–297.

Saltelli, A., Ratto, M., Tarantola, S. and Campolongo, F. (2005) Sensitivity analysis for chemical models, *Chemical Reviews*, **105**, 2811–2828.

Saltelli, A. and Tarantola, S. (2002) On the relative importance of input factors in mathematical models: safety assessment for nuclear waste disposal, *Journal of the American Statistical Association*, **97**, 702–709.

Saltelli, A., Tarantola, S., Campolongo, F. and Ratto, M. (2004) *Sensitivity Analysis in Practice: A Guide to Assessing Scientific Models*, Chichester: John Wiley & Sons, Ltd.*

Saltelli, A., Tarantola, S. and Chan, K. (1999) A quantitative model-independent method for global sensitivity analysis of model output, *Technometrics*, **41**(1), 39–56.

Saltelli, A., Chan, K. and Scott, E.M. (2000) (Eds) *Sensitivity Analysis*. Chichester: John Wiley & Sons, Ltd.

Sobol', I. (1967) On the distribution of points in a cube and the approximate evaluation of integrals, *Computational Mathematics and Mathematical Physics*, **7**, 86–112.

Sobol', I. (1993) Sensitivity analysis for non-linear mathematical models, *Mathematical Modelling and Computational Experiment*, **1**, 407–414.

Spear, R., Grieb, T. and Shang, N. (1994) Parameter Uncertainty and Interaction in Complex Environmental Models, *Water Resources Research*, **30**(11), 3159–3169.

Stein, M. (1987) Large sample properties of simulations using Latin hypercube sampling, *Technometrics*, **29** (2), 143–151.

Storlie, C.B. and Helton, J.C. (2008) Multiple predictor smoothing methods for sensitivity analysis: description of techniques, *Reliability Engineering & System Safety*, **93**(1), 28–54.

Tarantola, S., Gatelli, D. and Mara, T. (2006) Random Balance Designs for the Estimation of First Order Global Sensitivity Indices, *Reliability Engineering & System Safety*, **91** (6), 717–727.

Tarantola, S., Nardo, M., Saisana, M. and Gatelli, D. (2006) A new estimator for sensitivity analysis of model output: An application to the e-business readiness composite indicator, *Reliability Engineering & System Safety*, **91**(10–11), 1135–1141.

Wegman, E.J. (1990) Hyperdimensional Data Analysis Using Parallel Coordinates, *Journal of the American Statistical Association*, **85** (411), 664–675.

Xu, C. and Gertner, G. (2007) Extending a global sensitivity analysis technique to models with correlated parameters, *Computational Statistics & Data Analysis*, **51** (12), 5579–5590.

---

*Also suitable for beginners.

19

Presentation in a deterministic format

This chapter discusses a practical aspect of industrial uncertainty treatment. Most regulations or internal decision-making processes have to balance the pros and cons of determinism or probabilism, as a fully and systematic probabilistic approach is rarely feasible. The chapter will not add to the debate on the issue of the acceptability of probabilistic risk regulation, subject of a large literature (e.g. Bedford and Cooke, 2001); it is more practically a short summary of certain practices and a review of key questions which arise when using partial deterministic formats, such as safety factors. Three case studies from Part II use a (partial) deterministic representation of the results. In *Dyke reliability* this has to do with the traditional use of safety factors in dyke design. In *Spent nuclear fuel behaviour* and *Electromagnetic interferences in aircraft*, results involving extremely low exceedance probabilities have to be presented.

Partial safety factors have become a popular tool to deal with uncertainties without having to perform full probabilistic calculations (Ciria, 1977). The practitioner does not necessarily require a great deal of knowledge of the uncertainties in order to deal with them efficiently: if a convincing probabilistic model is available, it may be used, and otherwise an *a priori* penalized value will be employed (Section 19.1).

Beyond the particular domain of structural design, problems may arise in any uncertainty study involving very low exceedance probabilities as the quantity of interest. As it is difficult to communicate results including very low probabilities, or because they cannot easily be considered reliable (the relevance of the distribution tail model is important), the following options may be preferred:

- presenting the results in the form of an inequality (e.g. *probability* $< 10^{-9}$);

Uncertainty in Industrial Practice   Edited by E. de Rocquigny, N. Devictor and S. Tarantola,
© 2008 John Wiley & Sons, Ltd

- presenting the results in a deterministic format, without mentioning the calculated probability of failure, as, for example, in the safety factor (see below) – this avoids confusion about the accuracy and the applicability of the results;

- modifying the pre-existing model inputs by overestimating a load variable or underestimating a resistance variable – this approach is used in *Spent Nuclear Fuel Behaviour* and gives confidence in the result through the combination of a small probability and a deterministic safety margin.

19.1   How to present in a deterministic format?

Many (structural) codes are based on quantiles, with an additional deterministic margin to arrive at a 'safe' output variable of interest. This additional deterministic margin is called a *safety factor*; if deterministic margins are allocated to several model inputs, these coefficients are called *partial safety factors*. In practical implementation a number of complications have to be taken into account, this is why a number of generally accepted rules have been developed, which together form a 'code theory'. If a partial safety factor is defined for every random variable, the number of factors becomes far too large. The total number of factors has to be restricted by bundling variables and by calculating one partial safety factor for them all. The magnitude of the partial safety factors depends on the standard deviation of all of the base variables that occur in the reliability function. It is therefore not possible to determine a safety factor for a base variable which is independent of the system model. The factors in the regulations are therefore averages of a large number of reference cases (CUR, 1997). There is, however, no single way to translate the probabilistic results into safety factors, since different sets of factors may lead to the desired safety level. The 'optimal' set of factors is case-specific and may depend on historical and cultural factors.

19.1.1   (Partial) safety factors in a deterministic approach

A code defines the number of safety factors, their position in the design equations and the rules of the load combinations (permanent load, accidental load) which will be used in the checking stage. The various failure modes of the component are indexed, and each one is associated to a limit criterion G beyond which the component no longer satisfies the design requirements. Within the common framework of the book, $G(\underline{x})$ may be considered as the pre-existing system model and $G(\underline{x}) < 0$ may be viewed as a decision criterion under a deterministic setting; by convention $G(\underline{x}) < 0$ represents the failure domain. The checking of the design is done by computing the limit criterion value for the design values $\underline{x}_d$ and comparing the result with 0:

$$G(x_k^1 \cdot \gamma^1, x_k^2 \cdot \gamma^2, \cdots, x_k^N \cdot \gamma^N) \geq 0$$

where these design values have been assessed by assigning safety factors γ to the characteristic values $\underline{x}^k$ of the model inputs $\underline{x}_d = (x_k^1 \cdot \gamma^1, x_k^2 \cdot \gamma^2, \cdots, x_k^N \cdot \gamma^N)$.[1]

In the design case, the deterministic approach involves in finding the value of one (or several) fixed model inputs d of design (for example, pipe diameter, thickness, etc.), in such a way that the system model G equals 0 when all other model inputs equal their characteristic values multiplied by safety factors. In practice, it implies solving the following equation to obtain the value of dimensioning d_d of the model parameter d:

$$G\left(d_d, x_k^1 \cdot \gamma^1, x_k^2 \cdot \gamma^2, \cdots, x_k^N \cdot \gamma^N\right) \geq 0$$

Using safety factors in a deterministic approach underlies the assumption of a monotonic system model. In the operation of existing structures or systems, two types of studies are possible:

- The first consists in finding the value of one (or several) parameters d (often a flaw size or a load), in such a way that the system model G equals 0 when all other model inputs equal their characteristic values x_k multiplied by safety factors. Structure checking then consists in comparing this computed value with the values of the model inputs, observed on the in-service component, then taking corrective measures if necessary (for example, repairing a defect or decreasing the loading).

- The second consists in introducing the observed value of the model input into the rule and checking that this is strictly respected. The disadvantage of this method is that it does not estimate the limit value for the parameter and thus does not recognize the difference between the observed and the limit value; a probabilistic approach therefore seems more suitable and justifiable, owing to the operational feedback it provides.

Comments: The relevance of the use of 'quantiles' for the characteristic values depends on the quality of information on the probability distribution. When the data sample is small, the penalizing character of the value selected (also called the 'characteristic value') may be difficult to justify. Note that this so-called 'deterministic' approach may be seen in fact as a mixed deterministic-probabilistic setting when such probabilistic quantities are used for setting the characteristic values.

19.1.2 Safety factors in a probabilistic approach

Shifting the uncertainty setting to a more probabilistic one, which is the subject of safety factors in a probabilistic approach, appears more attractive to deal with the problem in many cases. This approach makes it possible, by means of parametric

---

[1]Usually, a partial safety factor is greater than 1. In this book the same convention is kept for a safety factor for a load variable and for a resistance variable, by using multiplication of the partial safety factor for load variables and division for resistance variables.

studies, to assess the impact of the choice of the probabilistic model on the risk. And if new information is available, Bayesian theory, which associates data and expertise (often referred to as 'objective' and 'subjective' information respectively), allows for a relevant updating of the probabilistic model and the results of the probabilistic study, as mentioned in Chapter 16.

Two kind of methods are usually used for assessing safety factors:

- The design point method based on FORM (see Chapter 17) described in detail by Thoft-Christensen and Baker (1982) and Ditlevsen and Madsen (1996);

- Global optimization methods more suitable for a rule design including a set of situations (Thoft-Christensen and Baker, 1982; Hauge, 1992; Sorensen, 1994; Ditlevsen and Madsen, 1996; Melchers, 1999).

Example: Application to piping

Consider the failure mode associated with tear instability of piping which has a flaw emerging in internal skin. The multiplicative margin with respect to the risk of instability is defined by:

$$M = \sqrt{J_{\Delta a}/J(a + \Delta a)}$$

where the tear strength $J_{\Delta a}$ of the material is a function of flaw size a and energy $J_{0.2}$, while the fissuring force $J(a+\Delta a)$ is function of a, of stress σ_∞, of elastic limit σ_Y and of Young modulus E. Failure by instability occurs for $M \leq 1$, and the associated limit function (i.e. the system model) is:

$$G = M(a, J_{0.2}, \sigma_\infty, \sigma_Y, E) - 1$$

Operating conditions lead to different values of the known stress, ranging between 200 and 250 MPa. Throughout those operating conditions, the failure probability should be lower or equal to 10^{-6}, which is translated by a reliability index equal to or higher than 4.75. In the code, a partial safety factor on the variables a, σ_∞, $J_{0.2}$ and σ_Y will be used. Table 19.1 gives the modelling of the uncertain model inputs.

Figure 19.1 shows the evolution of the partial safety factors as a function of the loading, obtained by the design point method; Figure 19.2 compares the reliability index, function of stress, as obtained by an optimisation method (see Curve B in Figure 19.2) and a method based on the set of partial safety factors obtained by the design point for the maximal load (see curve A in Figure 19.2). This last method leads to a lower reliability for the other situations This result is explained by the relative decrease of the influence of the defect height when the load increases. So the lower value of the factor obtained with 250 MPa means it is no longer possible to ensure the reliability necessary for the other loads. This example illustrates the need for an optimization approach when several situations (i.e. varying operating conditions or controlled scenario inputs encapsulated in $\underline{d}$) have to be covered.

Table 19.1 Uncertainty modelling of the uncertain inputs.

Model input	Variable	Unit	Distribution	Mean	Standard deviation or Coefficient of Variation (CV)	Characteristic values
Defect size	a	m	Log-normal	$\mu(a)$	CV = 20%	$\mu(a)(1+CV_a) = 1.2\mu(a)$
Stress	σ_∞	MPa	Log-normal	$\mu(\sigma_\infty)$	CV = 10%	$\mu(\sigma_\infty)(1+CV_{\sigma\infty}) = 1.1\mu(\sigma_\infty)$
Energy	$J_{0.2}$	MN/m	Log-normal	0,109	0,033	$7,710^{-2}$ (fractile 5%)
Elastic limit	σ_Y	MPa	Truncated Normal(+/− 2 standard deviation)	212	16	188,44 (fractile 15%)
Young modulus	E	MPa	Log-normal	191000	10000	No safety factor on this input.

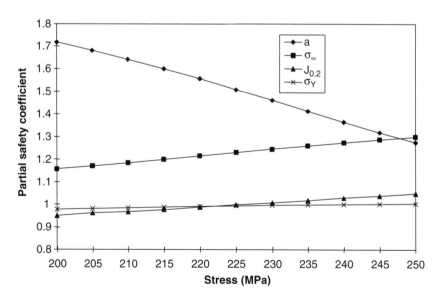

Figure 19.1 Partial safety factors according to the loading, by the design point method.

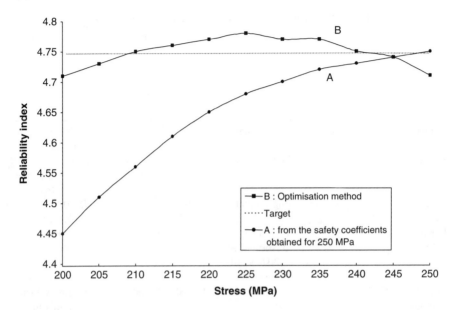

Figure 19.2 Comparison of the reliability indices obtained by various calibration methods according to the loading.

19.2   On the reliability target

In the deterministic approach, a component is designed by use of margins to avoid a failure, although lacking a full quantification of the reliability guaranteed. On the other hand, in a probabilistic approach it is necessary to accept a certain explicit risk level by associating to it a maximum acceptable failure probability P_f^t. From this probability, a target reliability index β^t is defined by:

$$\beta^t = -\Phi^{-1}(P_f^t)$$

The choice of the target reliability must take into account the characteristics of the system model and the nature of the phenomena involved: slow degradation or collapse, etc. A component likely to fail suddenly must be designed with a higher degree of reliability than a component having a ductile mode of failure for which corrective action could be taken in good time. The choice should also take into account the human and economic consequences of failure, as suggested by ISO 2394. The NKB (Nordic Committee on Building Regulations) recommends the use of various target probabilities according to the consequences of the failure (not very important, important, very important) and the failure mode (ductile, fragile, instable). For offshore platforms Thoft-Christensen and Baker (1982) propose, for the probability of failure targets P_f^t, taking the weighted average $\sum_{j=1}^{J} \omega_i P_f^i$ of the failure probabilities P_f^i of the components of the structure, in which the weights ω_i

represent the use frequency of each component included in calibration, such that $\sum_i \omega_i = 1$.

19.3   Final comments

Calibration methods determine a degree of freedom in the problem, the so-called pivot variable, in order to bring the component reliability closer to the target reliability. In a design study, the pivot variable may simply be a geometrical dimension of the component, a material property or a load variable. In in-service checking, choice of the pivot variable is more limited: often a flaw size, and sometimes a load variable. Consequently, the set of safety factors to be optimized in the deterministic code is limited to the choice carried out for the pivot variable. On the other hand, assuming that the safety factors were obtained through a realistic uncertainty model, conservatisms should not be introduced during the application of the code, for example, regarding the characteristic values.

Methods based on an optimization procedure do not usually have a unique solution. Therefore, different sets of partial safety factors could be obtained, a fact which grants a certain flexibility to the study. Using constraints in the optimization problem is a way to select certain solutions from among all the options. When FORM or SORM is used, it is necessary to validate the approximation of the probability, using, for instance, a Monte Carlo method as exemplified by *Dyke reliability*.

Regarding the pros and cons of deterministic or probabilistic presentations of uncertainty treatment, a conventional opinion, already mentioned above, is that the deterministic approach does not clearly give evidence to a full quantitative reliability estimate: it may be seen both as having practical advantages, in so far it avoids debates on explicit risk-level acceptability, and drawbacks, as it lacks any quantitative figure amenable to comparison with other competing risk or uncertainty situations. It is, however, worth bearing in mind that the explicit probabilistic reliability figure generated by a probabilistic approach may also be misleading in so far it completely depends on the uncertainty model developed on the inputs.

In relation to existing structures or systems it may be seen merely as a rectifiable *modelling* issue; concerning design codes for future systems it may involve uncontrollable deviations between the code input uncertainties and real material or loading values in future structures. Full control of the probabilistic reliability would then require a careful study of the variety of uncertainty models that may characterize the future structures, and a certain quantitative anticipation of the deviations in code application which may increase if more elaborate methods require more data and complex calculations. This may in fact generate multiple probabilistic levels, as in *Dyke reliability*, to ensure for instance that 95% of designed structures have a reliability greater than a given target. Note finally, as mentioned in Chapter 14 on the natures of uncertainty, that probabilistic or deterministic modelling of uncertainty also involves epistemological considerations; some may consider that the legitimacy of probabilistic modelling differs according to the nature of uncertainty sources. Care is therefore recommended in comparing the pros and cons of so-called 'deterministic' and 'probabilistic' approaches, particularly in codes applicable to future design practice rather than to existing facilities.

References

Bedford, T. and Cooke, R. (2001) *Probabilistic Risk Analysis – Foundations and Methods*, Cambridge: Cambridge University Press.

Ciria (1977) *Ciria Report 63: Rationalisation of Safety Factors in Structural Codes* (Available online at: www.ciria.org.uk).

CUR (1997) *Cur Report 190: Probabilistic in Civil Engineering*, CUR, Gouda, ISBN 90 376 0102 2.

Ditlevsen, O. and Madsen, H.O. (1996) *Structural Reliability Methods*, Chichester: John Wiley & Sons, Ltd.

Hauge, L.H. (1992) 'Optimal code calibration and probabilistic design', in *Proceedings of OMAE '92*.

Melchers R.E (1999) *Structural Reliability Analysis and Prediction* (2nd edition), Wiley.*

Sorensen, J.D. (1994) Code calibration as a decision problem, *Structural Safety & Reliability: Proceedings of ICOSSAR '93*, Innsbruck.

Thoft-Christensen, P. and Baker, M. J. (1982) *Structural Reliability Theory and Its Applications*, New York: Springer-Verlag.

---

*Also suitable for beginners.

20

Recommendations on the overall process in practice

After having reviewed the panel of applicable methods in previous chapters of Part III, this chapter aims to give some final recommendations on the choice of certain elements in the overall framework, bearing in mind factors such as data or expertise availability, computing budget, modelling manpower budget, time constraints, decision context and industrial processes.

20.1 Recommendations on the key specification step

The common framework presented in this book has given a great deal of attention to the pre-requisites of the specification step: the fact that a pre-existing system model $\underline{z} = G(\underline{x}, \underline{d})$ is available, the predominant final goal of a study, an explicit formulation of the uncertainty setting and the quantities of interest. In industrial practice, however, it often appears that those features are not so easy to identify: this is especially the case when initiating an uncertainty assessment process within an industrial sector where there is little or no tradition or regulatory guidelines regarding uncertainty. Formal decision criteria explicitly integrating uncertainty appear to be even rarer.

A complete process may then be necessary to specify clearly the final goal and quantities of interest, or even a final decision criterion. Two important practical features may guide the launch of the process:

Uncertainty in Industrial Practice   Edited by E. de Rocquigny, N. Devictor and S. Tarantola,
© 2008 John Wiley & Sons, Ltd

- simplicity of interpretation of the study for non-specialists;
- global cost affordability (data, computing, expertise, modelling, etc.).

Remember that the costs involved in an uncertainty study depend on many factors that are not often thoroughly considered: data collection (for uncertainty modelling), computing cost involved in propagation and sensitivity analysis; development, validation or simplification of the pre-existing model, which often encounters renewed criticism when the uncertainty assessment is launched, etc. Recommendations on the key specification steps of an uncertainty study will be given in the following, namely: choice of the system model; choice of the most convenient uncertainty setting; definition of the quantities of interest (q.i.); and separation of the model inputs into uncertain factors $\underline{x}$ and fixed factors $\underline{d}$.

20.1.1   Choice of the system model

Reference was made throughout the book to a *given* pre-existing model of the system: in practice there may be some flexibility at this initial step of specification of the study. If the analyst has to build a new model, has a choice of several existing models or can adapt the given model to his needs (e.g. by a close link to the model developer), some recommendations can be given. Especially in the case when the final goal of the study is Goal U (Understand) or Goal A (Accredit), a simple system model will be easier to handle. More complex models need higher computing costs and are often not easily made suitable for propagation and sensitivity methods. Moreover, complex models tend to have a higher dimension of x. As the amount of information available is often of a fixed and limited size, a smaller system model often allows for a better quantification of $\underline{x}$. During the study the model can be refined and may gain in complexity. The analyst should also keep in mind that although in the current situation the amount of information may be low, it may increase during the uncertainty study.

In fact, the study itself may even trigger further collection of information that will refine the model.

20.1.2   Choice of the uncertainty setting

The existing decision process may implicitly impose the setting; for example, asking for confidence intervals or for a 'worst-case value' in a variable of interest implies a probabilistic or deterministic setting, respectively. Otherwise, the most important choice in the uncertainty study is probably the selection of the uncertainty setting. Throughout Part II of this book, level-1 and level-2 settings with deterministic, probabilistic or DST representations were used. The selection is influenced by several aspects, however, the most obvious and most important is the following: *all study results are only correct conditional on the relevance of the uncertainty representation.* The first and foremost principle is therefore:

1. Select the setting which best represents the input information

If the choice of a distribution is straightforward or can easily be obtained from data, a probabilistic setting is adequate. If the input distribution is not known except in its mere range, a deterministic approach can be useful. If neither bounds nor distribution are known with an acceptable accuracy, an option may be to move to a level-2 setting (deterministic, probabilistic or possibly DST, for instance). If intervals are known for some sources, while distributions can be assumed for others, DST could be experimented as a combined representation, although it gets more complex.

If rule 1 does not lead to an obvious choice, it is important to bear in mind that results have to be computed and communicated. This leads to the second rule of thumb:

2. Select the simplest setting which represents the uncertainty adequately

A standard setting is easier to handle (data, computing cost) and to communicate than a two-level setting. Probabilistic settings are easier to communicate than extra-probabilistic (e.g. DST) settings.

Beyond these two main aspects, there are other, more specialized points that can support the decision. A common decision is the choice of a deterministic or a probabilistic setting. First, it is necessary to look at rule 1. In a probabilistic setting, all statistical confidence is conditional on the validity of the input distribution. Conversely, deterministic settings are only meaningful conditional on the accuracy of bounds of the inputs. *The setting to choose therefore depends mainly on the information available.*

Moreover, the choice of the setting depends on the characteristics of the pre-existing model. In the case of monotonic functions $G(\ldots)$, a deterministic or a DST setting is fast and produces exact results. However, if $G(\ldots)$ is a highly non-linear model with many local optima, obtaining the global optimum with a high level of confidence is a much more challenging task. While optimization algorithms are capable of finding a near-optimal result, it is not possible to express a degree of certainty in the fact that the global optimum has been discovered. In contrast to this, a probabilistic approach evenly explores the whole input space without particularly concentrating the sampling on the regions where the global maximum is to be found. However, confidence values in the quantity of interests and the uncertainty generated in the propagation and estimation step can generally be given.

Caution is recommended when using probabilistic inputs in a purely deterministic setting or vice versa. If, for example, 5%–95% fractiles on input variables are used for the bounds of a deterministic uncertainty model, the results at the output are conditional to a possibly-complex combination of input confidence levels represented by each individual fractile. Conversely, a deterministic input range is not completely equivalent to the inference of a uniform distribution inside this range.

Another practical aspect concerns the maturity of the available methods and the capabilities of further utilization of the results. Beyond the traditional deterministic setting, sensitivity analysis and decision theory are highly developed in probability theory. In DST, both fields are still virtually non-existent.

20.1.3   Choice of the quantity of interest

In some studies the decision process clearly defines the q.i.. For instance, if Goal C (Comply) is at stake, the uncertainty study may require a demonstration that a *probability of exceedance* or a *coefficient of variation* has stayed below a regulatory limit. But the situation is not always so clear. For Goal S (Select), several q.i. can be used in the comparison: for instance, two maintenance or design scenarios can be compared through the expectation, variance or quantile of a cost, or by computing the probability that one is cheaper than the other. This also applies to Goal U (Understand) and Goal A (Accredit), which are the most common goals encountered when dealing with design in early stages.

The choice of the q.i. should take into account some practical constraints. In a probabilistic setting, the assessment of a very low probability of exceedance or an extreme quantile often turns out to be strongly sensitive to the uncertainty model inputs. A good level of confidence in the uncertainty model inputs is then required, and consequently the available information (data samples and expertise) has to be sufficiently rich. This may be problematic, given the degree of knowledge available at early stages, in which information is scarce and the uncertainty model is often defined somewhat roughly (e.g. by a triangular probability distribution based on bounds and a most likely value provided by experts). Thus, the estimate of a low exceedance probability should be received with caution: the order of magnitude may be of interest, but the numerical value should not be thought of as accurate. On the other hand, a q.i. based on central dispersion – such as variance, coefficient of variation, mean or median – is generally less sensitive to the uncertainty model. Thus, a practical piece of advice would be to *start with a q.i. concerned with central dispersion if the available information is scarce, provided, of course, that it fits the decision process*.

Note that among the quantities of interest involving central dispersion, one should pay attention to the difference between two features that are often thought to be very similar: the mean of a probability distribution of the v.i., and its median value. These two quantities are equal in the very particular case in which the v.i. is symmetrically distributed (e.g. Gaussian distribution), but there may be a great gap between the two values (up to several orders of magnitude) when the probability distribution of the v.i. is non-symmetric. This is the case, for example, in *Nuclear waste repository*, including lognormal-like distributions. The choice between mean and median as the relevant q.i. then requires further analysis, involving decision theory considerations.

But even for a 'robust' quantity such as central dispersion, it is crucial to remember that its use at later stages should be cautious. For instance, taking the mean value plus 1.64 times the standard deviation is quite often erroneous if intending to assess a 95% quantile: it is only correct if the v.i's distribution is Gaussian, and may lead to totally inaccurate results otherwise. In other words, *mean and variance provide useful insights, but cannot generally be used to compute q.i. such as exceedance probabilities and quantiles* (at least without an arbitrary assumption on the probability distribution type, which is rather risky). Moreover, note that variance is meaningful mainly for close-to-symmetric distributions; otherwise, quantiles or higher-order moments should be used to handle dispersion (e.g. with a graphical representation such as 'box and whisker' plots).

20.1.4   Choice of the model input representation ('$\underline{x}$' and '$\underline{d}$')

Part I already mentioned that some pre-existing model inputs are considered uncertain ($\underline{x}$) while others can be fixed ($\underline{d}$). The borderline between these two categories is often subjective: *it is more a modelling choice than a theoretical distinction*, and even inputs that are acknowledged to be affected by uncertainty (such as complex scenarios or design characteristics) may sometimes be conventionally fixed, i.e. considered as part of $\underline{d}$.

In fact, the decision depends on the amount of data. For instance, there may be a point at which it is not even possible to model the uncertainty of a model input. This may be the case when no data is available and expertise is too imprecise. *If the timescale of the study is too short to gather more knowledge (data and/or expertise), then, rather than choosing an uncertainty model that might be quite unreliable and distort the final results, a reasonable approach is to choose a set of fixed values or scenarios.*

One of the values or scenarios may be designed to be 'penalized' when risk is at stake. External review is then clearly useful to validate these choices. But in any case, the model input considered, even though objectively uncertain, is afterwards considered as pertaining to a set of fixed values within vector $\underline{d}$.

The study is then carried out conditional to a given scenario. This has, of course, to be borne in mind in the interpretation of the study results: the order of magnitude of an exceedance probability does not have the same meaning when the conditioning scenario considered is 'penalized' rather than 'ordinary'. More generally, *results are tightly dependent on the values chosen for the vector $\underline{d}$, and to the delineation that has been chosen between $\underline{x}$ and $\underline{d}$.* This is true for all the q.i. considered (exceedance probabilities, minimum/maximum, quantiles, variance, etc.), the interpretation of which always change conditional on $\underline{d}$.

20.2   Final comments regarding dissemination challenges

As was stated in Chapter 13, new attitudes to uncertainty and a broadening of the scope of its analysis are now called for, beyond the traditional engineering approaches.

Decision-makers in industry or public authorities would benefit from acknowledging the importance of making uncertainty more explicit. Indeed, if uncertainty were adequately modelled and controlled, it would not be seen as necessarily negative, but rather as offering the possibility of positive outcomes: better designs, more accountable regulation or even market opportunities. The recognition that uncertainty management can be helpful – one of the fundamental lessons of financial option analysis – may signal a real move away from exclusively failure-driven risk management and from the idea that uncertainty is only connected to negative events. Indeed, at present, the rule is often 'the less we know, the more margins are taken'.

Dissemination of new ways of thinking certainly requires extensive communication and training, hopefully with the support of tools, concepts and methods presented in the course of this book. Favouring simple representations of uncertainty will obviously facilitate this dissemination. More thought about the value generated by uncertainty reduction, while bearing in mind the maximal investment that might be made, could go some way towards convincing decision-makers and ranking priorities. Reference could usefully be made to research on decision theory under uncertainty, and particularly on considerations on the expected value of information (see, for instance, Ang and Tang, 1984; Granger Morgan and Henrion, 1990).

References

Ang, A.H.S. and Tang, W.H. (1984) *Probability Concepts in Engineering, Planning and Design* (two volumes), Chichester: John Wiley & Sons, Ltd.

Granger Morgan, M. and Henrion, M. (1990) *Uncertainty – A Guide to Dealing with Uncertainty in Quantitative Risk and Policy Analysis*, Cambridge: Cambridge University Press.

Conclusion

The purpose of this book, the joint effort of a European network of industries and academics, was to develop an industrial reference for quantitative uncertainty management, so as to contribute to the dissemination of best practice in the worlds of engineering and industrial decision-making and regulation. The subject is currently gaining in strategic interest, due both to evolving regulatory demands (e.g. in safety, security or environmental control in certification or licensing processes) and to market needs (e.g. industrial process optimization or business development). Greater investigation is leading to a better understanding of the margins of uncertainty, possible ways of reducing them, and associated risks and opportunities.

Ten industrial case studies were developed to inspire thought (Part II), all structured according to a common methodological framework introduced in Part I and developed in Part III. While these studies, and most other similar undertakings, understandably use the domain-specific terminology and conventions of their real-world contexts, the common framework emerges as a generic foundation for the study of uncertainty. More importantly, the common framework should help to address the key questions that should be asked in any such study. The best practical trade-offs between the many competing methods and practical constraints prove to depend essentially on the initial step of the proper *specification of the study*, according to the real goals pursued (Understand/Accredit/Select/Comply). A variety of settings (i.e. mathematical paradigms to represent uncertainty), such as deterministic, probabilistic or extra-probabilistic, could be used within the same general framework. Note also that *combining* deterministic and probabilistic uncertainty settings generally proves to be more fruitful than putting them in opposition to each other.

In later stages, the appropriateness of various methods in the key steps of *uncertainty modelling, propagation and sensitivity analysis* depends essentially on the setting, the quantity of interest (a figure summarizing the output uncertainty), the information available (data or expertise) and the features of the system or pre-existing model (CPU per run, regularity, dimensionality of the model inputs), as explained in Part III.

Finally, the *feedback process* is indispensable to the industrial decision-maker and should not be neglected by the technical analyst. Could the uncertainty in the model output be reduced through better modelling and/or more data collection? *Should* it be reduced? Should the design be altered or the scenario modified? Could there be a simpler compliance process, beyond the reference study? Should the results be presented in a simplified format? and so on.

Further research, going beyond the rather mature methodologies and techniques explored in this book, would naturally be beneficial. In the light of the various cases studies and representations made by the authors, it would seem to be in the best interest of industry to support research on, for instance: high-performance computing (both through improved numerical algorithms and tools, and increased computing power), extended imprecise probability settings, model input dependency issues, sensitivity analysis adapted to large input dimensionalities and to quantities other than classical variance, etc.

The attempt to improve uncertainty treatment and management practices in a corporate manner poses a number of challenges, such as:

- An *information challenge*: a much more thorough data-collection process and cultivation of expertise will be necessary to qualify and assess the models (some of which have been designed on the basis of past experiments or by scientists no longer working in the field);

- A *scientific computing challenge:* high-performance computing and the coupling of dedicated uncertainty treatment tools to large multi-physics codes will soon be required;

- A *cultural and organizational challenge:* progress will depend on internal industrial capabilities and environments (and potentially on the external supply chain), while also remaining heavily dependent on regulatory developments.

Cultural awareness is generally the crucial limiting factor in the development of industrial best practice and possibly also in regulatory innovation. Extensive academic and industrial training and development may be required to facilitate the emergence of new disciplines and expertise to support evolving analytical needs.

To summarize, principles of good practice which may be derived from this book are captured in the following key messages:

1. Be conscious of uncertainty: it carries significant *risks and opportunities*.

2. Making uncertainty more *explicit* leads to more accountable decisions.

3. A *generic* approach is now available: the case studies illustrate its practical applicability.

4. *Mature* methods and tools can now address the majority of industrial needs.

5. Clarify the real *goals* and information available before beginning numerical modelling.

6. Keep it as *simple* as possible: for a given degree of information (data and expertise), overly-complex physical or statistical modelling reduces control over uncertainty.

7. A *multi-disciplinary* approach is necessary, combining statistics and numerical analysis, but also domain expertise in the system model.

8. The value of data collection and expertise-building *increases* in the presence of uncertainty.

9. Uncertainty management requires *corporate industrial investment* in training, high performance computing and strengthening of the knowledge base.

'Quantifying the unknown', a paradoxical activity perhaps, should ultimately conclude modestly . . .

Abbaye des Vaulx de Cernay

Appendices

Appendices

A

A selection of codes and standards

A limited number of codes and standards may be found regarding quantitative uncertainty treatment; there will be briefly reviewed in the following section, although the list is not extensive:

Eurocodes (http://www.eurocode1.com/en/index.html): set of ten European Standards that contain common structural rules for the design of buildings and civil engineering structures (EN 1990 Basis of structural design; EN 1991 Actions on structures; EN 1992 Design of concrete structures; EN 1993 Design of steel structures; EN 1994 Design of composite steel and concrete structures; EN 1995 Design of timber structures; EN 1996 Design of masonry structures; EN 1997 Geotechnical design; EN 1998 Design of structures for earthquake resistance; EN 1999 Design of aluminium structures)

ISO (1995) *Guide to the Expression of Uncertainty in Measurement* (G.U.M.), EUROPEAN PRESTANDARD ENV 13005 (published in 1999).

Comment: the Technical Advisory Group on Metrology of ISO assumed the responsibility to develop a guide on how to conduct measurement processes, as well as how to report on them and their related uncertainties. The result was the ISO 'Guide to the Expression of Uncertainty in Measurement', originally published in 1993 and reviewed in 1999. This guide is commonly known as GUM and was eventually published as a CEN standard (CEN standard ENV 13005:1999). The most relevant points are the adoption of the quantities of interest 'standard deviation' and 'enlarged uncertainty' as the main modes of expressing uncertainty about a measurement, the use of the Taylor approximation for the uncertainty propagation, and the classification of uncertainties as type A and type B, characterized by the use or lack of use of a statistical treatment for assessing the standard deviation.

Uncertainty in Industrial Practice   Edited by E. de Rocquigny, N. Devictor and S. Tarantola,
© 2008 John Wiley & Sons, Ltd

Owing to the limits of this method, the on-going revision of the GUM proposes to refer more often to Monte Carlo simulation in a forthcoming *Supplement*.

In fact, the approach described in GUM is consistent with the methodology introduced in this book. First, GUM's methodology is based on a pre-existing model (a metrological chain), and on quantities of interest defined in a given uncertainty setting. The 'standard deviation' is the q.i. in a standard level-1 probabilistic setting. The 'enlarged uncertainty' is another q.i. (which deals with coverage intervals), but which lies in fact in a level-2 probabilistic setting. The limited amount of information available for uncertainty modelling, which generates 'uncertainty about uncertainty', is taken into account in the enlargement factor, as was exemplified by the [CO_2 emissions] case study. Secondly, two of the major steps presented in this book – 'uncertainty modelling' and 'uncertainty propagation' – constitute the core of GUM, while the evolution of GUM on propagation, from Taylor approximation to MCS (as will presumably be presented in the *Supplement*), is consistent with the recommendations of Chapter 17 of this book. However, sensitivity analysis and, more generally, the feedback process are better highlighted in the present book. Some sensitivity measures should nevertheless be mentioned in the *Supplement*.

IPCC (2000), *Good Practice Guidance and Uncertainty Management in National Greenhouse Gas Inventories,* Intergovernmental Panel on Climate Change (Available online at: www.ipcc.ch).

Joint Committee on Structural Safety (2002), *JCSS probabilistic model code,* (Available online at: www.jcss.ethz.ch).

US (1997) Environmental Protection Agency. *Evaluating the uncertainty of emission estimates, Emission Inventory Improvement Project* (EIIP Volume VI – Chapter 4) (Available online at: www.epa.gov/ttn/chief/eiip).

USNRC (1990), *Severe Accident Risks. An assessment for Five US Nuclear Power Plants (Vols. 1, 2 and 3),* NUREG-1150.

USNRC (2002), *An Approach for Using Probabilistic Risk Assessment in Risk-Informed Decisions of Plant Specific Changes to the Licensing Basis*, Regulatory Guide 1.174, Revision 1.

Some specialized regulations give precise recommendations for the uncertainty treatment, for instance, the Directive 2004/107/EC of the European Parliament and the Council of 15 December 2004 relating to arsenic, cadmium, mercury, nickel and polycyclic aromatic hydrocarbons in ambient air, which refers explicitly to the G.U.M. EUROPEAN PRESTANDARD ENV 13005.

B

A selection of tools and websites

The following selection is subjective and not comprehensive; the tools for uncertainty and sensitivity analysis used by the authors have been retained:

- **@RISK**, a commercial add-on to Excel developed by Palisade (www.palisade-europe.com);
- **Crystal Ball**, a commercial add-on to Excel developed by Decisioneering (www.crystalball.com);
- **DAKOTA** (acronym for *Design Analysis Kit for Optimization and Terascale Applications*), developed by SANDIA Laboratories (www.cs.sandia.gov/DAKOTA); some components of this software is distributed under GNU General Public License;
- **LMS Optimus**, part of LMS.Virtual Lab dedicated to the robust design, developed by LMS International (www.lmsintl.com), a commercial software;
- **Open TURNS** (acronym for *Open Treatment of Uncertainties, Risks'N Statistics*), a software open source developed by a consortium from EDF, EADS and PHIMECA (www.openturns.org);
- **R** developed by the R Foundation for Statistical Computing (CRAN.R-project.org), and distributed under GNU General Public License v2;
- **Sunset**, developed by IRSN (www.irsn.fr), and distributed under an agreement with IRSN;
- **SUSA** (acronym for *A Program System for Uncertainty and Sensitivity Analysis*), developed by GRS (www.grs.de) and distributed under an agreement with GRS;

Uncertainty in Industrial Practice Edited by E. de Rocquigny, N. Devictor and S. Tarantola,
© 2008 John Wiley & Sons, Ltd

- **URANIE**, developed by CEA (www.cea.fr), and distributed under an agreement with CEA.

Two items of software dedicated to structural reliability, and also useful for the computation of an exceedance probability, have been added:

- **Phimeca Software**, commercial software developed by Phimeca (www.phimeca.com);

- **PROBAN** (acronym for Probabilistic Analysis), a commercial software developed by Det Norske Veritas (http://www.dnv.com/software/safeti/SafetiQRA/proban.asp);

- **Strurel**, commercial software developed by RCP GmbH (www.strurel.de).

The last item of software studied is the development framework provided for sensitivity analysis in the user working environment called **SimLab**, developed by the JRC (simlab.ec.europa.eu) and distributed under a Free License. This framework provides a set of Fortran, C++ and Matlab© subroutines dedicated to sensitivity analysis in a simple probabilistic framework. JRC offers also a forum dedicated to Sensitivity Analysis: http://sensitivity-analysis.jrc.cec.eu.int/.

The following three tables summarize the authors' understanding of the capabilities of these tools. For Tables B.2 and B.3, a case is marked if one method in the family is provided by the software. The reader should refer to the detailed documentation of the tool to obtain more information on the method implemented in the software.

For the extra-probabilistic framework, two other types of software exist:

- The **RAMAS Risk Calc** software, developed by Applied Biomathematics (www.ramas.com), provides some uncertainty and sensitivity tools in several frameworks: probabilistic, probability bounds analysis, fuzzy arithmetic, interval analysis, probability box, fuzzy Dempster-Shafer Theory. It is commercial software available in the Windows environment.

- The **Imprecise Probability Propagation (IPP) Toolbox**, developed by the Duisburg-Essen University, free for use (available at www.uni-due.de/il/software).

Because of a lack of practice with this tool, it could not be included in the following tables.

The following websites provide additional information on extra-probabilistic frameworks:

- Sandia workshop: http://www.sandia.gov/epistemic/;

- Society for Imprecise Probability: Theories and Applications: http://www.sipta.org/index.html.

Another interesting website on risk and uncertainty management is proposed by Matthew Leitch: http://www.managedluck.co.uk/basics/index.html.

Table B.1 General information on the selected tools.

	@RISK	Crystal Ball	DAKOTA	Open TURNS	LMS Optimus	Phimeca Software	PROBAN	R	SimLab	Strurel	Sunset	SUSA	URANIE
Release	4.5.6	7.2.1	4.0	1.0		2.5	4.4	2.4.1	3.0.8	7.0		3.5	1.0
Plateform availability	Win.	Win.	Win. Linux	Linux	Linux	Win. Linux	Win. Linux	Win. Linux	Win. Linux	Win.	Linux	Win.	Linux
Uncertainty framework	Probabilistic	Probabilistic	Probabilistic + hybrid	Probabilistic	Probabilistic	Probabilistic	Probabilistic	Probabilistic	Probabilistic	Probabilistic	Probabilistic + hybrid	Probabilistic	Probabilistic
Dependency modeling (correlations/ n-dimensional models …)	Ranks corr.	Ranks corr.	Rank corr./ Nataf/n-dimensional models	Rank corr./ Nataf/n-dimensional models		Nataf/ functional dependence	Rank corr./ Nataf/	Rank corr./ Nataf/n-dimensional models	Rank corr./ dependency Tree/ Elliptical copulae	Nataf et Hermite/ functional dependence	Ranks corr./ n-dimensional models	Pearson, Kendall and Rank corr./ functional, conditional, inequality	Ranks corr.
Coupling methods	DLL (as for Excel)	DLL (as for Excel)	Coupling by text files.	Dynamic library coupling or coupling by text files	Text files	Dynamic library coupling or coupling by text files	Coupling by text files/ Fortran language	With C and Fortran codes.	Library of Functions	Coupling by text files/ Fortran language	Coupling by text files/ Fortran language	Coupling by text files/ Fortran language	Dynamic library coupling or coupling by text files
Gradient computation provided			Yes by finite difference	Yes by finite difference		Yes by finite difference	Yes by finite difference	Yes by finite difference		Yes by finite difference	Yes by finite difference	Yes by finite difference	Yes by finite difference
Metamodel			Large possibilities via packages	Polynomial with degree ≤ 2	Polynomial	Polynomial for SORM	Polynomial	Large possibilities via packages		Polynomial for SORM		Polynomial	Polynomial + neural networks

Table B.2 Capabilities for uncertainty analysis (grey filling indicates the capability).

	@RISK	Crystal Ball	DAKOTA	Open TURNS	LMS Optimus	Phimeca Software	PROBAN	R	SimLab	Strurel	Sunset	SUSA	URANIE
Classical statistics													
Graphical method (histogram)*													
Sample fitting and Statistical tests													
Deterministic method (DOE)													
Numerical integration (4th moments)													
Taylor approximation													
Monte-Carlo simulation and V.R.T.†													
V.R.T. for exceedance probability													
Methods based on a stochastic development													
FORM/SORM													
Hybrid methods (FORM+V.R.T.)													

*OpenTurns proposes also Henry-line, pp-plot and qq-plot, and the capabilities of URANIE are the same than ROOT.

†LHS is not available for Phimeca Software and Strurel.

Table B.3 Capabilities for sensitivity analysis (grey filling indicates the capability).

	@RISK	Crystal Ball	DAKOTA	Open TURNS	LMS Optimus	Phimeca Software	PROBAN	R	SimLab	Strurel	Sunset	SUSA	URANIE
Elaborated Graphical method*		●							●		●	●	●
Differential methods			●	●	●			●					
Differential methods (From FORM)			●	●	●	●	●			●			
Screening					●								●
Sampling-based	●	●	●	●			●		●		●	●	●
Non-parametric statistics													
Variance-based Decomposition (correlations)								●	●		●	●	●
Variance-based Decomposition (Sobol, FAST)			●					●	●				●
Distribution sensitivity techniques													

*Cobweb for SimLab and SUSA, and capabilities from ROOT for URANIE.

C

Towards non-probabilistic settings: promises and industrial challenges

One way is to describe the uncertainty around the model input through a *probabilistic* distribution, assigning probability masses to points in this neighbourhood. Uncertainty calculation is then conducted in a probabilistic framework. A second, competing way is to describe the model input uncertainty using an interval, representing the range in which this model input can vary. In this *deterministic* framework, no distribution is assumed and the results are obtained by inferring the best and worst cases using optimization algorithms. Besides their distinct theoretical foundations and decision-theory interpretations, both approaches have some practical advantages and disadvantages.

In a probabilistic setting, the shape of the distribution has a strong influence on the result. In a deterministic setting, a distribution is not necessary, but the results may prove very conservative. The Dempster-Shafer Theory of Evidence (DST) can be considered in the light of an attempt to develop synergy between the two frameworks, allowing uncertainty quantification with probabilistic methods, distribution-free intervals and mixtures of both (Ferson *et al.*, 2003). This Appendix aims only to give a brief introduction to DST, a relatively new uncertainty framework still under development but perhaps promising for some future applications (Helton *et al.*, 2004).

This Appendix is divided into three parts. The first presents some general aspects of DST. As DST is not as well-known as probability theory, a short introduction by means of a notional example is given in Section C.1. Section C.2 discusses the differences between DST and a probabilistic or deterministic framework regarding practical applications, and on this basis, Section C.3 gives some

Uncertainty in Industrial Practice   Edited by E. de Rocquigny, N. Devictor and S. Tarantola,
© 2008 John Wiley & Sons, Ltd

suggestions on when DST could be an alternative to a probabilistic or a deterministic framework. However, as DST is one of the newer methods in uncertainty modelling, it is not as widely used as purely probabilistic models. The comments are therefore based rather on research than on a broad overview of completed studies.

C.1   A notional example

Some principles of uncertainty modelling in DST are very similar to those found in probabilistic modelling, while others are quite different. DST techniques require some mathematical instrumentation which is different to probability calculus. While this section will summarise some of the features, the reader is invited to refer to (Yager, 1986, Klir, 2005) and (Ferson *et al.*, 2003) for a complete introduction of the formulation. Throughout this section, a notional example will help to illustrate the DST approach. The example is kept as simple as possible: a reliability block diagram (RBD) with two components. The system modelled is a sensor with two failure modes. The corresponding components are 'sensor works properly' and 'connection to controller is not interrupted'. The RBD belonging to this example is shown in Figure C.1.

RBDs are Boolean functions which are mostly monotonic for all inputs. The pre-existing system model G for the example is given as:

$$G : \{0,1\}^2 \rightarrow \{0,1\}$$
$$G(x) = x^1 \wedge x^2 \tag{C.1}$$

If an input variable $x^i = 0$, then the component i is in a state of failure; if $x^i = 1$, it is in its working state. A probability $\theta_i = P(x^i = 0)$ is estimated, which represents the component's probability of being in a state of failure. From these probabilities, the system failure probability $p_{sys} = P(G(x) = 0)$ can be obtained. This is a probabilistic representation, with the parameters θ_i describing the uncertainty induced by *variability* in time or over a population that could be observed on frequency records. However, the value of θ_i is rarely well known, being rather a quantity estimated by experts, especially when considering early design stages with

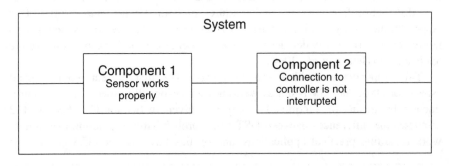

Figure C.1 Reliability block diagram of a sensor.

poor reliability or dependability records. Therefore, a second level of uncertainty representing *lack of knowledge* of the variability probabilities is a reasonable way of coping with the experts' relative ignorance or disagreement. To illustrate the differences between the settings that may be considered on the second level, as listed in Chapter 14, we reformulate the level 1 model as a black box and look on the following function G':

$$G' \begin{pmatrix} \theta_1 \\ \theta_2 \end{pmatrix} = p_{sys} = P\left[G(X) = 0 \middle| \begin{pmatrix} \theta_1 \\ \theta_2 \end{pmatrix} \right] \qquad (C.2)$$

The calculation of G' is not subject of this section, see e.g. Chapter 17.

C.1.1   Representation (modelling)

At least four different settings are available to quantify the uncertainty on this second level. On one hand, a point value could be used to model the parameters θ_1 and θ_2. A best/worst-case scenario, a probabilistic and a DST approach are other uncertainty settings that could be used. These four settings on level 2 are compared to illustrate the working principle of DST.

Setting 1: Point value

If uncertainty on level 2 is neglected (setting 1), a first simple solution is to estimate θ_1 and θ_2 as point values. In the notional example, θ_1 is estimated as $1.5 \cdot 10^{-7}$ and θ_2 as $4 \cdot 10^{-7}$ (see Table C.1, row 1).

Setting 2: Best/worst-case scenario (i.e. level-2 deterministic setting)

A second approach is to consider the level 1 probabilistic setting with a level 2 best/worst-case scenario: select optimistic and pessimistic values for θ_1 and θ_2 and calculate point-value results for $p_{sys} = P(G(x) = 0)$. In this setting no distribution is assumed in θ_1 and θ_2. The only assumptions on θ_1 and θ_2 are intervals, i.e. that $\theta_1 \in [\underline{\theta}_1, \bar{\theta}_1] = [10^{-7}, 2 \cdot 10^{-7}]$ and $\theta_2 \in [\underline{\theta}_2, \bar{\theta}_2] = [2 \cdot 10^{-7}, 5 \cdot 10^{-7}]$ (see Table C.1, row 2). While only two calculations are necessary for such best or worst cases if G is completely monotonic, propagating the intervals through a system function is generally done by finding the lower/upper system function bounds through optimization algorithms, not via sampling. This may be quite costly in computing time. Figure C.2 (left) shows the estimated interval value for θ_2.

Setting 3: Probabilistic scenario (i.e. double probabilistic setting)

A double probabilistic setting may be here classically interpreted as a 'probability of frequency approach' as mentioned in Part I. Failure probability θ_1 in this example is given by a weighted mixture of three uniform distributions[1],

---

[1] $U(a,b)$ represents a uniform distribution between a and b.

Table C.1 Four different uncertainty settings on level 2 given a level 1 probabilistic setting.

Level 2 uncertainty setting	Estimated values for θ_1	Estimated values for θ_2
Setting 1: Point value	$1.5 \cdot 10^{-7}$	$4 \cdot 10^{-7}$
Setting 2: Best/worst-case scenario	$[10^{-7}, 2 \cdot 10^{-7}]$	$[2 \cdot 10^{-7}, 5 \cdot 10^{-7}]$
Setting 3: Probabilistic scenario	$U(1 \cdot 10^{-7}, 1.5 \cdot 10^{-7})$ with weight 0.3 $U(1.3 \cdot 10^{-7}, 1.7 \cdot 10^{-7})$ with weight 0.4 $U(1.7 \cdot 10^{-7}, 2 \cdot 10^{-7})$ with weight 0.3	$Tr(2 \cdot 10^{-7}, 4 \cdot 10^{-7}, 5 \cdot 10^{-7})$
Setting 4: DST	$m(\theta_1 \in [1,1.5] \cdot 10^{-7}) = 0.3$ $m(\theta_1 \in [1.3,1.7] \cdot 10^{-7}) = 0.4$ $m(\theta_1 \in [1.7,2] \cdot 10^{-7}) = 0.3$	$Tr(2 \cdot 10^{-7}, [3,4] \cdot 10^{-7}, [5,5.5] \cdot 10^{-7})$

representing three expert interval estimates: $U(1 \cdot 10^{-7}, 1.5 \cdot 10^{-7})$ with weight 0.3, $U(1.3 \cdot 10^{-7}, 1.7 \cdot 10^{-7})$ with weight 0.4, $U(1.7 \cdot 10^{-7}, 2 \cdot 10^{-7})$ with weight 0.3. θ_2 is quantified by a triangular distribution[2] $Tr(2 \cdot 10^{-7}, 4 \cdot 10^{-7}, 5 \cdot 10^{-7})$. Setting 3 is a reasonable extension of setting 1. The distribution on θ_2 is given in Figure C.2 (left, solid line).

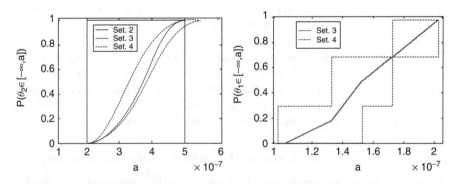

Figure C.2 Left: Estimation for θ_2 in settings 2, 3, 4; right: Estimation for θ_1 in settings 3 and 4.

[2] $Tr(a,b,c)$ is a triangular distribution with lower limit a, mode b and upper limit c.

In the case of setting 3, θ_I is describable by a distribution. To show the differences to the DST approach, it is helpful to illustrate the case that θ_I is described by a discrete distribution, which means that a mass $m(a)$ can be put on each value that this variable can take, corresponding to $P(\theta_i = a)$. It is important to highlight this equality, because it will not hold in the case of DST. From m, the probability $P(\theta_i \in [\underline{a}, \bar{a}])$ can be obtained by summing, over all mass values:

$$P(\theta_i \in [\underline{a}, \bar{a}]) = \sum_{a \in [\underline{a}, \bar{a}]} m(a) \qquad (\text{C.3})$$

Setting 4: DST

But what if there were a need to represent not only one, but a set of best/worst-case scenarios? There are several more elaborate probabilistic settings (including Bayesian), which could be considered. The Dempster-Shafer Theory of evidence (DST), one of the options, is considered hereafter. Failure probability θ_I is given as a so-called mass function, a DST mathematical concept that is distinct to that of a probability distribution, and which may be understood as representing three weighted best/worst-case scenarios/intervals (Figure C.2, right). Each of these scenarios is propagated through the system using optimization (setting 2), rather than sampling. The triangular distribution is maintained but with uncertain parameters. The mode b and upper limit c are specified by intervals (Figure C.2, left, dashed lines). In this way the uncertainty in the distribution parameters of the triangular distribution can be represented.

As informally introduced in this short example, DST may be seen as a synthesis of the deterministic best/worst-case view and the probabilistic perspective – informally, it is a 'probabilistic best/worst-case analysis'. As such it utilizes methods from both worlds for the modelling and propagation of uncertainty.

An original set of axioms is necessary to properly define this approach, quite different to that of probability calculus. While not reproducing it here (cf. for instance : Helton *et al.*, 2004), some simplified comments will be given to help understanding the approach. In DST, a mass $m([\underline{a}, \bar{a}])$ is expressed by an interval $[\underline{a}, \bar{a}]$, for each best/worst-case scenario. The function m is often referred to as 'basic probability assignment' (BPA) or possibly better as 'basic belief assignment' since it is not truly equivalent to a probability. Inside the interval $[a, \bar{a}]$, no distribution (i.e. uniform or Gaussian) is assumed. It can be seen as an 'elementary best/worst-case scenario', called a 'focal element'. Figure C.3 (left) shows a BPA on θ_i. Focal elements are drawn as rectangles with their height representing their mass. It can be seen that, as a difference with probabilistic representations, a given value of θ_i could be included in several focal elements.

According to the DST uncertainty paradigm, it is not possible to give statements that are fully equivalent to probability ones such as $P(\theta_i \in [\underline{a}, \bar{a}])$ or even $P(\theta_i = a)$. However, lower and upper probability bounds may be calculated, in a certain sense in analogy to the best- and worst-case perspective. These bounds are referred

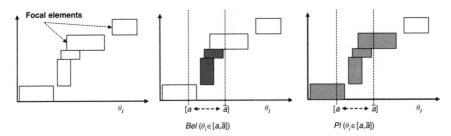

Figure C.3 (Left) Exemplary plot of a BPA; (centre) Focal elements contributing to $Bel(\theta_i \in [\underline{a}, \bar{a}])$; (right) Focal elements contributing to $Pl(\theta_i \in [\underline{a}, \bar{a}])$.

to as 'Belief' Bel and 'Plausibility' Pl of an event and are defined as:

$$Bel(\theta_i \in [\underline{a}, \bar{a}]) = \sum_{A \subseteq [\underline{a}, \bar{a}]} m(A) \tag{C.4}$$

$$Pl(\theta_i \in [\underline{a}, \bar{a}]) = \sum_{A \cap [\underline{a}, \bar{a}] \neq \emptyset} m(A) \tag{C.5}$$

In Figure C.3 (centre), it can be seen that only the mass of the dark grey focal element contributes to $Bel(\theta_i \in [\underline{a}, \bar{a}])$, since it is the only one which is completely enclosed by $[\underline{a}, \bar{a}]$. Regardless of which points in this focal element the mass is distributed over, it will be included in $[\underline{a}, \bar{a}]$. Figure C.3 (right) illustrates the plausibility of an event. The masses of all light grey focal elements contribute to $Pl(\theta_i \in [\underline{a}, \bar{a}])$ because of the non-empty intersection with $[\underline{a}, \bar{a}]$. It is possible to distribute the mass of each focal element on points which are enclosed in $[\underline{a}, \bar{a}]$.

Informally, the belief function represents the lowest possible value for $P(\theta_i \in [\underline{a}, \bar{a}])$ (e.g. the worst case) and the plausibility function represents the highest possible value for $P(\theta_i \in [\underline{a}, \bar{a}])$ (e.g. the best case). Figure C.4, below, shows this probabilistic interpretation of a BPA. A common way of visualizing the BPA of a variable is to plot $Bel(-\infty, a)$ and $Pl(-\infty, a)$ in analogy to a *cdf*. Included are some possible *cdfs* bounded by belief and plausibility functions.

C.1.2 Propagation

Depending on the uncertainty representation, the propagation methods vary. To illustrate how uncertainty is propagated in DST, it is helpful to look again at how uncertainty is propagated in the other frameworks. The four settings are different regarding the uncertainty representation of θ_1 and θ_2, and this uncertainty needs to be propagated through $G'(\boldsymbol{\theta}) = p_{sys}$.

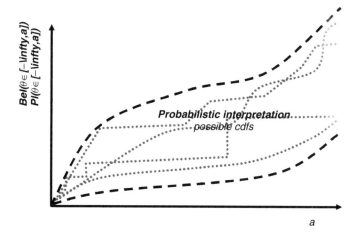

Figure C.4 Bel/Pl ($[-\infty, a]$) bound all possible cdfs covered by a BPA.

Setting 1: Point value

In this simplest case, only the point values $\theta_1 = 1.5 \cdot 10^{-7}$ and $\theta_2 = 4 \cdot 10^{-7}$ are propagated. The function $G'(\theta) = p_{sys}$ needs to be calculated only once for these specific values. The result is a sharp value, $p_{sys} = 5.5 \cdot 10^{-7}$.

Setting 2: Best/worst-case scenario

The chosen uncertainty model is a two-dimensional set generated by intervals with limiting best- and worst-cases on θ_1 and θ_2, $\theta_1 = [\underline{\theta}_1, \bar{\theta}_1] \in [10^{-7}, 2 \cdot 10^{-7}]$ and $[\underline{\theta}_2, \bar{\theta}_2] \in [2 \cdot 10^{-7}, 5 \cdot 10^{-7}]$. Propagation is done by a deterministic optimization approach. The following optimization problems need to be solved:

$$p_G = \left[min\, G'\left(\frac{[\underline{\theta}_1, \bar{\theta}_1]}{[\underline{\theta}_2, \bar{\theta}_2]} \right), \quad max\, G'\left(\frac{[\underline{\theta}_1, \bar{\theta}_1]}{[\underline{\theta}_2, \bar{\theta}_2]} \right) \right] \tag{C.6}$$

In other words: find the minimal and maximal values of p_{sys} for all possible values of θ_1, and θ_2. The result is an interval: $p_{sys} \in [3 \cdot 10^{-7}, 7 \cdot 10^{-7}]$.

Setting 3: Probabilistic scenario

In the probabilistic setting, the uncertainty is modelled by distributions describing θ_1 and θ_2 (see Figure C.2). Monte Carlo sampling strategies and other probabilistic propagation methods can be used to propagate the uncertainty through G'. In the case of independence, marginals can be sampled directly and propagated through the system, otherwise e.g. dependence models can be applied (see Chapter 16). The result is a pdf (or a cdf) on p_{sys}, as shown in Figure C.6 (the 'Set. 3' curve representing the cdf).

Setting 4: DST

In the DST setting, the BPAs describing θ_1 and θ_2 contain the uncertainty to propagate (see Figure C.2). Uncertainty propagation is a combination of the methods for setting 2 and setting 3. Marginal BPAs can be sampled in a similar way (Ferson et al., 2003) than marginal probability. In probability theory a joint distribution exists which can be used for sampling in the case of dependent inputs. DST contains a similar concept: the joint BPA m. There are several ways of obtaining m from marginal BPAs, e.g. through the use of copulas when additionally assuming a dependence structure (Ferson et al., 2004). From a joint BPA m, arbitrary sampling strategies can be used to sample focal elements from the joint distribution.

In the propagation step the component failure probabilities with joint BPA m must be propagated through G'. The resulting is a BPA on p_{sys}, which will be noted m_{sys}, that is the analogue of the output probability distribution in a double probabilistic setting. Propagating a sampled focal element through the system function is not identical to propagating a sampled point value from a distribution. As a mixture of probabilistic and interval arithmetic, DST relies on optimization (as was mentioned for setting 2) to propagate focal elements through the system function. Remembering the analogy of weighted best/worst-case scenarios, every time one specific best/worst-case scenario is sampled from m either by standard Monte Carlo sampling or more sophisticated methods as in (Helton et al., 2007), this best/worst-case scenario needs to be calculated via optimization methods. Assuming the following specific best/worst-case scenario $[\underline{\theta}, \bar{\theta}]_i$ with mass 0.001 corresponding to the i-th sample has been selected from the joint BPA:

$$[\underline{\theta}, \bar{\theta}]_i = \begin{pmatrix} [\underline{\theta}_1, \bar{\theta}_1]_i \\ [\underline{\theta}_2, \bar{\theta}_2]_i \end{pmatrix} = \begin{pmatrix} [1 \cdot 10^{-7}, 1.5 \cdot 10^{-7}] \\ [2.3 \cdot 10^{-7}, 2.7 \cdot 10^{-7}] \end{pmatrix}, \quad m([\underline{\theta}, \bar{\theta}]_i) = 0.001 \quad \text{(C.7)}$$

This specific scenario $[\underline{\theta}, \bar{\theta}]_i$ (i.e. the focal element) is propagated through the system function, and the corresponding focal element of the BPA m_G is obtained by minimization/maximization (obtaining the best and worst cases of propagating).

$$m_{sys}\left(\left[\min G'\begin{pmatrix} [\underline{\theta}_1, \bar{\theta}_1]_i \\ [\underline{\theta}_2, \bar{\theta}_2]_i \end{pmatrix}, \max G'\begin{pmatrix} [\underline{\theta}_1, \bar{\theta}_1]_i \\ [\underline{\theta}_2, \bar{\theta}_2]_i \end{pmatrix}\right]\right)$$

$$= m_{sys}([3.3 \cdot 10^{-7}, 4.2 \cdot 10^{-7}]) = 0.001 \quad \text{(C.8)}$$

The propagation is carried out for each of the samples in a BPA. Thus, if a sample size of 1000 is chosen, 1000 outputs of minima/maxima must be calculated. The amount of computing power required in this step varies widely with the complexity of the function representing the system model. Obviously, functions which tend to be (at least locally) easy to solve with optimization methods would not increase the calculation effort substantially.

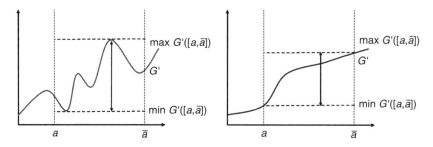

Figure C.5 Bounds on $G'([\underline{\theta}, \bar{\theta}]_i)$ depending on the shape of G (1-D schematic representation). If G is monotonic, the optima are at the borders of $G'([\underline{\theta}, \bar{\theta}]_i)$.

Figure C.5 illustrates the finding of the lower and upper bounds of a 1-D system model. On the left-hand side, the bounds of $G'([\underline{\theta}, \bar{\theta}]_i)$ must be obtained by optimization. The function of the right-hand side is monotonic, representing a system reliability model. $G'([\underline{\theta}, \bar{\theta}]_i)$ reaches its maximal values at the interval boundaries.

C.1.3   Interpretation of results

This section discusses the interpretation of results and how these may differ depending on the setting chosen. Table C.2 compares various quantities of interest for the four settings.

Setting 1: Point value

In setting 1, where no uncertainty on level 2 was modelled, the result is a single value for p_{sys}. This value is used for further evaluation of the goals, within the potential decision criteria and the associated feedback process.

Setting 2: Best/worst-case scenario Representation (modelling)

In the best/worst-case scenario, lower and upper bound on p_{sys} are obtained. Therefore, it can be safely stated that p_{sys} lies between $3 \cdot 10^{-7}$ (best case) and $7 \cdot 10^{-7}$ (worst case). This approach is not probabilistic, hence no distribution is assumed in the interval $[3 \cdot 10^{-7}, 7 \cdot 10^{-7}]$. Therefore it is not possible to state whether p_{sys} lies more likely in $[3 \cdot 10^{-7}, 4 \cdot 10^{-7}]$ or in $[5 \cdot 10^{-7}, 6 \cdot 10^{-7}]$.

Setting 3: Probabilistic scenario

In the probabilistic case, the measure of uncertainty is a (joint) pdf (or cdf) on p_{sys}. Figure C.6 shows the resulting BPA of the system failure probability p_{sys}. The solid line shows the result of setting 3, the dashed lines the result of setting 4. In setting 3, popular *quantities of interest* are, for example, the probability of exceeding

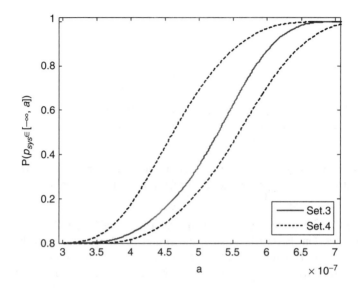

Figure C.6 Resulting BPA of p_{sys} after uncertainty propagation (settings 3 & 4).

a threshold probability (or frequency), or the expected value of this probability (or frequency). Table C.2 gives values for the *quantities of interest* resulting in the example.

Setting 4: DST

The measure of uncertainty in a DST approach is a BPA on p_{sys}. From a BPA, quantities of interest corresponding to that of the probabilistic case can be obtained. It is important to note that all these quantities are single numbers in setting 3, while they become intervals in a DST setting. The interval given for $E(p_{sys})$, for example, could be interpreted as enclosing all possible expected values that would result from probability distributions bounded by the BPA.

After the results of the uncertainty study have been obtained, they need to be interpreted and communicated. Here lies a fundamental difference between DST and probabilistic modelling. DST makes it possible to quantify uncertainty by interval width and by the distribution of focal elements. It is possible to infer statements on quantities of interest such as the expected value, similarly to probability theory. However the results are intervals, between best and worst cases of the expected value, instead of a point value for the expectation in a probabilistic approach.

The results in Table C.2 can be interpreted as follows. Suppose the exceedance of a threshold is tested, such as requiring $P(p_{sys}>6 \cdot 10^{-7})<0.05$ which could be formulated as 'there is less than 5% credibility (or 'confidence') in the fact that $p_{sys}>6 \cdot 10^{-7}$'. Credibility is then bounded by the belief $Bel(p_{sys}>6 \cdot 10^{-7})$ and the plausibility $Pl(p_{sys}>6 \cdot 10^{-7})$. In the example, $Bel(p_{sys}>6 \cdot 10^{-7}) = 0.03$ and

Table C.2   Quantities of interest of the system failure probability.

Quantity of Interest	Setting 1	Setting 2	Setting 3	Setting 4
p_{sys}	$5.5 \cdot 10^{-7}$	$[3 \cdot 10^{-7}, 7 \cdot 10^{-7}]$	–	–
$E[p_{sys}]$	–	–	$5.2 \cdot 10^{-7}$	$[4.7 \cdot 10^{-7}, 5.6 \cdot 10^{-7}]$
$Med[p_{sys}]$	–	–	$5.2 \cdot 10^{-7}$	$[4.6 \cdot 10^{-7}, 5.6 \cdot 10^{-7}]$
$Q_{95}[p_{sys}]$	–	–	$6.2 \cdot 10^{-7}$	$[5.9 \cdot 10^{-7}, 7.6 \cdot 10^{-7}]$
$P(p_{sys} > 6 \cdot 10^{-7})$	–	–	0.10	$[0.03, 0.29]$
$Bel(p_{sys} > 6 \cdot 10^{-7})$	–	–	–	$0.03 / 0.29$
$Pl(p_{sys} > 6 \cdot 10^{-7})$				

$Pl(p_{sys} > 6 \cdot 10^{-7}) = 0.29$. Therefore, the conclusion is that the available information is not sufficient to establish with the desired 'level of credibility' either the exceedance of the threshold or the shortfall of the threshold.

One point should be noted, since it represents a trap in the application of DST: averaging intervals of resulting quantities of interest into point values is questionable. In the quantification phase, uncertainties that could not be described by distributions were described by intervals, i.e. 'best/worst-case' estimates. This interval uncertainty is propagated and preserved in the results. Averaging intervals would render the uncertainty quantification useless. This can be illustrated in comparison to a best/worst-case study. An analyst trying to obtain the worst and the best case for an input range would never average best and worst case at the end, but rather provide both values as results. This is a clear difference to double probabilistic settings where a full expectation may be computed, averaged over the two levels (say aleatory/variability and epistemic/lack of knowledge, as illustrated in Table C.2 for setting 3), and retained for the decision-making.

C.2   Preliminary comments on comparing DST and probabilistic settings

Several case studies applying DST have been documented (Tonon, 2004; Kriegler, 2005; Démotier *et al.*, 2006). However, there is still a need for practical recommendations on when DST may present an interesting alternative to other uncertainty settings. Two opposing basic representations of uncertainty have been illustrated: the deterministic (setting 2) and the probabilistic (setting 3). DST is, to a certain extent, a mixture of both representations and therefore can be considered as a rather flexible approach. By introducing interval uncertainty (*e.g.* by a wide focal element), a controversial or hard-to-quantify source of uncertainty may be represented without the need to argue over the choice of a distribution. On the other hand, the analyst may utilize the concepts of a probabilistic approach where these are more adequate than best/worst-case scenarios. Different best/worst-case scenarios can be weighted by masses. As DST has been formalised as a mathematical generalization of the probabilistic representation, distributions, intervals and a mixture of both can be represented in a DST framework (Dempster, 1967). This section attempts to give some hints in comparing DST to both the deterministic and the probabilistic settings from a practical point of view. As outlined, this comparison is partly based on practical experience and partly on research, since DST is not yet commonly applied.

Figure C.7 illustrates the various settings at level 2 and their application. Five different settings are grouped according to their use, depending on the robustness requirements in the study. If the need for a robust uncertainty model is small, quite often a fixed value is assumed. If an uncertainty model is required, the practitioner may move to a probabilistic setting. DST can be chosen if the need for a robust uncertainty model is even greater than this. Distributions may be enclosed within bounds, making the model more robust against a possibly inadequate choice of the distribution. The most conservative approach would theoretically be a best/worst-case analysis, providing that the entire *possible* model input space could be accurately described. However, in practice, only a limited best/worst-case analysis is applied. The largest *'plausible'* range of the model input values is retained using limited 'best and worst' cases, making the approach conservative only with respect to the bounds.

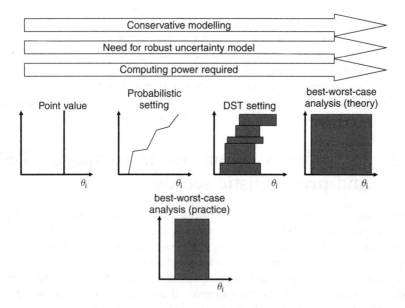

Figure C.7 Settings 1–4 arranged according to their conservativeness, robustness and required computing power. This is a very approximate ranking as the real effort varies with the problem-specific uncertainty and system models.

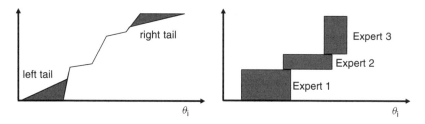

Figure C.8   Examples of uncertainty quantification in DST. Left: Quantifying distribution tails. Right: Aggregation of three best/worst-case estimates.

C.2.1   Quantification of uncertainty

To illustrate possible uses of the extended modelling flexibility of DST, two cases are illustrated (Figure C.8). The first case is a distribution on θ_i in which the tail behaviour is not well known. In a probabilistic approach the tails can be estimated, but there will be uncertainty involved if this estimation is true. In a DST approach, the known part of the distribution may be defined, but the tails of the distributions can be bounded without specifying the exact distribution. The second case shows the aggregation of three expert estimates of θ_i. In DST it is possible to aggregate interval estimates of several experts, and there are several ways of merging estimates to a combined BPA (Sentz *et al.*, 2002).

C.2.2   Computation for propagation

Enhanced modelling freedom in DST does not come for free. With the merging of interval uncertainty and probabilistic uncertainty models, the computational effort rises. Whether this rise is large or small depends on the shape of the system model (see Figure C.5) which has an impact on the complexity of the resulting optimization task. If the computational costs constitute a constraint, there are two adjustment screws in DST: on one hand, reducing the samples of the joint BPA, on the other, reducing the accuracy of the optimization algorithm involved. Use of a meta-model to replace the full pre-existing system model would of course be an alternative solution, but the same limitations as in probabilistic propagation will be encountered.

C.2.3   DST and sensitivity analysis

While uncertainty modelling and propagation in the DST setting are supported by significant theory, sensitivity analysis has yet to be fully developed. Although there are several existing works, such as (Hall, 2006) and (Helton *et al.*, 2004), the inventory of case studies covering imprecise sensitivity analysis is rather short. More detailed coverage of this topic can be found in (Ferson *et al.*, 2006).

C.3   Possible areas of application

This section gives some suggestions as to the type of situation in which DST can be applied. A drawback in DST is that it is not as standardized nor as widespread as probabilistic modelling. Therefore, the analyst planning to apply DST may face some difficulty in finding adequate methods and tools for the purpose. However, some circumstances exist under which the application of DST is anticipated to be valuable. Several factors that could guide the choice between DST and probabilistic settings are listed below.

C.3.1   Uncertainty model

As discussed in Part I of this book, an uncertainty study may aim at several goals and be influenced by several factors. One possible target for criticism in a study is the choice of the appropriate distribution. The maximum entropy principle, although controversial, may be an answer to this criticism while remaining within a probabilistic approach. If this argumentation is not possible, questions regarding the distribution choice could be circumvented altogether. With DST a whole range of distributions may be bounded without preferring one over the other, spanning the range over all adequate distributions. Therefore a DST study can be thought of as a possible response to this argument.

The same holds if the analyst is not sure how to quantify the uncertainty sources or if there is a very sensitive variable whose modelling by a given distribution is critical. In this case, the analyst may refer to a best/worst-case estimate (or a weighted best/worst-case estimate such as a BPA), thus levering the whole study into a DST setting. If some sort of 'deterministic tradition' exists, in the sense that the target audience would prefer a deterministic setting, but the analyst has good reasons to prefer a probabilistic setting, DST can be used as a 'third way'.

In DST, it is possible to aggregate expert estimates which are given in a probabilistic, interval-valued or a combined form. DST has some natural capacities to fuse data and expert estimates to deal with these different representations. The resulting BPA still contains the characteristics of interval and probabilistic uncertainty.

C.3.2   System model and interpretation of results

DST has both the advantage and disadvantage of the interval paradigm. If the system model involves some localised extreme values which may be important but are poorly-known beforehand, a probabilistic approach tends to minimise their relative importance in the sense that their likelihood would be low (as a proportion of the localised space they represent to prior knowledge). That type of issue may be circumvented by DST. However, in consequence it means that if the compliance of a threshold such as 'failure probability $p_{sys} > 6 \cdot 10^{-7}$ with credibility lower than 5%' is tested, the outcome could be:

- $Pl(p_{sys} > 6 \cdot 10^{-7}) < 0.05$: Compliance with the threshold criterion despite all interval uncertainties.

- $Bel(p_{sys} > 6 \cdot 10^{-7}) \geq 0.05$: The threshold criterion is not fulfilled when including all interval uncertainties.

- $Bel(p_{sys} > 6 \cdot 10^{-7}) < 0.05$, but $Pl(p_{sys} > 6 \cdot 10^{-7}) \geq 0.05$: It cannot be concluded whether the criterion has been reached or not, because the amount of interval uncertainty is too large.

Therefore, even more with DST than in a double probabilistic setting, the amount of epistemic (or level 2) uncertainty may be very large on top of aleatory uncertainty (level 1). Demonstrating compliance with a threshold in DST may hence be a very strong supporting argument. On the other hand, a great deal of discipline is necessary, because an excessively pessimistic use of interval uncertainty may lead even more quickly to ambiguous results. Note, indeed, that eliciting interval bounds on the basis of expertise may be tricky and controversial (are the bounds certain?), while using purely physical bounds may conversely be over-conservative. Whether a probabilistic, deterministic or DST approach proves to be the optimal choice therefore depends heavily on the objectives of the analyst and the type and level of information available.

References

Bae, H-R., Grandhi R.V. and Canfield R.A. (2004) Uncertainty Quantification and Optimization of Structural Response using Evidence Theory, *Reliability Engineering & System Safety*, **86**(3), 215–225.

Dempster A.P. (1967) Upper and lower probabilities induced by a multivalued mapping, *Annals of Math. Statistics*, **38**, 325–339.

Démotier, S., Schön, W. and Denoeux, T. (2006) Risk assessment based on weak information using belief functions: a case study in water treatment, *IEEE Transactions on Systems, Man and Cybernetics*, **36**(3), 382–396.

Denoeux, T. (1999) Reasoning with imprecise belief structures, *International Journal of Approximate Reasoning*, **20**(1), 79–111.*

Dubois, D. and Prade, H. (1992) 'On the combination of evidence in various mathematical frameworks'. In: J. Flamm and T. Luisi (Eds) *Reliability Data Collection and Analysis*, 213–241.

Ferson, S., Kreinovich, V., Ginzburg, L., Myers, D. S. and Sentz, K. (2003) *Constructing Probability Boxes and Dempster-Shafer Structures*, Sandia Report SAND2002-4015, Sandia National Laboratories.*

Ferson, S., Hajagos, J., Berleant, D., Jianzhong, Z., Troy, T., W., Ginzburg, L. and Oberkampf, W. (2004) *Dependence in Dempster-Shafer theory and probability bounds analysis*, Sandia Report SAND2004-3072, Sandia National Laboratories.

Ferson, S. and Tucker, W. T. (2006) *Sensitivity in Risk Analyses with Uncertain Numbers*, Sandia Report SAND2006-2801, Sandia National Laboratories.

---

*Also suitable for beginners.

Hall, J.W. (2006) Uncertainty-based sensitivity indices for imprecise probability distributions, *Reliability Engineering & System Safety*, **91**(10–11), 1443–1451.

Helton, J. C., Johnson, J. D. and Oberkampf, W. L. (2004) Sensitivity Analysis in Conjunction with Evidence Theory Representations of Epistemic Uncertainty, *Proceedings of the 4th International Conference on Sensitivity Analysis of Model Output (SAMO 2004)*, Santa Fe.

Helton, J. C., Johnson, J. D. *et al.* (2004), An exploration of alternative approaches to the representation of uncertainty in model predictions, *Reliability Engineering and System Safety*, **85**(1–3), 39–71.

Helton, J.C., Johnson, J., Oberkampf, W. and Sallaberry, C. (2006) Sensitivity analysis in conjunction with evidence theory representations of epistemic uncertainty, *Reliability Engineering & System Safety*, **91**(10–11), 1414–1434.

Helton, J. C., Johnson, J.D. *et al.* (2007) A sampling-based computational strategy for the representation of epistemic uncertainty in model predictions with evidence theory, *Computer Methods in Applied Mechanics and Engineering*, **196**(37–40), 3980–3998.

Klir, G.J. (2005) *Uncertainty and Information: Foundations of Generalized Information Theory*, Chichester: John Wiley & Sons, Ltd.

Kriegler, E. (2005) *Imprecise Probability Analysis for Integrated Assessment of Climate Change*, PhD Thesis, Universität Potsdam.

Sentz, K. and Ferson, S. (2002) Combination of Evidence in Dempster-Shafer Theory, *Sandia Report SAND2002-0835*, Sandia National Laboratories.*

Shafer, G. (1976) *A Mathematical Theory of Evidence*, Princeton University Press.

Tonon F. (2004) Using random set theory to propagate epistemic uncertainty through a mechanical system, *Reliability Engineering & System Safety*, **85**(1–3), 169–181.

Walley, P. (1991) *Statistical Reasoning with Imprecise Probabilities*, London: Chapman & Hall.

Yager, R. R. (1986) Arithmetic and other operations on Dempster-Shafer structures, *International Journal of Man-Machine Studies*, **25**, 357–366.

Zadeh, L.A. (1978) *Fuzzy Sets as a Basis for a Theory of Possibility*, Philadelphia: Society for Industrial and Applied Mathematics.

---

*Also suitable for beginners.

Index

Uncertainty in Industrial Practice   Edited by E. de Rocquigny, N. Devictor and S. Tarantola,
© 2008 John Wiley & Sons, Ltd